INTRODUCTION TO ENGINEERING

edited by

Robert M. Glorioso

and

Francis S. Hill, Jr.

introduction to engineering

Prentice-Hall, Inc., Englewood Cliffs, New Jersey

Library of Congress Cataloging in Publication Data

GLORIOSO, ROBERT M
    Introduction to engineering.
    Includes bibliographical references.
    1. Engineering. I. Hill, Francis S., joint author.
II. Title.
TA145.G57 1974   620   74-3430
ISBN 0-13-482398-2

© 1975 by Prentice-Hall, Inc.
Englewood Cliffs, New Jersey

All rights reserved. No part of this book may be reproduced in any form or by any means without permission in writing from the publisher.

10 9 8 7 6 5 4 3 2

Printed in the United States of America

PRENTICE-HALL INTERNATIONAL, INC., *London*

PRENTICE-HALL OF AUSTRALIA, PTY. LTD., *Sydney*

PRENTICE-HALL OF CANADA, LTD., *Toronto*

PRENTICE-HALL OF INDIA PRIVATE LIMITED, *New Delhi*

PRENTICE-HALL OF JAPAN, INC., *Tokyo*

*To
Dee and Scott
and
Susan and Jessica*

contents

contributing authors — xiii

preface — xiv

what is engineering? — 1

1.1 INTRODUCTION	2
1.2 THE ROLE OF ENGINEERING IN SOCIETY	4
1.3 THE ROLE OF THE ENGINEER IN THE TECHNICAL COMMUNITY	5
1.4 ENGINEERING AND THE SCIENCES	13
1.5 SPECTRUM OF JOBS FOR ENGINEERS	14
1.6 DESIGN	21
1.7 CONCLUSION	30

man, the bridge builder — 2

2.1 INTRODUCTION	33
2.2 FORCES AND EQUILIBRIUM	36
2.3 FORCES AND DEFORMATION	41
2.4 DEFORMATION OF BRIDGE FORMS	44
2.5 TYPICAL BRIDGE GEOMETRY	48
2.6 STATICS OF BEAMS	50
2.7 STATICS OF SUSPENSION CABLES	54
2.8 STATICS OF ARCHES	60
2.9 EVOLUTION OF THE TRUSS	63
2.10 STATICS OF TRUSSES	63
2.11 HISTORICAL AND MODERN EXAMPLES	69
2.12 DESIGN PROJECT	83

mass transportation 3

3.1 INTRODUCTION	*88*
3.2 HISTORY OF MASS TRANSPORTATION	*90*
3.3 DEMAND AND PLANNING	*96*
3.4 CRITERIA	*98*
3.5 SYSTEMS CONCEPTS	*106*
3.6 ENVIRONMENTAL CONSIDERATIONS	*121*
3.7 SUMMARY	*123*
EXERCISES AND PROJECTS	*125*

energy and pollution 4

4.1 INTRODUCTION	*128*
4.2 ENERGY RESOURCES AND USES	*128*
4.3 ENERGY CONVERSION SYSTEM ANALYSIS	*134*
4.4 ENERGY CONVERSION SYSTEMS	*139*
4.5 ENVIRONMENTAL ASPECTS OF ENERGY CONVERSION SYSTEMS	*147*
4.6 SUMMARY	*154*
EXERCISES	*154*

air pollution concepts 5

5.1 INTRODUCTION	*158*
5.2 EFFECTS OF POLLUTION AND CLIMATE	*163*
5.3 EFFECTS OF POLLUTION ON PLANTS AND ANIMALS	*171*
5.4 SPECIFIC ATMOSPHERIC POLLUTION	*174*
5.5 ATMOSPHERIC SCAVENGING PROCESSES	*187*
5.6 AIR POLLUTION LEGISLATION	*189*
EXERCISES	*199*

engineering materials

6.1 MATERIALS IN ENGINEERING DESIGN	202
6.2 MECHANICAL TESTING	203
6.3 EXAMINATION OF MATERIALS	220
6.4 SELECTION OF MATERIALS	224
6.5 MATERIALS OF THE FUTURE	233
6.6 SUMMARY	

fundamentals of electrical networks

7.1 DEFINITIONS, CONCEPTS, AND CIRCUIT LAWS	236
7.2 THE BASIC CIRCUIT FORMS	242
7.3 TIME-VARYING SIGNALS	251
7.4 NONDISSIPATIVE PASSIVE ELEMENTS—INDUCTORS AND CAPACITORS	260
7.5 TRANSIENT BEHAVIOR IN SERIES RL AND RC CIRCUITS	264
7.6 SUMMARY	268
EXERCISES	269

computer engineering

8.1 HISTORICAL PERSPECTIVE	273
8.2 GENERAL-PURPOSE DIGITAL COMPUTER	276
8.3 COMPUTER CODING AND NUMERICAL REPRESENTATIONS	283
8.4 SIGNALS IN LOGICAL SYSTEMS	291
8.5 DIGITAL LOGIC	292
8.6 DE MORGAN'S THEOREM	295
8.7 BINARY FUNCTIONS	296
8.8 THE FLIP-FLOP AND ITS APPLICATION	298
8.9 COMPUTER SYSTEMS	299
8.10 SUMMARY	301
EXERCISES	301

communications in the modern world 9

9.1 INTRODUCTION	305
9.2 A BASIC COMMUNICATIONS SYSTEM	309
9.3 SOME FUNDAMENTAL CONCEPTS IN COMMUNICATIONS	322
9.4 CONCLUSION	350
EXERCISES	350

introduction to biomedical engineering

10.1 INTRODUCTION	356
10.2 NERVES	357
10.3 THE ELECTROENCEPHALOGRAM (EEG)	361
10.4 THE HEART	364
10.5 ELECTROCARDIOLOGY	367
10.6 AUTOMATIC ELECTROCARDIOGRAPHIC ANALYSIS	370
10.7 POPULATION GENETICS	370
10.8 SUGAR BALANCE	375
10.9 THE KIDNEY	378
10.10 THE BALANCE ORGANS	379
10.11 SUMMARY	381
EXERCISES	382

introduction to experimental technique 11

11.1 INTRODUCTION	*384*
11.2 ESTIMATION OF ERRORS	*387*
11.3 SYSTEMATIC ERROR AND INSTRUMENT CALIBRATION	*391*
11.4 PRESENTATION OF RESULTS	*394*
11.5 REDUCTION OF VARIABLES	*401*
EXERCISES	*411*

chemical and thermal processes 12

12.1 INTRODUCTION	*414*
12.2 CONSERVATION OF MATTER	*414*
12.3 CONSERVATION OF ENERGY AND THE FIRST AND SECOND LAWS OF THERMODYNAMICS	*420*
12.4 KINETICS AND CHEMICAL EQUILIBRIUM	*426*
12.5 HEAT TRANSFER	*432*
12.6 MASS TRANSFER AND DIFFUSION	*436*
12.7 CONSERVATION OF MOMENTUM (MOMENTUM BALANCES)	*440*
12.8 PROCESS DESIGN	*441*
12.9 PROCESS CONTROL	*442*
12.10 SUMMARY	*442*
EXERCISES	*442*

appendix A

TRIGONOMETRIC IDENTITIES *445*
OFTEN-USED TRIGONOMETRIC FUNCTIONS *446*
APPROXIMATIONS FOR SMALL ANGLES *447*

appendix B

PREFIXES AND SYMBOLS *448*

appendix C

CONVERSION FACTORS *449*

index *453*

CONTRIBUTING AUTHORS

Professor Lawrence L. Ambs
Department of Mechanical and Aerospace Engineering
University of Massachusetts, Amherst

Professor Geoffrey Boothroyd
Department of Mechanical and Aerospace Engineering
University of Massachusetts, Amherst

Professor Duane E. Cromack
Department of Mechanical and Aerospace Engineering
University of Massachusetts, Amherst

Professor Robert M. Glorioso
Department of Electrical and Computer Engineering
University of Massachusetts, Amherst

Professor William P. Goss
Department of Mechanical and Aerospace Engineering
University of Massachusetts, Amherst

Professor Francis S. Hill, Jr.
Department of Electrical and Aerospace Engineering
University of Massachusetts, Amherst

Professor James R. Kittrell
Department of Chemical Engineering
University of Massachusetts, Amherst

Professor Jon G. McGowan
Department of Mechanical and Aerospace Engineering
University of Massachusetts, Amherst

Professor Melton M. Miller
Department of Civil Engineering
University of Massachusetts, Amherst

Professor Richard V. Monopoli
Department of Electrical and Computer Engineering
University of Massachusetts, Amherst

Professor John E. Ritter, Jr.
Department of Mechanical and Aerospace Engineering
University of Massachusetts, Amherst

Professor Donald E. Scott
Department of Electrical and Computer Engineering
University of Massachusetts, Amherst

Professor W. Leigh Short
Department of Chemical Engineering
University of Massachusetts, Amherst

preface

preface

Freshman engineering curricula have been undergoing considerable change through the past decade. Initially, many schools introduced computer programming into the freshman program while sacrificing some material on drafting, graphics, or descriptive geometry. This trend continued, and in many schools graphics has been completely eliminated from the curriculum.

The question then arises, What material should properly replace graphics, while providing the student with a clear perspective of and a heightened motivation toward an engineering career?

After several semesters of experimentation, the following program has evolved at the University of Massachusetts, Amherst. The freshman year is divided into six modular periods, with three modules to each semester. Each student is required to take two modular mini-courses in computer programming: Introductory Programming and Advanced Programming. During each of the remaining modules the student is free to elect any one of several mini-courses that are offered by the various departments.

This book has grown out of the mini-courses that have been given in these modular periods, but it has been designed to be used in standard semester, trimester, or quarter-type courses as well. The material contained in this book can be covered in approximately one and three quarter semesters. Thus, in a one semester course, for example, the instructor can cover material which is appropriate for his institution and students. This is easily accomplished since each chapter is designed to stand alone.

The material covered in the text represents a broad cross section of topics in contemporary engineering disciplines. It is not the intent of this book to be all-inclusive, but rather to provide subject areas that are of particular interest to today's engineering students. For example, contemporary problems of energy consumption and its concomitant generation of pollution are treated by Professors Ambs and McGowan in Chapter 4, Energy and Pollution. Specific air pollution problems and their analysis are covered by Professors Kittrell and Short in Chapter 5.

On the other hand, techniques of computer programming have not been included, since each school has individual requirements and circumstances, and several good texts already exist on this subject. The text is organized roughly in order of increasing difficulty where the first six chapters require less mathematical sophistication than the remaining chapters.

The author of each chapter is an expert in his area and has taught this material to freshmen for at least two modules. Furthermore, each author is presently doing research and is teaching upper-division and graduate courses in his area of specialization.

We wish to acknowledge the School of Engineering of the University of Massachusetts for the encouragement received in the preparation of this text. Also, a large portion of the manuscript was splendidly typed by Ms. Rhea Cabin and Ms. Helga Ragle.

Amherst, Massachusetts

ROBERT M. GLORIOSO
FRANCIS S. HILL, JR.

INTRODUCTION TO ENGINEERING

1
what is engineering?

R. M. Glorioso

F. S. Hill, Jr.

1.1 INTRODUCTION

In this chapter we shall present an overview of the engineering profession and describe many of the roles of the engineer in society and in the scientific community. We shall also discuss the various opportunities available to an engineering student in choosing a career in the engineering professions.

A rich host of possibilities awaits the graduating engineer. Most graduates apply their training to problems in areas of society such as industry, government, and education. However, graduate engineers also go into other fields; for example, medicine, law, and management. We shall discuss these various opportunities and show how engineering training provides the technical background and intellectual discipline needed for these professions.

Does a bachelor's degree in engineering directly prepare one for all this? Obviously not. A bachelor's degree provides only a strong foundation upon which an individual must build. After graduation an engineer has several choices: He may go into industry, continue his engineering education in graduate school, or move into another profession. In industry an engineer must learn the technical and procedural processes of his company. Industry does not expect a newly graduated engineer to be able to contribute a great deal to his assigned projects or tasks immediately. It is generally understood that several months on the job are needed for orientation and training. Thus, it can be said that a bachelor's degree in engineering *prepares one to become* an engineer in industry. Although an engineer's formal education may end with a bachelor's degree, he is expected to continue learning the technological details of his job as well as other scientific advances. This is accomplished by reading the journals of his profession and attending conferences. He must be up to date.

On the other hand, one may wish to go to graduate school. An advanced degree in engineering prepares one for more complex and analytical technical work, thereby allowing him to command a higher salary. It is also true that in government and industry those who have advanced degrees generally progress faster through the job hierarchy. This is true for both the master's degree and the Doctor of Engineering or Ph.D. degrees. However, a doctorate is essential for a career in education or advanced research. The third alternative, that of moving to another field, is also frequently chosen by graduate engineers and will be discussed later in this chapter.

Assuming that one does stay in the field of engineering, what will be expected of him? The engineer is basically a *problem solver*. He must be capable of taking a problem (sometimes only vaguely defined), analyzing it, reducing it to its most fundamental ingredients, and formulating a collection of alternative solutions. The formulation of this set of solutions calls upon his resourcefulness and creativity. The selection of the best solution from among the alternatives depends on many, sometimes conflicting, factors. Of these factors, some of the most important are technical feasibility, cost, producibility, and reliability. Solving these problems is not usually a nine-to-five affair. Some of an engineer's most effective thinking goes on while he is driving, in the shower, etc. Indeed, some preliminary designs are often drawn on the backs of envelopes, paychecks, or scraps of toilet paper.

Let us now examine how the technical and intellectual skills developed through an engineering education relate to problem solving. The goal of the engineering curriculum is to develop basic technical skills, intellectual discipline, and the ability to communicate ideas lucidly. The basic courses in mathematics, chemistry, and physics lay the foundation for the study of their application in later engineering subjects. A second but equally important purpose of these courses is to develop intellectual discipline and the power to reason clearly. These basic attributes (knowledge of fundamentals, discipline, and analytical ability) enable the engineer to do creative problem solving. Some students feel that many of their courses are not directly relevant to the areas of their greatest interest. However, it is important for the student to bear in mind the above objectives of the engineering curriculum. In addition, as one becomes more involved with these courses, one often finds interesting relationships between seemingly unrelated subjects. For example, the mathematical relationships that describe the motions of mechanical systems with springs, masses, etc., are identical to those describing currents and voltages in electrical networks. You will discover many other similarities in your course work. The insights provided by these relationships broaden the engineer's perspective, thereby providing him with the ability to apply his background in new and different areas. This flexibility and diversity of background allow the engineer to solve new problems in society as they emerge.

Technical background and discipline go a long way toward preparing the student to be a successful engineer. However, he must also be able to communicate his ideas. To whom? First, to his coworkers, so that they can share new ideas and help develop them into practical solutions. Second, to the management of the company he works for, so that they will understand his ideas well enough to support and encourage him to follow up on them. And third, to other engineers around the world—through journal papers and conference talks. These talks and papers spread new ideas to the entire profession and help to enhance an engineer's reputation. Basically then, an engineer must continually convince others of the value of his ideas and abilities. As in any profession, he must be able to sell himself and his ideas.

1.2 THE ROLE OF ENGINEERING IN SOCIETY

Our society is highly technological and is rapidly becoming more so. We use dozens of sophisticated devices every day: automobiles, television sets, telephones, elevators, and bridges, to mention only a few. In addition there are thousands of less visible devices and methods. The equipment we use could not exist were it not for the remarkable assortment of materials and manufacturing techniques technology has spawned. It is very unlikely that our civilization will ever turn around and become less technological: People want comfortable living, easy communications and travel, health care, and recreation too much to turn back.

Why have all these technological advances appeared, and what role does the engineer play in developing them? Almost all invention begins with a *need*. An engineer, scientist, or the lone inventor in his basement recognizes a need and searches for a way to satisfy it. Transportation is one example. The whole history of transportation from the horse-and-buggy days to the present is full of cases where a new device or technology has come along to replace the old. In each case the new development answers the need for safe, inexpensive, and reliable travel better than the old one. Better roads came along because automobiles were beginning to move at higher speeds and because smoother all-weather roads were needed. Another example is in communications. The need for instantaneous communication over large distances brought the telegraph, which replaced the pony express. Then the need for direct voice conversations gave rise to the telephone, and communications has been blossoming ever since.

The engineer's job is to always come up with better ways of satisfying a need: to satisfy it more safely, reliably, and at a lower cost to the user. This may require a totally new technique, as in the case of radar, or it may result in a long series of small improvements, as in the development of the automobile over the last 50 years.

Today in our society there are many needs still awaiting the emergence of workable solutions. Two that have become prominent in recent years are energy production and pollution removal. These two demands seem to be on a collision course. On the one hand, everyone is using energy (in heating, lighting, travel, entertainment, etc.) at an ever-increasing rate, and very few people are seriously willing to give up the devices that consume this energy. On the other hand, there is no truly pollution-free way of generating energy on the large-scale basis required. Some methods such as those used in coal-burning power plants, dump pollutants into the air, while others require the disposal of residual pollutants such as radioactive wastes. This is, to a large extent, a technological problem requiring new techniques and devices, although there are important social, economic, and political problems to be solved here also.

Another growing need in our society is that of creating and maintaining cities that are safe and comfortable for human use. Since, at present, there are many things that motivate people to settle in certain geographical areas, it is too easy a solution to say that people should spread out. To respond to the need as it exists today, engineers, social scientists, and other professionals must combine forces to solve the many difficult problems that arise when people are crowded together. Some of these are primarily technological problems. We must learn how to provide power, water, and food in ample quantities and to remove wastes from the air and water. In addition, housing, crime, and intracity transportation problems must be attacked and solved.

In developing devices and methods for solving these problems, engineers must always be aware of the effects of their ideas. In a society as complex as ours new devices can have both good and bad uses, and an engineer must consider how a new idea will be used. One very prominent example of this is the development of nuclear devices. This technology has led not only to radiological treatment of cancer and potential sources of energy but also to atomic bombs. Sometimes an engineer will feel he must urge his company not to manufacture a certain product even when it may be profitable to do so. The engineer may foresee that the product will be harmful in some ways to the consumer or that it will lead to misuse and hence endanger others. Presently, engineers can have a great deal of trouble convincing their employers to halt production of profitable but potentially harmful items. They may even lose their jobs if management is unsympathetic.

Other pressures exist that influence what is produced and marketed. These involve private groups that report on the value and safety of consumer products. Consumer's Union is a well-known example. It publishes information on brand-name items in many categories, evaluates the items, and warns the public about dangerous or poorly made brands.[1] Ralph Nader and his "raiders" have become powerful voices as consumer advocates, especially in the area of automotive products.[2] Groups such as these can have a tremendous effect on manufacturers, who know that good products will be publicly praised and bad ones denounced. The net effect is a trend toward responsible and farsighted production of merchandise, for the benefit of the entire society.

1.3 THE ROLE OF THE ENGINEER IN THE TECHNICAL COMMUNITY

A practicing engineer also has a specific responsibility to both himself and his colleagues in the scientific and engineering community. This responsibility lies at the heart of any profession, and engineering is no exception. The most obvious example of this professional role is the consulting engineer, perhaps self-employed, who deals directly with the public. Most engineers in this category are concerned with the public works type of project such as the design and construction of a highway bridge or sewage disposal system or the specification of a traffic control or police communications network. These engineers must be registered professionals in the states in which they practice. This is not unlike other professions such as law or medicine. An engineer may become a licensed professional engineer (P. E.) by passing the state examinations and by obtaining the proper practical experience.[3]

Another aspect of a profession is a code of ethics for the practitioner. A code of ethics for the engineer also exists and is expanded and communicated by the National Society of Professional Engineers, the Engineers Joint Council for Professional Development, and the other engineering societies. These organizations are also concerned with standards of professional excellence and enforce them by periodically examining and accrediting engineering schools. In recent years these organizations have been involved with the presentation of the position and image of engineering to the various sections of the federal government.

CANONS OF ETHICS FOR ENGINEERS AND RULES OF PROFESSIONAL CONDUCT

Foreword

Honesty, justice and courtesy form a moral philosophy which, associated with mutual interest among men, constitutes the foundation of ethics. The engineer should recognize such a standard, not in passive observance, but as a set of dynamic principles guiding his conduct and way of life. It is his duty to practice his profession according to these Canons of Ethics.

As the keystone of professional conduct is integrity, the engine will discharge his duties with fidelity to the public, his employers and clients, and with fairness and impartiality to all. It is his duty to interest himself in public welfare and to be ready to apply his special knowledge for the benefit of mankind. He should uphold the honor and dignity of his profession and avoid association with any enterprise of questionable character. In his dealings with fellow engineers he should be fair and tolerant.

Professional Life

CANON 1. *The engineer will co-operate in extending the effectiveness of the engineering profession by interchanging information and experience with other engineers and students and by contributing to the work of engineering societies, schools and the scientific and engineering press.*

Rule 1. He will be guided in all his relations by the highest standards.

Rule 2. He will not lend his name to any enterprise about which he is not thoroughly informed and in which he does not have a positive belief.

Rule 3. He should seek opportunities to be of constructive service in civic affairs and work for the advancement of the safety, health and well-being of his community.

Rule 4. He will not offer to pay, either directly or indirectly, any commission, political contribution, or a gift, or other consideration in order to secure work, exclusive of securing salaried positions through employment agencies.

CANON 2. *He will not advertise his work or merit in a self-laudatory manner and he will avoid all conduct or practice likely to discredit or do injury to the dignity and honor of his profession.*
Rule 5. Circumspect advertising may be properly employed by the engineer to announce his practice and availability. The form and manner of such advertising shall satisfy in all respects the dictate and intent of the Canons. Only those media shall be used as are necessary to reach directly an interested and potential client or employer, and such media shall in themselves be dignified, reputable and characteristically free of any factor or circumstance that would bring disrepute to the profession or to the professional using them. The substance of such advertising shall be limited to fact and shall contain no statement or offer intended to discredit or displace another engineer, either specifically or by implication.
Rule 6. Telephone listings shall be limited to name, address and telephone number under each branch listing in which he qualifies.
Rule 7. He will not allow himself to be listed for employment using exaggerated statements of his qualifications.

Relations with the Public

CANON 3. *The engineer will endeavor to extend public knowledge of engineering, and will discourage the spreading of untrue, unfair and exaggerated statements regarding engineering.*
Rule 8. He will avoid belittling the necessity for engineering services.

CANON 4. *He will have due regard for the safety of life and health of public and employees who may be affected by the work for which he is responsible.*
Rule 9. He will regard his duty to the public welfare as paramount.

CANON 5. *He will express an opinion only when it is founded on adequate knowledge and honest conviction while he is serving as a witness before a court, commission or other tribunal.*

CANON 6. *He will not issue ex parte statements, criticisms or arguments on matters connected with public policy which are inspired or paid for by private interests, unless he indicates on whose behalf he is making the statement.*
Rule 10. He will not advocate or support enactment of community laws, rules, or regulations that he believes are not in the public interest.

CANON 7. *He will refrain from expressing publicly an opinion on an engineering subject unless he is informed as to the facts relating thereto.*

Relations with Clients and Employers

CANON 8. *The engineer will act in professional matters for each client or employer as a faithful agent or trustee.*
Rule 11. He will not undertake or agree to perform any engineering service on a free basis.
Rule 12. He will be conservative and honest in all estimates, reports, statements and testimony.
Rule 13. He will advise his client when he believes a project will not be successful.

Rule 14. His plans or specifications will not be such as to limit free competition, except with his client's consent.

Rule 15. He will associate himself only with projects of a legitimate character.

Rule 16. He will not solicit or accept employment to the detriment of his regular work or interest.

Rule 17. An engineer in private practice may be employed by more than one party when the interests and time schedules of the several parties do not conflict.

Rule 18. While in the employ of others, he will not enter into promotional efforts or negotiations for work or make arrangements for other employment as a principal or practice in connection with a specific project for which he has gained particular and specialized knowledge without the consent of all interested parties.

CANON 9. *He will act with fairness and justice between his client or employer and the contractor when dealing with contracts.*

Rule 19. He will insist on contractor compliance with plans and specifications.

CANON 10. *He will make his status clear to his client or employer before undertaking an engagement if he may be called upon to decide on the use of inventions, apparatus or any other thing in which he may have a fianancial interest.*

Rule 20. Before undertaking work for others in connection with which he may make improvements, plans, designs, inventions or other records which may justify copyrights or patents, the engineer should enter into a positive agreement regarding the ownership.

Rule 21. When an engineer or manufacturer builds apparatus from designs supplied to him by a customer, the designs remain the property of the customer and should not be duplicated by the engineer or manufacturer for others without express permission.

Rule 22. A clear understanding should be reached before the beginning of the work regarding the respective rights of ownership when an engineer or manufacturer and a customer may jointly work out designs and plans or develop inventions.

Rule 23. Designs, data, records and notes made by an employee and referring exclusively to his employer's work are his employer's property.

Rule 24. A customer, in buying apparatus, does not acquire any right in its design but only the use of the apparatus purchased. A client does not acquire any right to the ideas developed and plans made by a consulting engineer, except for the specific case for which they were made.

CANON 11. *He will guard against conditions that are dangerous or threatening to life, limb or property on work for which he is responsible, or if he is not responsible, will promptly call such conditions to the attention of those who are responsible.*

Rule 25. He will not complete, sign, or seal plans and/or specifications that are not of a design safe to the public health and welfare. If the client or employer insists on such unprofessional conduct, he shall call building authorities' attention to the case and withdraw from further consulting business or service on the project.

CANON 12. *He will present clearly the consequences to be expected from deviations proposed if his engineering judgment is overruled by nontechnical authority in cases where he is responsible for the technical adequacy of engineering work.*

Rule 26. He will not apply his signature of approval or seal on plans that do not meet accepted engineering standards.

CANON 13. *He will engage, or advise his client or employer to engage, and he will cooperate with, other experts and specialists whenever the client's interests are best served, by such service.*

Rule 27. He will not undertake responsible engineering work for which he is not qualified by experience and training.

CANON 14. *He will disclose no information concerning the business affairs or technical processes of clients or employers without their consent.*

CANON 15. *He will not accept compensation, financial or otherwise, from more than one interested party for the same service, or for services pertaining to the same work, without the consent of all interested parties.*

CANON 16. *He will not accept commissions or allowances, directly or indirectly, from contractors or other parties dealing with his clients or employer in connection with work for which he is responsible.*

Rule 28. He will not accept financial or other considerations, including free engineering designs, from material or equipment suppliers for specifying their product.

CANON 17. *He will not be financially interested in the bids as or of a contractor on competitive work for which he is employed as an engineer unless he has the consent of his client or employer.*

Rule 29. He will not accept personal consideration in any form. This assures that his recommendations for the award of a contract cannot be influenced.

CANON 18. *He will promptly disclose to his client, or employer any interest in a business which may compete with or affect the business of his client or employer. He will not allow an interest in any business to affect his decision regarding engineering work for which he is employed, or which he may be called upon to perform.*

Relations with Engineers

CANON 19. *The engineer will endeavor to protect the engineering profession collectively and individually from misrepresentation and misunderstanding.*

Rule 30. The engineer will insist on the use of facts in reference to an engineering project or to an engineer in a group discussion, public forum or publication of articles.

CANON 20. *He will take care that credit for engineering work is given to those to whom credit is properly due.*

Rule 31. Whenever possible, he will name the person or persons who may be individually responsible for designs, inventions, writings or other accomplishments.

Rule 32. He will not accept by voice or silence, credit rightfully due another engineer.

Rule 33. He will not sign or seal plans or specifications prepared by someone other than himself or an employee under his supervision.

Rule 34. He will not represent as his own the plans, designs, or specifications supplied to him by a manufacturer or supplier of equipment or material.

CANON 21. *He will uphold the principle of appropriate and adequate compensation for those engaged in engineering work, including those in subordinate capacities, as being in the public interest and maintaining the standards of the profession.*

Rule 35. He will not undertake work at a fee or salary that will not permit professional performance, according to accepted standards of the profession.

Rule 36. He will not accept work in the geographic area in which he practices or intends to practice at a salary or fee below that recognized as a basic minimum in that area.

Rule 37. He will not accept remuneration from either an employee or employment agency for giving employment.

Rule 38. When hiring other engineers, he shall offer a salary according to the engineer's qualifications and the recognized standards in the particular geographical area.

CANON 22. *He will endeavor to provide opportunity for the professional development and advancement of engineers in his employ.*

Rule 39. He will encourage attendance at professional or technical society meetings by his engineer employees.

Rule 40. He should not unduly restrict the preparation and presentation of technical papers by his engineer employees.

Rule 41. He will encourage an employee's efforts to improve his education.

Rule 42. He will urge his engineer employees to become registered at the earliest possible date.

Rule 43. He will assign a professional engineer duties of a nature to utilize his full training and experience, insofar as possible, and delegate lesser functions to subprofessionals or to technicians.

Rule 44. He will not restrain an employee from obtaining a better position with another employer by offers of short-term gains or by belittling the employee's qualifications.

CANON 23. *He will not directly or indirectly injure the professional reputation, prospects or practice of another engineer. However, if he considers that an engineer is guilty of unethical, illegal or unfair practice, he will present the information to the proper authority for action.*

Rule 45. He will report unethical practices of another engineer with substantiating data to his professional or technical society, and be willing to appear as a witness.

CANON 24. *He will exercise due restraint in criticizing another engineer's work in public, recognizing the fact that the engineering societies and the engineering press provide the proper forum for technical discussions and criticism.*

Rule 46. He will not review the work of another engineer for the same client, except with the knowledge or consent of such engineer, or unless the connection of such engineer with the work has been terminated.

CANON 25. *He will not try to supplant another engineer in a particular employment after becoming aware that definite steps have been taken toward the other's employment.*

Rule 47. He will not attempt to inject his services into a project at the expense of another engineer who has been active in developing it.

CANON 26. *He will not compete with another engineer on the basis of charges for work by underbidding, through reducing his normal fees after having been informed of the charges named by the other.*

Rule 48. The practice of engineering is a learned profession, requiring of its members sound technical training, broad experience, personal ability, honesty and integrity. The selection of engineering services by an evaluation of these qualities should be the basis of comparison rather than competitive bids.

Rule 49. Competition between engineers for employment on the basis of professional fees or charges is considered unethical practice by all professional engineering groups. Hence, the announced intent of an owner or governmental body to request such competitive bids removes from consideration many engineers who may be the best qualified to be entrusted with the proposed work.

Rule 50. It shall be considered ethical for an engineer to solicit an engineering assignment, either verbally or written. Such solicitation may be in the form of a letter or a brochure setting forth factual information concerning the engineer's qualifications by training and experience and reference to past accomplishments and clients.

Should the engineer be asked for a proposal to perform engineering services for a specific project, he should set forth in detail the work he proposes to accomplish and an indication of the calendar days required for its accomplishment. The engineer's qualifications may be included if appropriate. A statement of monetary remuneration expected should be avoided if possible. Should such a statement be deemed necessary, the proposed fee shall be equal to or more than the fees recommended as minimum for the particular type of service required, as established by fee schedules or practice in the geographical area where the work is to be done. Where a fee cannot be established in this manner, the ethical fee shall be equal to actual cost plus overhead plus a reasonable profit.

Rule 51. He will take a professional attitude in negotiations for his services and shall avoid all practices which have a tendency to affect adversely the amount, quality or disinterested nature of professional services; such as charging inadequate fees for preliminary work or full services, competing for an engineering assignment on a price basis, spending large amounts of money in securing business or consenting to furnish monetary guaranties of cost estimates.

CANON 27. *He will not use the advantages of a salaried position to compete unfairly with another engineer.*

Rule 52. While in a salaried position, he will accept part-time engineering work only at a salary or fee not less than that recognized as standard in the area.

Rule 53. An engineer will not use equipment, supplies, laboratory or office facilities of his employer to carry on outside private practice without consent.

CANON 28. *He will not become associated in responsibility for work with engineers who do not conform to ethical practices.*

Rule 54. He will conform with registration laws in his practice of engineering.

Rule 55. He will not use association with a non-engineer, a corporation, or partnership, as a "cloak" for unethical acts; but must accept personal responsibility for his professional acts.

Miscellaneous

Rule 56. An engineer who is in sales or industrial employ is entitled to make engineering comparisons of the products offered by various suppliers, but will avoid aspersions upon their character, standing or ability.

Rule 57. If in sales employ, he will not offer, or give engineering consultation, or designs, or advice other than specifically applying to the operation of the equipment being sold.

Rule 58. No engineer in the employ of equipment or material supply companies will tender designs, plans, specifications, advice or consultation to operations beyond the limits of a machine or item of material or supply, except as is required for proper functioning of the particular item.

Rule 59. He will not use his professional affiliations to secure the confidence of other engineers in commercial enterprise and will avoid any act tending to promote his own interest at the expense of the dignity and standing of the profession.

Rule 60. He will admit and accept his own errors when proven obviously wrong and refrain from distorting or altering the facts in an attempt to justify his decision.

Rule 61. Any firm offering engineering services must, in conformance with the laws of the state in which it operates, have its operations under the direction and responsibility of registered professional engineers.

Rule 62. He will not attempt to attract an engineer from another employer by methods such as offering unjustified salaries or benefits.

These *Canons of Ethics* for Engineers were developed and promulgated by the Engineers' Council for Professional Development and have been adopted by the Board of Directors, National Society of Professional Engineers, October 28, 1946. The *Rules of Professional Conduct* were developed by the Ethical Practices Committees of NSPE as a means of expanding and furthering understanding of the Canons of Ethics.

As mentioned above, one of the interesting aspects of the role of the engineer and his relation to the profession is the communication of new technical ideas to his colleagues. The media for these communications take several forms, such as technical conferences, professional journals, and seminars. It is necessary that an engineer be aware of the latest developments and techniques for proper performance of his job, and these media become an important part of an engineer's professional life.

Similarly, it is part of an engineer's responsibility to participate in the communication of his ideas to his profession by participating in the affairs of his professional societies. This participation may be manifested by giving papers, taking part in panel discussions, writing papers and articles, and working in the local or national administration of the society. Most employers look favorably upon and encourage the activity of engineers in professional societies. A list (by no means complete) of some of the professional societies that function in the engineering profession follows[4]:

American Institute of Civil Engineers
American Institute of Chemical Engineers
American Institute of Industrial Engineers
American Society of Mechanical Engineers
Institute of Electrical and Electronics Engineers
Association for Computing Machinery

A set of Greek-letter honor societies also functions in engineering both to recognize achievement in the profession and to foster the profession's interests. A list of some of these societies and the interests which they serve follows:

ΤΒΠ: Tau Beta Pi, Engineering Honor Society
HKN: Eta Kappa Nu, Electrical Engineering
ΑΠΜ: Alpha Pi Mu, Industrial Engineering
ΠΤΣ: Pi Tau Sigma, Mechanical Engineering
ΧΕ: Chi Epsilon, Civil Engineering
ΣΞ: Sigma Xi, Research Society

Most of the organizations listed above also have student chapters that are active on many campuses, and student participation in these professional groups can be an interesting and rewarding part of college life.

1.4 ENGINEERING AND THE SCIENCES

It is apparent from an examination of an engineering curriculum that engineering is firmly based in the sciences, especially physics, chemistry, and mathematics. All areas of engineering require this background before the specific disciplines can be studied. What, then, is the relationship between the sciences and engineering at the advanced levels? For example, what is the difference between the work of the physicist and that of the engineer?

This difference can best be described as follows: The scientist is primarily concerned with the study and understanding of basic principles and relationships, whereas the engineer is mainly concerned with the application of these principles to the solution of problems. Thus, the difference is *applied* science versus *basic* science. However, the difference between these two areas is not always easily discernable, and many engineers do indeed get involved with work in the basic sciences. Similarly, many people trained in the basic sciences find themselves working in areas of applications.

In the past several years there has been an increased interest in applications of technology toward specific problems of society, and engineers are now solving these problems. This has led the engineer into areas of science heretofore alien to him. For example, the need for better medical services and increased hospital efficiency has led engineers into biomedical engineering. Here, the engineer needs a background in biology, zoology, and physiology as well as in physics and chemistry. These sciences are referred to as the "hard" sciences, since they are based on direct experimental investigation, theoretical verification, and mathematical description. What, then, are the "soft" sciences, and do they relate to engineering? First, the soft sciences, such as sociology, economics, and psychology, refer to those areas that are based on observations of the behavior of a system so complex that only indirect experimental investigations can be used. Further, theoretical verification is generally difficult if not impossible to obtain, and mathematical description is generally limited to statistical relationships. The soft sciences are playing an increasing role in the training and practice of engineering. The problems that engineers must solve today must be examined in a different framework. The implications of technological developments must be examined with respect to their long-term impact on society and the environment. Also, engineers are getting involved in problems of the environment, transportation, law enforcement, and the cities. An understanding of the results of the work in the soft sciences is needed before the engineer can effectively cope with these problems. Finally, the engineer must be versatile and willing to work with "soft" scientists in interdisciplinary teams in order to solve these problems.

1.5 SPECTRUM OF JOBS FOR ENGINEERS

Engineers can move from school into a wide variety of endeavors, some thoroughly immersed in the engineering profession and others seemingly far removed. The background an engineering student receives is excellent preparation for study and work in many fields, because the intellectual discipline he has acquired carries over faithfully to almost any job. In this section we shall discuss some of the areas an engineer might wish to enter and attempt to give a feeling for what he can expect.

First, what are the distinctions among an engineer, a technician, and a draftsman? Many people confuse these roles, mainly because, in the recent past, an engineer was something of all three. Today, however, the training required and the outlook of the three positions are quite different. An engineer's main job is in problem solving and design. To do this well requires a broad background in mathematics and the physical sciences as well as up-to-date knowledge of available materials and devices. An engineer will take an idea or project goal, analyze it, and come up with a preliminary design for the required prototype device. A technician will then work closely with the engineer in the laboratory to *implement* the design in terms of hardware. Good technicians are an essential part of a project team, relied upon by everyone to work out the myriad details of constructing a given device. Experienced technicians have a wealth of practical knowledge that is invaluable in avoiding pitfalls and in translating a sometimes rough design into a working piece of equipment.

Once the prototype has been thoroughly developed and tested, the agreed upon design must be recorded in a clear and unambiguous language. This is where the draftsman is so valuable. The engineer's sketches of the device layout are given to the draftsman, who produces accurate and detailed drawings of all parts of the device. These drawings are the permanent record of the project, in a form that can be interpreted accurately during the manufacturing phase of the project.

(a) Industry

There are many types of industries active today. Among the most prominent are the electronics, automotive, steel, and petroleum industries. All industries require a great deal of engineering to keep new ideas and developments coming and to refine and improve manufacturing techniques. We can subdivide the roles of engineers in industry roughly as follows:

1. *Research and Development.* As discussed in Section 1.2, these are the primary tasks of an engineer: to explore new avenues, invent new solutions to problems, and design new devices. In the research phase of a project, the engineering team usually has found a new way of doing a job and is analyzing it (using mathematics and computers) to see how feasible the idea is and how well it will work ideally. The development stage then follows. Here the idea is implemented, usually as a prototype model, in the laboratory. These processes vary markedly among different projects, but the basic thrust is the same: Turn the idea into a working reality. The development stage lasts as long as it needs to, until the working device (along with improvements discovered en route) has been constructed and tested. Then the manufacturing phase begins, during which it may be necessary to alter some plans in the light of practical manufacturing constraints.

2. *Production Engineering.* Following the development of a new device or product, it must be manufactured, usually in large quantities. There is plenty of room for creativity in this area, for the cost of manufacturing the device will strongly influence its final price and hence its marketability. Any manufacturing technique that lowers the cost even a few cents per item can be of great value to the company. In the last few decades a whole field of automated manufacturing techniques has been developed, requiring new engineering skills to invent and improve machines that automatically construct other machines efficiently and reliably.

3. *Testing.* During production and at the end of the assembly line, a product must be carefully tested to determine if it will perform its job properly and reliably. Frequently this testing procedure must be done automatically as well, to avoid a production bottleneck, and engineers whose specialty is testing are now being trained. They develop procedures and machines to carry out what can be a very intricate sequence of tests. For example, consider the enormous problem of testing a large computer to see if it will perform *all* its tasks correctly.

In a large-scale production operation, be it many identical small items such as radios of cars or a single item such as an oil refinery, there are so many problems or scheduling and planning that a brand of engineer, the *industrial engineer*, has grown up to tackle them. These engineers study such topics as logistics, work simplification, optimum routing of manpower and materials, and computer-oriented scheduling schemes. In an industrial environment their basic job is to plan the sequence of steps necessary for the successful completion of a task at minimum cost. The complexities of such work are tremendous, often requiring computer analysis to determine which steps are potential bottlenecks in the process.

4. *Marketing and Sales:* Except in special circumstances a company must sell large quantities of its products in order to realize desired profits. (The exceptions are in contracting companies that are paid, say by the government, to develop and produce a limited number of special items.) Consequently a high proportion of company resources are used in analyzing the potential market for a product and in promoting its sale. In products of a technical nature, such as computers, laboratory equipment, and some types of machinery, an engineer can be a very valuable salesman, for he can communicate on a sophisticated level with his potential customers. He has the technical knowledge needed to explain his product to a technical audience and can demonstrate the product's value in the customer's terms. Many engineers who like business and enjoy talking with people move into sales and make it their specialty.

There is also a need for the engineer in the marketing end of a company. He can be used as a consultant in many aspects of marketing: to estimate the future need for a device, to size up potential competition, and to predict profitable applications for a new idea.

(b) Engineering in Government

Engineers respond to government needs just as they do to society's needs. The federal government recognizes that it needs high-quality engineering expertise just as much as does private industry. Consequently it supports a large amount of engineering work in many different forms. The largest effort is usually in engineering for military applications. The Department of Defense and the branches of the armed forces support hundreds of projects. The usual procedure is to contract with an independent company in industry to do the engineering, so that although the engineers are ultimately supported by the government, they are hired by a private company. However, many engineers also work directly within the armed forces, such as in the Signal Corps or the Naval Research Laboratory. Many of these men oversee the projects under contract, writing specifications for them and carefully examining their progress and quality.

There are other engineers in the government whose principal job is to judge the technical needs of the nation and to make recommendations (sometimes to the President or to Congress) as to where the national resources ought to be focused in future years. The National Science Foundation

(NSF) reviews proposals for scientific research projects and determines which projects it will support. The criteria for choosing deserving projects are both the excellence of the researchers and the present technical needs of the nation. Engineers in government also develop plans for responding to national emergencies such as floods and droughts. Many of these problems are technical as well as social and political, so the engineer's technical talent is very important in the decision-making process.

There are also regulatory groups within government that oversee the operation of large enterprises and enforce fair practice laws. Two of the more technical of these are the Federal Communications Commission (F.C.C.) and the Federal Aeronautics administration (F.A.A.). The F.C.C. oversees the use of radio and television signals (see Chapter 9) so that broadcasting stations, the police, private users, etc., can communicate by radio without significantly interfering with one another. As more and more people want to communicate and as new devices and techniques come along, there are enormous problems of fair allocation of signal frequencies that must be solved by the F.C.C. The job of the F.A.A. is somewhat similar, only here the ultimate goal is safety of air passengers and materials. The F.A.A. regulates the use of air routes and sets policies at airports for the ever-increasing congestion of air traffic. In addition, it sets standards for the safety of aircraft and for pilot qualifications. Here again there are many highly technical problems to be solved (to design better automatic landing systems, for instance), and engineers are urgently needed in this area.

In the areas of pollution control and conservation the government is now very active, and engineers have much to contribute. Better techniques for sewerage treatment, for heat removal from power plants, and for reducing automotive pollution are basically technical problems to which engineers must respond and for which governments will most likely have to provide funds.

There are several primarily scientific organizations within the government that employ many engineers. One is the National Bureau of Standards (NBS), which performs research on measurement techniques and maintains legally accepted standards of weights, lengths, and volumes. These standards are metal rods or blocks that have been made with the greatest precision (at tremendous cost) to serve as references against which any other items can be compared. In some intricate legal or business cases it is essential to have a reference that everyone agrees is exactly 1 m long or weighs exactly 1 lb. Another highly technical organization is the National Aeronautics and Space Administration (NASA), which organized and oversaw the impressive space program of the 1960s; today it continues on a somewhat reduced scale to develop techniques for manned space platforms, planetary surveyor satellites, and space shuttles. Projects such as these are extremely costly, requiring gigantic resources of manpower and money; they are possible only when a large nation deems them sufficiently important investments.

There are many other places in government where engineers are employed. The CIA and FBI are highly technical organizations, requiring much scientific planning and development. The Internal Revenue Service is another area where a great deal of data processing is necessary, requiring computer engineers to develop systems and programs to perfect the processing routines.

(c) Engineering in Law

There are two ways in which engineers may become involved with law. The first is the case of the engineer who goes on to earn a law degree, thus becoming a lawyer-engineer, and the second is the situation where an engineer spends part of his time advising lawyers in technical matters. The most common form of the engineer-lawyer is the patent attorney, who reviews patent applications, searches through the existing patents to find any overlap with existing patent ideas, and follows through on many extremely intricate legal problems that arise in patent disputes.[5,6] He must be well trained both legally and scientifically, so that he can understand the purpose, technical details, and ramifications of the invention to be patented.

The other situation, where an engineer does not become a lawyer but becomes involved with law, is by far the more common. Every engineer must be familiar with the basic aspects of patent law and laws in general, because these laws will affect the fruits of his work. Whenever an engineer invents a new technique or device, questions arise: Is this patentable, has it already been patented, should I publish the idea or will this jeopardize its value? Most often an engineer works for a company, and the company will follow through the details of patent searching and application. Frequently an engineer signs an agreement with his company to assign all patent rights to the company, recognizing that the company supported the work that led to the idea. If the engineer works alone, however, he must pursue on his own the rather tortuous route to acquire a patent.

Engineers also act as consultants in legal disputes that require technical expertise. Just as doctors clarify medical points in a trial, so do engineers clarify technical points. In this case the engineer probably is a licensed professional.

Whatever the situation, today's engineer must have at least a fundamental knowledge of law, especially business law, as he will frequently come up against certain legal issues and should know how to obtain and interpret the required information.

(d) Engineers in Management

Engineers working in industry often rise in the company to positions of responsibility where they must direct groups of people. It is fairly common for a good engineer to become a project manager, leading a group of perhaps a dozen engineers and technicians toward a project goal. (In fact, it has been found that 5 years after graduation 50% of all engineers are in some form of management.) He must be very familiar with all the technical details of the project; he guides the direction of the work, making difficult technical decisions along the way. In addition, he must have a broad overview of the manner in which the project fits into the overall company program, for he must report his progress to his management and be prepared to defend his requests for more funds or equipment. He is also held primarily responsible if the project fails and must be able to explain what happened. Project supervisors who successfully lead a number of projects are usually promoted into a higher level of management, perhaps ultimately to the role of company director.

There is another route for an engineer to take after graduation. If he desires to be primarily a manager in his career, he can go to business school, and earn a master's degree in business administration (MBA). With his engineering background and this advanced degree he will be eagerly sought after by technical companies and will quickly be brought into the management part of the business. This higher level of management, whether reached directly with an MBA or through the project-leader route mentioned above, is usually somewhat removed from the technical details of the various company projects. A manager oversees the developing projects from the point of view of the whole company, its direction and growth. He must see the whole picture, technical as well as financial. Sometimes a strong and promising project in one area must be canceled to make room for another project, and management must make the difficult decision when necessary.

An engineering background is a great asset to the man who wants to go into management, for the disciplines developed while studying and practicing engineering are directly transferable to management skills. Many engineers choose to avoid the business side of their company's efforts, wishing to concentrate wholly on the technical problems. But for those who think they might enjoy management, an engineering degree is an excellent first step.

(e) Engineers in the Medical, Biological, and Social Sciences

Some engineers become doctors, entering medical school immediately after earning their bachelor's degrees. An engineering background is very valuable to a doctor, since so much of medicine today is involved with highly technical equipment and procedures.

We want to stress here, however, the role of the engineer who remains primarily an engineer but who finds work in the medical, biological, or social sciences. Hospitals have recently been hiring an increasing number of engineers as part of their staff or as consultants. The engineer is no longer considered a glorified technician who will just fix the doctor's equipment. He becomes involved in large-scale programs, developing new instruments and techniques in operating rooms, laboratory analysis, and patient monitoring. Large computer facilities are becoming more common in hospitals, and they require skilled engineers to develop programs and the attendant sensing instruments. For instance, patient-monitoring systems are being developed that keep track of the heartbeat, respiration, and other critical functions of hundreds of patients simultaneously. This information is fed into a computer, which monitors the data and instantly alerts the staff when dangerous irregularities are detected. Systems like this are very sophisticated, requiring the cooperation of medical personnel and engineers.

In the broader area of the biological and social sciences, engineers are finding their skills highly sought after, especially in the areas of *modeling* and *simulation*. Engineers are learning how to construct mathematical models of very large and complex systems, such as the workings of a city or the development of large animal populations. A simulation program is then developed that takes the important variables involved in the study and predicts how they will interact as a function of time. These efforts permit scientists to forecast the effects of various plans. For instance, a governor may want to know the effects of putting a highway through the middle of a city, or a conservationist may want to know the effects of an oil pipeline placed across the migration routes of caribou. Without large-scale simulation programs these real-world "systems" are much too complicated to attempt any detailed predictions.

(f) Self-Employed Engineers

There is also the free-lance consultant (frequently a civil or industrial engineer), who does not work for any one company on a permanent basis but who is paid to act as a source of ideas and information on specific projects. Often he must make important recommendations on materials to use in bridge construction or suggest a better way to process a difficult waste material. He is hired to do specific jobs just as a building contractor or architect is. Police departments sometimes hire electrical engineers to help solve difficult crime-detection problems or to design surveillance devices; chemical engineers are hired to find new processing techniques; and so forth.

The important factor here is that the independent consulting engineer must continually sell himself, for he must convince each potential customer that he is the best man for the job. He must demonstrate his ability and availability. This puts additional pressure on the engineer, but for those who have a salesman's flair and a great deal of energy, it can be a very rewarding route to take.

(g) Engineers in Education

Some engineers have a compelling desire to teach, and they go into high schools, colleges, and technical schools as full-time teachers. If an engineer chooses to go into high school teaching he will most likely not be doing actual engineering but rather will be using his background to teach more basic courses in mathematics, chemistry, and physics. Beyond the bachelor's degree in engineering, most high schools require a certificate in education, which is commonly a one-year graduate program.

For teaching in a technical school, it is usually necessary to have had some experience as a professional engineer. Such experience, which is valuable in any teaching position, is essential here because the technicians being trained must acquire the very practical outlook they will need in their jobs. Some teachers in technical schools also consult part time, and some do engineering research, although most of their daily efforts are in teaching.

At the university level an engineering educator must almost without exception have a Ph.D. degree today. Special training in teaching is not required, but the professor's background must be both broad and deep, and this usually requires extra years of graduate school, doing research, and taking additional courses. The job of a professor is normally multifaceted. Besides pursuing vigorous teaching activity, he is expected to do independent research in his specialty and to publish new results. Universities want their faculty to become well known in their fields, as this attracts good students, which in turn attracts better faculty, etc., upgrading the whole university system. Faculty members also must seek financial support for their projects, usually from the government (e.g., through NSF or NASA). This research support also helps to finance graduate students. A professor is also expected to do a share of public service, either within the university serving on committees and giving talks or out in the community helping to inform the public of the role of the university.

1.6 DESIGN

If there were one word to describe engineering, it would be design. A great deal of the work of engineers revolves around design and the design process. Thus, it is in order here to explore the design process. We shall first examine the general nature of this process and then describe two case studies.

Why do we design something? Generally, a machine, gadget, building, system, bridge, or what have you is designed in response to some real or assumed need. The initial step in the design process is the *recognition of a need*. In some cases the recognition is a simple process of responding to the competition or fulfilling a requirement imposed by a governmental agency. For example, the competition for the small foreign automobile market encouraged the domestic manufacturers to design and market small cars of their own (Colt, Pinto, Vega, Gremlin, etc.). On the other hand, the automotive industry has been forced to design new safety and pollution-control equipment by government regulations in order to remain in business.

The recognition of a need may also originate from other sources. Often, the natural evolution to products within a company's product line leads directly to a new version of an old product or even an entirely new product. In other cases a company may respond to some cited societal need or fear. For example, the development of nonpolluting washing products and of environment-monitoring equipment was in response to societal need. The flood of new auto and home burglar alarms that are being marketed is a response to the increase in crime and peoples' natural fear of being robbed. These examples are somewhat alien to the rather glamorous image of the clever lone inventor cloistered in his basement and saying things like "wouldn't it be nice if..." or "Eureka!" Don't despair, these people still do exist and often come up with ideas that satisfy some real or anticipated need. The specific problems associated with the basement inventor/designer are related to his general lack of resources or his inability to carry out the complete design process. A great idea that cannot be manufactured at a marketable price is not such a great idea after all. The evolution of a product from conception to market is part of the design process. Let us now examine the design process in detail.

The development and design of a new product is an iterative process. That is, one goes through the same sequence of steps several times before the process is complete. The particular steps outlined below are representative parts of the design process; many variations are possible.

As stated above, the first step in the design process generally requires recognition of a need. In addition to the methods mentioned above, the marketing department or administrative units of a company often identify a need and refer it to the engineering department for the next step. At this point, however, a list of specifications may have been generated that lays out the basic nature of the object under consideration.

More often than not the specification reads "Can we build something that does *this* and sells for no more than X dollars?" The object of the second step then is to establish a relationship between the need and the feasibility of the idea, which requires a detailed analysis of the problem and the formulation of alternative solutions. These alternative solutions often take the form of competing rough designs. For example, one alternative for the drive system of a vehicle is electric motors and batteries, whereas another alternative is a conventional transmission and an internal combustion engine. In another case, one alternative implementation of an electronic system makes use of transistors, while another uses integrated circuits. The analysis of these competing designs often takes many forms. In some cases simple models that demonstrate the practicality of the different approaches are constructed; in other situations, sophisticated computer simulations of complex systems are developed and run in order to gain insight into the opposing alternatives.

The third major step in the design process is to select one or more of the alternative designs for further development. At this point detailed designs are worked out and prototypes are constructed. Here the first iterations begin as the prototype is tested, evaluated, and redesigned in the context of the original design philosophy. This step in the design process lays the foundation for the future of a product, and any weakness in the design here can doom a product. Thus, good engineering at this stage is mandatory if a product's chances for success are to be maximized. The result of this step is a decision as to whether or not the design approach taken is technically feasible. If it is not feasible, another decision must be made: Is it worth trying another iteration, or should this product be abandoned? This decision requires the application of "engineering judgment," which is one of the abilities that an engineer gains with on-the-job training and experience. If the decision is made to continue, another design is tried. The flow of these processes and decisions is illustrated in Figure 1.1.

On the other hand, if the design is technically feasible, then the fourth major design step and a change in the complexion of the design process take place. The manufacturability, cost of manufacturing, potential selling price, and predicted market size must be studied in detail. In many cases

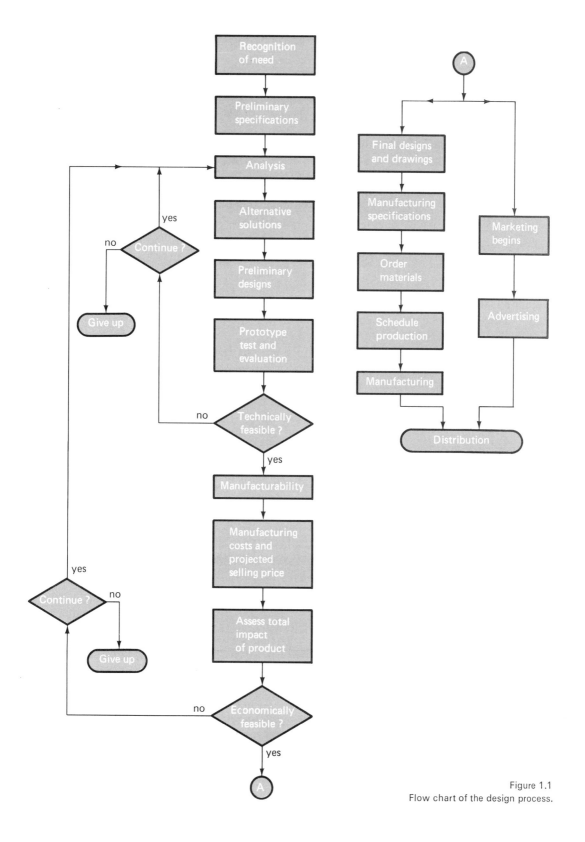

Figure 1.1
Flow chart of the design process.

the manufacturability and cost of manufacture of a product are paramount to the design engineer. For example, to manufacture a product it is often necessary to design or purchase expensive or complex processing equipment, and these costs are necessarily reflected in the price of the product. This in turn must be weighed against the potential market size and projected selling price. It is imperative that the product be salable and fit into the market projected by the marketing surveys. Another factor creeps into the process at this time. There is considerable pressure on the designer to reduce costs, and he is often urged to reduce some specification or tolerance in an effort to reduce costs. It is important that the engineer maintain his integrity and not sacrifice the quality or safety of the product to the great god Cost. It is also the engineer's responsibility to attempt to assess the total impact of the product on the environment and its potential for harm in light of the usefulness of the product. For example, questions such as "Is it biodegradeable?," "Can it be easily maintained and repaired?," "Are there special precautions that must be taken in using the product?," and "Are there any specific dangers to children who come in contact with the product?" must all be reconciled at this time.

If this step shows promise, marketing of the product begins in earnest. It is necessary to determine if the market exists or if it must be developed. What are the best ways to reach this market? Is direct mail an alternative, or is a network of distributors needed? What packaging scheme will advantageously display the product? What kind of advertising is needed to reach the projected market? The engineer's role in the marketing effort depends on the particular company that he is working for. In any event, the engineer must be capable of communicating the technical and salable features of the product to those who have the responsibility of selling it.

On the other hand, it is possible that the results from the fourth step are not satisfactory. Then a decision must be made whether to continue or to abandon the product. If the decision is made to continue, the specifications must be adjusted, based on the knowledge gained so far.

Let us now assume that the fourth stage is successful and that the product is to be marketed. Final drawings must be prepared, and manufacturing specifications must be drawn up. Facilities must be scheduled, and parts and raw materials must be ordered. It is crucial that production schedules be met, because once a production line is committed, a delay can cost thousands of dollars. Thus, adequate lead times (the time between placing an order and actual delivery) with a margin for error must be considered. The final step in the design process is the manufacturing and distribution of the product.

Some observations of the design process are now in order. The amount of engineering design and creativity in this process may seem a bit sparse. Most of this process is involved with assessment of technical, manufacturing, and economic feasibility. This observation led Thomas Edison to say that "invention is one percent inspiration and 99 percent perspiration." Another factor not specifically programmed into the design process described here is the role of patents in the design process. Patents often provide a helpful background in the initial formulation of alternatives. Also, new ideas developed at any stage of the design process may be worth patenting even if the product is not developed at the time. Patents that are not used by a company may have value in the future, or they may be sold to another company. It is also important that a company assure itself that it is not infringing on someone else's patent, thereby exposing itself to a potential lawsuit.

The development of two different products is described below. The first was developed and patented by lone inventors working in a basement laboratory and finally manufactured and marketed by a small company. The other was developed, manufactured, and marketed by a large, well-established company.

1.6.1 Design Study 1: Guitar Tuner

The design and development of a device that tunes each string of a guitar or bass guitar is the subject of this case study. The inventors of this device did not work for a large company but rather can be classified as lone inventors.

The recognition of the need was simple and probably representative of that for many other products. In this case the inventors were part of a musical group with four guitars and a bass guitar, which meant that 28 strings had to be tuned before the group could play. Unless an individual has particularly sharp musical acuity, this tuning process can be difficult and time consuming. At this point the two inventors said, "Wouldn't it be nice if we could tune each string perfectly and easily by means of some device?" In the next breath they said, "And it's got to be simple and inexpensive!" Thus, the rough specifications were developed.

The next step was to examine the available technology and define the type of indicating device that would allow the unit to be easy to use. The first alternative was to use a flashing light that responded to the difference between the frequency of the string and the frequency of a reference signal. A second alternative was to drive the light at the frequency of the string; this would illuminate a disc (with alternate black and white marks on it) spinning at a constant rate past the light, thus creating a stroboscopic effect. A third alternative was to use a meter, where the center of the meter indicates the correct frequency. A meter reading below center indicates that the string frequency is low and must be tightened, while one above center indicates that the string frequency is high and must be loosened. Further, there were two alternatives for realizing this meter readout. One uses built-in standard reference frequencies and the other uses accurately precalibrated digital timing circuits.

The analyses and evaluation of the three alternatives was carried out by first sketching rough designs and then building "breadboards" (rough physical embodiments), as shown in Figure 1.2. Since the alternatives differed mainly in the method of readout, these aspects of the system were studied first. The first two alternatives were quickly eliminated because they proved difficult to read and could cause eyestrain. Also, the first approach gave no

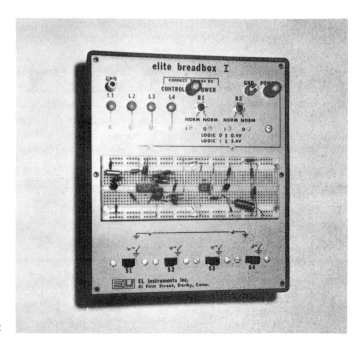

Figure 1.2

indication whether the string was higher or lower than the reference frequency.

A simple electronic circuit that could be used to drive the meter was designed first, and then the amplifiers and filters needed to process the signals from an electric or an acoustic guitar were designed. These designs were then put into the crude package shown in Figure 1.3. The device could then

Figure 1.3

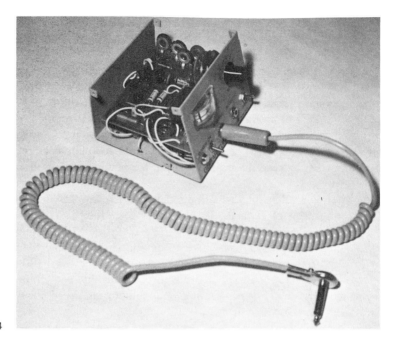

Figure 1.4

be tested with a variety of instruments and modified accordingly. This phase of the design was very successful, and the inventors decided at this point that an investigation of the patentability of this design was worthwhile. Thus, a patent attorney was secured, a patent search was carried out, and on March 4, 1970, an application was filed with the U.S. Patent Office. The filing of a patent application afforded the inventors some protection for their ideas. However, several other versions of the tuner were designed that incorporated new solid-state technologies; these were tested with respect to thermal and mechanical stability. Once these designs were proved, several companies that the inventors thought might be interested in manufacturing this device were contacted. Care was taken not to disclose too many details of the invention before the patent was issued so as to ensure the integrity of the patent.

On January 25, 1972, Patent 3,638,113 was issued on this invention and the rights to manufacture and market the product were licensed to EL Instruments on February 2, 1973. At this time production engineering began and the final package shown in Figure 1.4 resulted.

It is important to note that the rights to an idea or patent are not easy to sell if the inventor has not reduced the idea to the point where production engineering, manufacturing, and marketing are the next step. Many companies cannot or will not invest money in only an idea but prefer to invest in a well-developed product. That is not to say that an undeveloped idea or patent cannot be sold; it is just more difficult.

1.6.2 Design Study 2: Automatic Pleating Machine*

The motivation for designing a piece of equipment does not always come from a simple obvious problem. Often the problem is recognized indirectly, as shown in this example.

As is true with many consumer goods industries, the textile or needle-trade industry works with a very low profit margin and must compete with foreign manufacturers as well. Here, labor costs constitute 50% or more of the total cost of producing sewn goods. A significant problem is the high turnover rate of a relatively low-skilled labor force—it is not unusual to hire 100 workers per year in order to maintain a 100-man work force. Thus, it is important to keep the training period for new workers as short as possible.

The motivation for this project developed in the following sequence. The management of company A was eager to reduce costs to maintain a competitive edge in the market place. The plant industrial engineers were called upon to identify areas where automation would make the most significant contribution to lowered costs. Their studies revealed that the pleating and sewing operations in the manufacture of curtains and draperies were the single most costly operation in the manufacturing process and that any help here would be significant.

To appreciate the problem, consider the process of pleating a drapery. The top of each drape must be pleated and sewn properly to make it look full when hung. The pleats must be parallel, and each set of pleats must be properly spaced from its neighbors. The process used at the time was as follows: The drapes were creased by a machine with hot "fingers" and pressure to mark each pleat. This machine is illustrated in Figure 1.5(a). Next, the drapes were sent to a sewer who felt each crease, formed the pleats by hand, and sewed them in a heavy-duty sewing machine, as illustrated in Figure 1.5(b). Although the training time for the first operation was relatively short, the training time for the pleating and sewing operation was significant. It took 16 weeks before an operator was producing enough to cover his salary, and, in an industry where the average employee stays only 52 weeks, this is costly.

At this time the industrial and mechanical engineers of the company started tackling the problem. Initially an attempt was made to carry out both operations with only one operator. Thus, a method for forming a pleat was devised, using strips of bent sheet metal fingers and hinges much like the ones in the final product shown in Figure 1.6. The material is placed in the machine by the operator, and the top fingers come down and mesh into the bottom fingers. The fingers are then pressed together, thereby forming the pleat, and the operator removes the finished pleat from the machine ready for sewing. Several versions of the pleater were built and tested, some for over a year, until the final unit evolved. This pleater has several simple adjustments that are used to accommodate different weight materials and the various pleat sizes of several different products.

---

*This study was developed with the aid and cooperation of Louis Hand, Inc. (a Division of Aberdeen Mfg. Corp.) of Fall River, Mass.

Figure 1.5 (a)   (b)

Figure 1.6

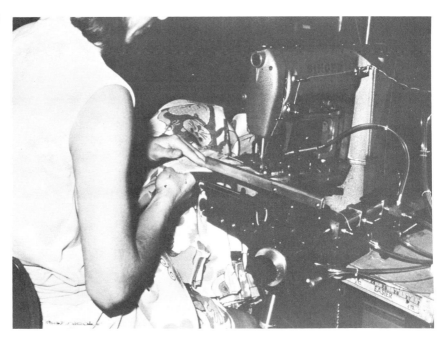

The first versions of the machine used foot-pedal-operated holding devices that held the pleat in the semiautomatic sewing machine while the final pleat was stitched. Also, these early designs required that the operator measure the distance between pleats on a scale. However, the machine shown in Figure 1.6 has an automatic holding bar, an automatic measuring mechanism that sets the distance between pleats, and sensors that set the machine to sew the beginning and ends of the drapery.

At the end of this development phase it was decided that patent protection should be secured in order to maintain an edge over the competition. This edge was significant—the automatic pleating machine eliminated one operation and one operator, decreased training time from 16 to 8 weeks, and increased production. Company A was subsequently awarded Patent 3,331,345 on this machine. Before the patent was issued, it was necessary to find a manufacturer who was capable of producing these machines in quantity. An agreement was ultimately reached with company B which would make machines both for company A and for other drapery companies. The other companies would, of course, pay royalties to company A. In this way the cost to company A of design and development would be defrayed still further.

Engineers are called upon in many capacities to solve problems in manufacturing and production. Here, industrial engineers helped define the problem and develop specifications, while mechanical engineers designed and developed the final product. More sophisticated equipment design often requires the use of electrical, chemical, and computer engineers as well.

1.7 CONCLUSION

The thrust of this chapter has been to describe the role of the engineer in the modern world and the breadth of opportunities available to him. Engineers are needed in many different areas of our society and are continually moving into new areas as the need for them develops. The training an engineer receives serves as an excellent foundation for whatever career he ultimately chooses, be it as an engineer per se or in some other technical field. In years to come the image of an engineer and the roles he fills will surely change, just as they have dramatically changed in the last 30 years. This is one important sign of a healthy and responsive profession.

REFERENCES

1. *Consumer Report*. Mt. Vernon, N.Y.: Consumers Union of United States, Inc.
2. NADER, RALPH, *Unsafe at Any Speed*. New York: Grossman Publishers, Inc., 1965.
3. CONSTANCE, JOHN D., *How to Become a Professional Engineer*. New York: McGraw-Hill Book Company, 1958.

4. KEMPER, J. D., *The Engineer and His Profession*. New York: Holt, Rinehart and Winston, Inc., 1967.
5. HOLLOMAN, J. H., "The U. S. Patent System," *Scientific American*, June 1967, 19–27.
6. ARNOLD, T., and F. S. VADEN II, *Invention Protection for Practicing Engineers*. New York: Barnes & Noble, Inc., 1971.

1. BEAKLEY, G. C., and H. W. LEACH, *Careers in Engineering and Technology*. New York: The Macmillan Company, 1969.
2. WALLS, H., *Copyright Handbook*. New York: Watson-Guptill Publications, Inc., 1963.

GENERAL BACKGROUND READINGS

2

man, the bridge builder

M. M. Miller

2.1 INTRODUCTION

Bridges, like everything else made and used by man, have a history. The history of bridges is inseparable from the history of other activities of man. There has been a complex interaction among all the activities of man as he evolved through the ages. This interaction and evolution are probably as important to his activity as a bridge builder as to any other activity. The history of man and his bridges has its beginnings in the prehistoric era. As a result we can only make educated guesses as to how man the hunter and food gatherer discovered, used, and eventually imitated those bridges that nature made available.

In this study of bridges we shall carefully examine how a bridge supports the loads we place on it. Any bridge is created to support the weight of people and their animals or vehicles over an opening. This weight represents a force acting down on the bridge, and it is balanced by forces acting up at the points where the bridge is supported. Just how the bridge accomplishes this task is the major concern of this chapter.

There are four distinctly different bridge forms: the beam, the arch, the suspension, and the truss. Of these, the first three can be found in nature. The beam, the arch, and the suspension bridge all have their beginnings in naturally occurring situations, which were available to primitive man for his use.

The beam-type bridge can be seen wherever a tree has fallen across a stream. Primitive man surely used such tree bridges to cross rivers that were too deep to ford. The location of the fallen tree may not have been very convenient for him, which meant he had to go out of his way to get from one place to another. When early man had acquired the use of cutting tools and developed some of the impatience of modern man, he built his bridges where he wanted them by cutting or burning down a tree. Figure 2.1 is an example of this crude bridge form. Until that time he simply looked for a bridge or remembered where one was located and used it.

The simple beam bridge made of a crude log was only the beginning of this beam form. As man's reasoning powers expanded, he learned that by placing piles of stones in a stream he could span wider streams. Shorter logs, or even a crude flat stone, could be made to reach from one pile to another in succession, making a multiple-span beam bridge. There are other types of beam bridges that differ from the simple beam only in the manner in which they are supported. The distinction between simple beams and other types of beams will be made clear later in the chapter.

Figure 2.1

Many of the so-called natural arch bridges that are so spectacular are not really arches at all. The openings give the appearance of arches, but the material spanning the opening acts much like a beam. The archlike openings were probably caused by the erosion of soft material from beneath a more durable material. The result of this erosion can result in huge dramatic structures such as those found in the western part of the United States. Erosion may also cause the formation of a simple cave. The cave that provided shelter for primitive man is closer to the true arch than its more dramatic counterpart in nature.

A natural arch can also be formed as the result of a rock slide that occurs in such a way that rocks are piled over a stream bed, as illustrated in Figure 2.2. The rocks may wedge together tightly enough to provide a reasonably strong arch.

Figure 2.2

Just how man accomplished his first true arch bridge is not known. As early as 4000 B.C. the people of Mesopotamia were using the true arch, or *vault*. Certainly these were preceded by many attempts to imitate the inspiring arch shapes in nature.

Suspension bridges had their beginnings in the tropical climates where vines grow to enormous lengths. A suspension bridge today is not fundamentally different from the crude system developed by early man. All modern man has done is make his suspension bridges longer, stronger, and more comfortable.

The first use of the vine to cross a river was most likely accomplished by swinging from one bank to another. Unfortunately, once he arrived on the other side of the river and let go of the vine, the man lost his means of transportation back. It was not long before some bright person thought to tie his vine to a tree and have it there when he came back to recross the river. The next thing to occur to him was that by securing each end of a vine to a tree on each bank of the river he could get himself back and forth quite safely by a hand-over-hand technique. This finally was the fundamental form of the suspension bridge.

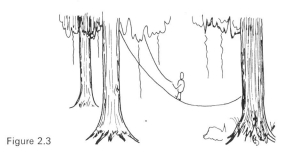

Figure 2.3

By placing two vines parallel to each other and spanning the opening between them with short beams of small branches, early man was able to improve the suspension bridge by providing a floor. This was further improved by attaching two more vines above the floor level to provide sides for the bridge, as shown in Figure 2.3. Each improvement to the suspension bridge would seem to be in a direct evolutionary line to the modern version of this exciting bridge form.

The fourth bridge form, the truss, came into being much later in man's history. The gable or trussed roof had been used for some time earlier, but it was not until the Renaissance that the principle of the truss was understood and used extensively.

This brief description of the early interaction of man and bridges is, of course, highly speculative. The details and the time involved in the evolution of these bridge forms are obscure. However, the facts are that prehistoric man did use, and eventually imitated, the bridge forms found in nature, and these are the fundamental forms that modern man uses today.

2.2 FORCES AND EQUILIBRIUM

The bridge as a structure is subjected to the push and pull of forces. The loads imposed on the bridge by the man, animal, or vehicle using the bridge and the weight of the bridge itself are forces that ultimately push *down* on the points where the bridge is supported. These points, known generally as *piers* or *abutments*, provide for forces that push *up* on the bridge to balance the forces pushing down. Assuming that the bridge does not break under the loads, we can say that the bridge is in *equilibrium*. The notions of *force* and equilibrium are important to the understanding of how the four bridge forms work.

A force is felt by the human body in terms of effort expended to push or pull an object. A more useful definition of force is that it is a magnitude (pounds) and a direction, that is, a line of action in space. The force may be represented graphically by an arrow, the length being proportional to the magnitude and the arrowhead pointing in the proper direction along the line of action.

We shall concern ourselves with a very restricted system of forces in our discussion, namely, forces acting in one plane only. If a group of forces is restricted to acting in a plane (obviously the plane containing the bridge), it is called a coplanar force system.

Equilibrium implies balance. Balance may be illustrated by the action of two people on a seesaw, as shown in Figure 2.4. If one of the people weighs more than the other, he must sit a different distance from the balance point in order that the two people balance each other. When this happens, we say that the moments of the forces (the weight of the person times his distance from the balance point) about the balance point is algebraically zero. The person on the left exerts a counterclockwise moment about the balance point, while the person on the right exerts a clockwise moment. Assuming that the plank (a beam?) does not break, the two people have achieved equilibrium.

Figure 2.4
Equilibrium.

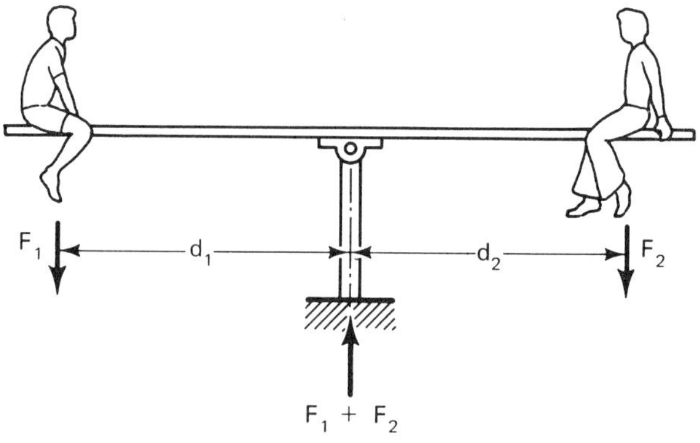

One other condition must be satisfied in order to completely assure equilibrium. The pipe that furnishes the balance point must be strong enough to provide an upward force equal to the combined weights of the two people. In other words, the algebraic sum of the forces acting on the plank must be zero.

Our seesaw involves only forces in the vertical direction. We must consider one more possibility to describe the requirements of equilibrium of forces in a plane, that is, the possibility of horizontal forces acting in the plane. In a real bridge, horizontal forces may be exerted by the action of vehicles as they start and stop on the bridge roadway. In the cases of the arch and suspension forms, as we shall see later, the structure itself exerts horizontal forces on the foundations.

The nature of forces is such that their total effect on a bridge may be separated into vertical and horizontal effects. We do this by breaking the force into two components, one with a horizontal orientation and one with a vertical orientation. Actually, forces may be broken up into any pair of components associated with directions other than vertical and horizontal, but these are the most convenient for our purposes.

The calculations needed for determining the vertical and horizontal components involve only the trigonometry of right triangles. In Figure 2.5 the inclined force F, represented by the line $0A$, exerts an influence on the block (B). This influence is a combination of three effects: one, the tendency to push the block down against the plane, and two, the tendency to push the block to the left along the plane. The third effect is not so obvious. It is the tendency of the force to tip the block over. Whether or not the block will tip over depends on the actual dimensions of the block and the magnitude of the frictional resistance to slipping along the plane. This third effect is easier to deal with in the case of actual bridges that do not rely on the notion of friction. We shall deal with this idea in more detail later.

The first two effects, however, can be separated from each other if we break the force into two components. The vertical component, F_v, will produce the same effect vertically as the original force F; likewise, F_h accounts

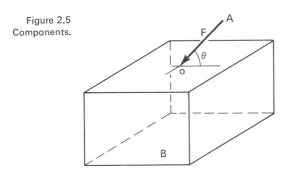

Figure 2.5
Components.

for the total horizontal effect of the original force. The magnitudes of these components are simply the lengths of the sides of the right triangle that has 0A as the hypotenuse. The trigonometry of the right triangle will give the following relationships:

$$F_v = F \sin \theta$$

$$F_h = F \cos \theta$$

For example, in Figure 2.5 if $\theta = 60°$ and the force $F = 100$ lb,

$$F_v = 100 (\sin 60) = 100 (.707) = 70.7 \text{ lb} \downarrow$$

and

$$F_h = 100 (\cos 60) = 100 (.500) = 50.0 \text{ lb} \leftarrow$$

The arrows $\downarrow$ and $\leftarrow$ are used to indicate the directions of the force components. Their meaning is obvious, and they are easier to use than some sign convention. The same results for F_v and F_h would have been obtained if a line 100 units long were drawn at 60° with the horizontal and if the vertical and horizontal sides of the corresponding right triangle were measured.

The next idea to consider is the equilibrium of a ring acted upon by a group of forces as shown in Figure 2.6. This system is known as a *concurrent force system*, which simply means that the forces go through a common point.

The equilibrium requirements for this system can be summarized by the following:

1. The algebraic sum of the horizontal components of the forces acting at zero must be equal to zero.
2. The algebraic sum of the vertical components of the forces must be equal to zero.

In Figure 2.6(a) the forces F_1 and F_2 are given, while the forces F_3 and F_4 are assumed to be known in direction but unknown in magnitude.

Figure 2.6
Concurrent force system.

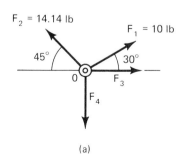
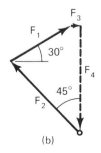

(a)   (b)

Since there are two equilibrium conditions to be satisfied, we can solve for these two unknown magnitudes, *assuming* that the ring is in equilibrium.

For example,

$$F_{2,h} = 14.14 \, (\cos 45°) = 10.00 \text{ lb} \leftarrow$$

$$F_{1,h} = 10.0 \, (\cos 30°) = 8.66 \text{ lb} \rightarrow$$

Therefore F_3 must be 1.34 lb → in order to satisfy equilibrium in the horizontal direction. In a similar manner,

$$F_{2,v} = 14.14 \, (\sin 45°) = 10.00 \text{ lb} \uparrow$$

$$F_{1,v} = 10.0 \, (\sin 30°) = 5.00 \text{ lb} \uparrow$$

Therefore $F_4 = 15.0$ lb ↓ in order to satisfy equilibrium.

The subscripts 2,*h* and 1,*h* indicate the horizontal components of forces F_2 and F_1, respectively. Likewise, 2,*v* and 1,*v* refer to the vertical components of F_2 and F_1.

The graphical solution to this problem is illustrated in Figure 2.6(b). The known forces are plotted one following the other, with the polygon being completed by drawing the two closing sides parallel to the unknown forces F_3 and F_4.

Equilibrium problems of concurrent force systems can be posed in many different ways. For our purposes the previous example is typical of the problems we shall solve later in the analysis of bridge trusses.

We shall consider next the equilibrium of a nonconcurrent, coplanar force system. The system is illustrated by the group of forces acting on a rigid block as shown in Figure 2.7. You will notice that the forces F_1 and F_2 are the same as in the previous example but are applied at two distinctly

Figure 2.7
Equilibrium of nonconcurrent force system.

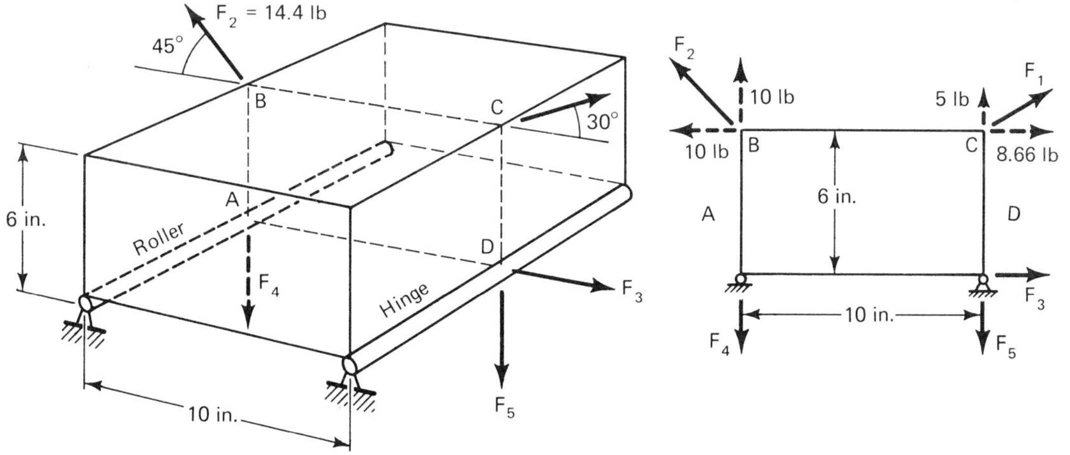

different points on the block. The block is supported at A by a device capable of resisting a vertical force only; the force may be either up or down. At B the hingelike device will resist both vertical and horizontal forces. The forces F_3, F_4, and F_5 are shown in the assumed correct directions. Notice also that the block has definite dimensions.

As in the case of concurrent force systems the horizontal and vertical equilibrium requirements must be satisfied. In this example, horizontal equilibrium requires

$$F_2 \overleftarrow{\cos} 45° + F_1 \overrightarrow{\cos} 30° + \vec{F}_3 = 0$$

or

$$F_3 = 1.34 \text{ lb} \rightarrow$$

This is the same result as in the example of the concurrent force system. The requirements of vertical equilibrium are

$$F_2 \sin 45° \uparrow + F_1 \sin 30° \uparrow + F_4 \downarrow + F_5 \downarrow = 0$$

Writing this relationship algebraically yields

$$10.0 + 5.0 - F_4 - F_5 = 0$$

or

$$F_5 + F_4 = 15.0$$

There are two unknowns in this equation; therefore we must find another relationship between the unknowns in order to solve for F_4 and F_5.

So far our equilibrium equations have accounted for the tendency of the forces to slide the block horizontally and vertically. By saying that the forces in these two component directions add up to zero we mean that the net tendency to slide the block is zero. However, we have not said anything about the tendency of the block to rotate. This tendency, to which each individual force contributes, must also be zero if the *total* equilibrium of the block is to be assured.

The measure of the tendency of a force to cause a body to rotate about a point is called the *moment* of the force about that point. The moment is the product of the force and the distance from the force to the point measured along the direction perpendicular to the force. In this particular example we shall use point D as a reference point for computing the moments of the forces.

Assuming that moments in the clockwise direction about D are positive,

$$(F_1 \cos 30° \times 6) + (F_2 \sin 45° \times 10) - (F_2 \cos 45° \times 6) - (F_4 \times 10)$$
$$= 0$$

Notice that the vertical component of F_1 as well as F_3 and F_5 go through point D and do not appear in the equilibrium equation. Simplifying this relationship yields

$$(8.66 \times 6) + (10 \times 10) - (10 \times 6) - (10 \times F_4) = 0$$

or

$$F_4 = 9.2 \text{ lb} \downarrow$$

From the previous equilibrium equation

$$F_4 + F_5 = 15.0 \text{ lb}$$

we obtain

$$F_5 = 15.0 - 9.2 = 5.8 \text{ lb} \downarrow$$

We might check this result by using point A as the reference point for the moments of the forces. The arithmetic yields $F_5 = 5.8$ lb $\downarrow$.

To summarize briefly, the conditions of equilibrium for a coplanar force system acting on a body are

1. The summation of the horizontal components of *all* the forces applied to the body is zero.
2. The summation of the vertical components of *all* the forces applied to the body is zero.
3. The summation of the moments of *all* the forces applied to the body about *any* point in the plane of the body is zero.

These three conditions express the fact that the tendency of the individual forces to move the body add up to zero when all the forces act together. These three conditions permit us to solve for as many as three unknown forces in any problem of coplanar equilibrium. If the force system is a concurrent-coplanar system, then the third equilibrium condition is satisfied automatically, and as a result only two unknowns may be involved in any problem.

2.3 FORCES AND DEFORMATION

It was convenient in our discussion about equilibrium not to concern ourselves with the deformations of the bodies on which the forces act. In fact, normally the deformations are so small compared to the dimensions of the body that they are negligible. Assuming the deformations to be zero means that the original geometry of the body may be used in the equilibrium calculations without any significant error.

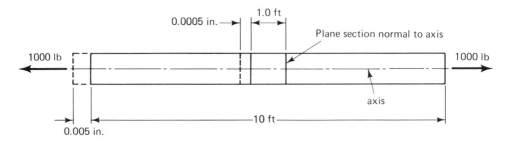

Figure 2.8
Pure axial load.

As small as they are, it is useful for us to examine these changes in shape as the bodies are loaded. Each of the types of forces that may be applied to a body causes a distinctly different type of deformation. Each of the bridge forms we want to study is characterized by one of these deformations. It is convenient to establish this terminology now.

The easiest deformation to visualize is simple stretching. A bar being pulled by a force at each end will increase in length. This increase in length will be small but uniform. That is, if the bar has a uniform cross section, we can say that each unit length of the bar is stretched an equal amount.

For example, consider the bar in Figure 2.8. The bar is originally 10 ft long and is pulled by forces equal to 1,000 lb. Assume that the bar stretches an amount equal to .005 in. If the bar has a uniform area, we can say that each 1-ft section of the bar stretches .0005 in. Another very important aspect of these deformations is that each longitudinal element will be stretched .005 in. This means that any plane section perpendicular to the axis of the bar will move a certain distance but *will remain undistorted.*

Bars subjected to a longitudinal compressive load will behave in the same way except that the stretching will be replaced by shortening. These two types of deformations are considered together as one, namely, pure longitudinal deformation.

One thing more should be said about axial or longitudinal compression before we leave the subject. If the compressive force reaches a certain limit, the bar will buckle (see Figure 2.9). If you push on the end of a yardstick, you can cause this to happen. The tendency for long, slender bodies under compression to buckle is important when you consider the design of bridge elements.

The next most visible deformation to consider is the bending deformation. The seesaw mentioned in a previous example would show significant bending deformation if two rather large people were to use it. The characteristic of the bending deformation is the curvature of the plank.

The plank shown in Figure 2.10(a) has changed from an initially straight element to a curved element. Imagine that we had marked the edges of the plank with two parallel lines, *a-c* and *b-d*, perpendicular to top and bottom surfaces. These two lines will still be straight when the plank is bent, but

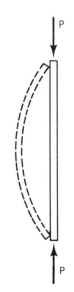

Figure 2.9
Buckling of a slender compression member.

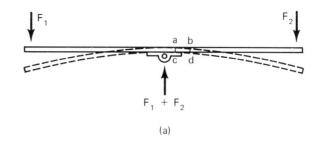

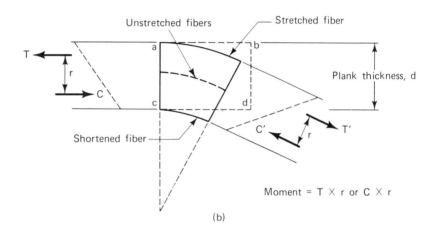

Figure 2.10
Bending deformations.

they will no longer be parallel. The distance between the lines at the top of the plank will increase, while the distance at the bottom surface will decrease. Figure 2.10(b) shows this clearly. The fibers at some point between the top and bottom surfaces would be unstretched. This deformation causes tension or stretching forces to develop in the upper portion of the plank, while compression forces develop in the lower part. The total tensile force T must be equal to the total compressive force C. The two forces are separated by the distance r, which is a fraction of the thickness of the plank d. The net result of this pair of forces is a pure moment or *couple*.

We note two major differences in comparing the beam with the bar subjected to an axial load. First, the loads on the beam are perpendicular to the long dimension of the beam, while the bar loads are at the end and parallel to the long dimension. The second difference is that the beam is subjected to *both* tensile and compressive forces, while the bar is subjected to *either* tensile or compressive forces. The latter difference is particularly important when we select materials with which to build our beams.

The last deformation to consider is perhaps the most difficult to visualize. The change in shape of a body due to pure shear forces is illustrated

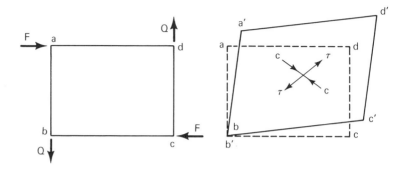

Figure 2.11
Shear deformations.

in Figure 2.11. The block has been subjected to a pair of equal and opposite shear forces, F and Q, on the two faces. The block changes from a cube into a rhombic shape. The result is that the diagonal b-d stretches and the diagonal a-c shortens. This shear type of deformation is important in the design of structures precisely because it gives rise to tensile and compressive deformations along these diagonals. A material that is inherently strong in shear may have a weakness in tension that will ultimately limit its usefulness in shear.

In real structures we rarely encounter the pure version of any one of these deformations. What is most likely to occur is a situation where one of these deformations dominates and the others are present to a minor degree. The tension bar may have a small amount of bending deformation because we cannot grip it properly to cause pure tension. The same can be said about our seesaw example. There is some shear deformation present along with the bending deformation but it is usually very small.

We shall make reference to these different types of deformations as we begin to distinguish one bridge form from another. Two bridges may have a very similar geometric characteristic but be of different classes because they develop quite different deformations under load.

2.4 DEFORMATION OF BRIDGE FORMS

The major differences in the behavior of the various types of bridges are most easily understood by observing how each form changes shape under a load. Of course we assume that the bridges we build will not deform drastically while we are using them, but they do deform slightly. This deformation, however small, is the key to understanding the differences in their behavior.

The beam deforms by changing from a more or less straight member to a definitely curved member. The experience of crossing a ditch filled with water using a beam that sags is not particularly comforting. The plank is a beam bridge and serves to illustrate beam behavior. In Figure 2.12 the beam axis is shown both in its unloaded shape and in its loaded shape.

The suspension bridge is best illustrated by a slack string connected to points on either side of the opening to be spanned. In Figure 2.13 this string is shown in its normal shape, a smooth curve. By attaching a weight at a point in the span, the shape is changed into two straight lines. The string has no resistance to bending; therefore the force in the string must be directed along the string. The applied weight is held in equilibrium by the two string forces. The string shape is determined by the requirements of equilibrium on this concurrent force system. The chief characteristics of the simple suspension system are as follows: (1) the string or cable in a real bridge resists only tension, and (2) the cable changes shape depending on where the cable is loaded. The second characteristic is not very desirable in modern suspension bridges. Anyone who has crossed a suspended footbridge is well acquainted with the sensations of motion under foot. This motion is not unlike that of a small ship in a heavy sea.

2.4
deformation of bridge forms

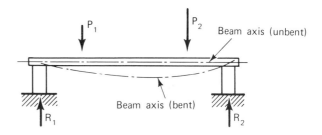

Figure 2.12
Simple beam.

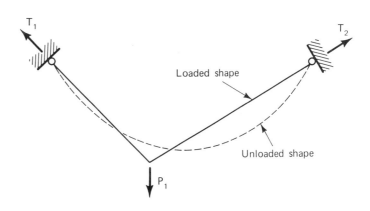

Figure 2.13
Simple suspension cable.

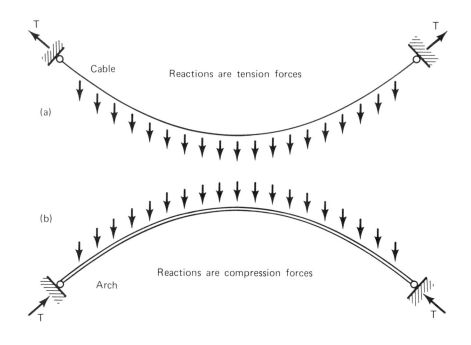

Figure 2.14
Cable and equivalent arch.

The arch is the mirror image of the suspended cable. If we take a flexible cable, as shown in Figure 2.14(a), and place loads on it in the same pattern with which an arch is to be loaded, the cable will assume an equilibrium configuration. The cable will be in a state of pure tension (no bending deformations present). Now imagine that the loaded cable is sprayed with a material that will cause it to become stiff and able to resist compression. This stiffened cable may then be turned over and used as an arch, as shown in Figure 2.14(b). The shape of the arch will ensure that only compression forces will exist in the arch. Each segment of the arch will undergo a small amount of shortening, but no bending deformations will occur. Equilibrium is possible at every load point because the internal forces are oriented along the direction of the arch segment as they were when the arch was a cable.

This description of arch behavior is for a perfect arch. A perfect arch has no bending if its shape is determined by the load distribution that the arch is to carry. However, a real arch must carry a traffic load, which obviously changes the original load pattern. Unlike the cable, the arch cannot change shape freely to accommodate the new load pattern and maintain a pure compression deformation. The arch accommodates the changed load pattern by combining a small amount of beam action with the fundamental arch action to maintain equilibrium. A well-designed arch will accommodate traffic loads (as opposed to the dead load of the arch itself) with only a minimum of bending action.

The truss is another entirely different form. A simple truss is shown in Figure 2.15. The geometry of the triangle provides its rigidity when loaded. The ideal truss is a network of members connected so as to form a series of triangular figures. Each member is connected at a joint by an assumed frictionless pin, and the loads are placed on the truss only at the joints. Frictionless pins are not really used in the actual fabrication of trusses, but the assumption simplifies the analysis of a truss. The configuration of a truss bridge, particularly the arrangement of the beams carrying the roadway, ensures that the traffic loads do reach the truss only at its joints.

The fundamental behavior of the truss depends on the ability of the triangular shape (made up of three truss members) to resist changes in shape. Consider the center panel (*a-b-c-d*) of the truss in Figure 2.15 to be isolated without member *b-c*. The resulting rectangular configuration is shown with the forces from the adjacent parts of the truss and the loads applied to that panel. This configuration is unstable; that is, it will not be able to resist the loads because it will simply fold up.

If the member *b-c* is inserted in the rectangular panel, the panel will maintain its shape under load. For the panel to collapse, the member *b-c*

2.4 deformation of bridge forms

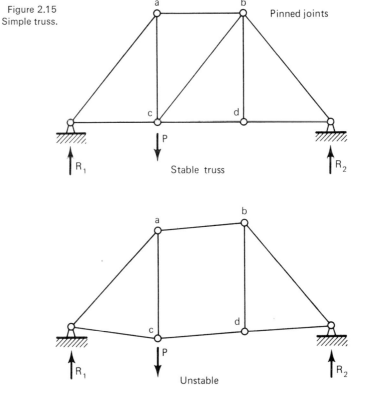

Figure 2.15
Simple truss.

would have to be compressed or stretched. There is a tendency for all the panels to change shape in a truss. In fact, all the members are stretched or compressed just enough to develop the required internal resistance needed to maintain equilibrium. Again, the details of the truss analysis will be treated in a later section.

2.5 TYPICAL BRIDGE GEOMETRY

All bridges are made up of some combination of these basic forms: the beam, the truss, the arch, and the suspension cable. All bridges have elements in common. In this section we shall review the overall geometry of a typical bridge before we concentrate on the details of specific forms.

The span length is the most obvious element common to all bridges. A particular bridge may cross a river with a single span. If the river is not too wide, the bridge may not require any intermediate piers between the abutments on the shore. Each of the bridge forms have different capabilities with respect to span length. In general, the suspension bridge and the arch bridge are used where long spans are needed. Beams are limited generally to shorter spans, and trusses satisfy the requirements of intermediate spans. What constitutes a long or short span depends on the material as well as the form of the bridge. Steel beams may be used where a wooden truss would be necessary. Stone arches may be limited to 100 ft or so, whereas a steel arch may easily span 200 to 800 ft.

Bridges may be made of a series of short spans by the use of intermediate piers, and such bridges are called *multiple-span bridges*. They may utilize a single form throughout or a combination of different forms. Conditions,

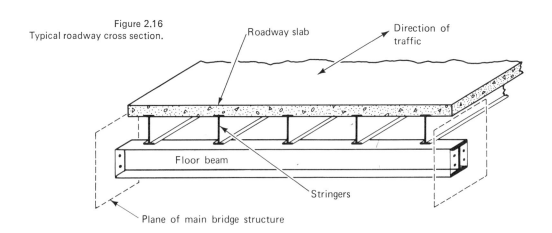

Figure 2.16
Typical roadway cross section.

such as the requirements of navigation channels, usually determine the combination to be used in any particular bridge location.

For our detailed discussion of the behavior of bridge forms we shall assume a single-span bridge. We shall also assume that the roadway is supported between two main structural bridge forms. Figure 2.16 shows a typical roadway. The roadway slab is supported on longitudinal beams called *stringers*. The stringers are oriented parallel to the direction of traffic and are supported at various points by crossbeams. These crossbeams are supported on each end by the main structural form. Figure 2.17 illustrates a few common bridge forms. If the bridge is a truss, the crossbeams are connected at the joints of the truss at the level of the roadway. If the bridge is an arch, the

2.5 typical bridge geometry

Figure 2.17

Deck Type Arch

Through Type Arch

Deck Type Truss

Through Type Truss

Typical Suspension Bridge

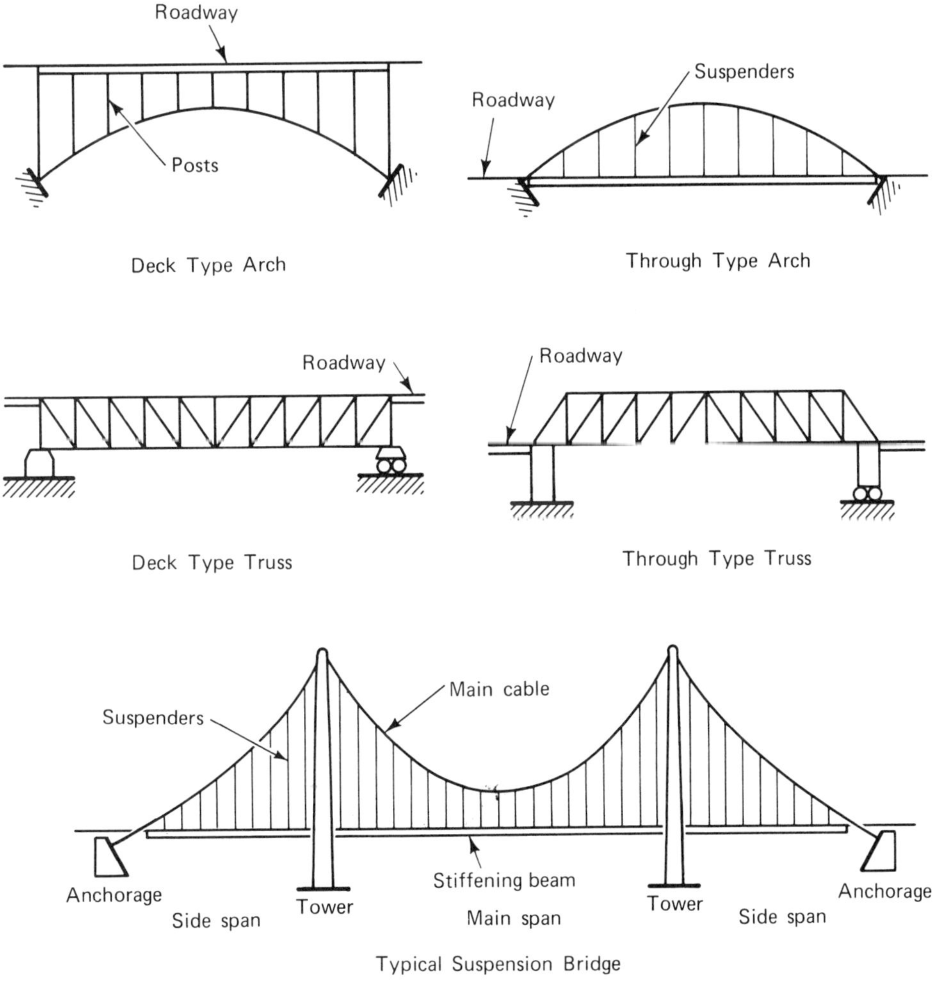

crossbeams may be supported by posts or hangers connected to the arch. The suspension bridge supports the roadway in a manner quite similar to the others except that the floor beams are supported by a stiff girder (parallel to the traffic), which in turn is supported by hangers from the main suspension cable.

We shall now return to specific bridge forms and the manner in which they respond to loads and examine the beam in more detail. Every bridge form must be in equilibrium. The overall support provided for the bridge must be strong enough to resist the forces exerted by the bridge. We shall assume here that the design of the abutments is adequate.

2.6 STATICS OF BEAMS

In Figure 2.18a we see a symbolic drawing of a beam. The horizontal line A-B represents the longitudinal axis of the beam. The symbols at A and B represent a roller support and a fixed support, both with theoretical knife edges. The roller support (at A) cannot resist any horizontal load.

This particular example has a span length of 40 ft with two loads applied to the beam at 10 ft and 25 ft from A. The supports provide the reactions necessary to maintain this beam in equilibrium. Because of the roller at A, the reaction R_A must be vertical. The horizontal loads applied to the beam are resisted by the support at B. The support at B also provides a vertical reaction R_B.

The beam is to be treated as a rigid body subjected to a nonconcurrent coplanar force system made up of the loads and reactions. The beam is in equilibrium (hopefully), and the techniques of Section 2.2 may be used.

Consideration of the equilibrium of the horizontal components leads directly to

$$\vec{H}_B = \overleftarrow{P}_{2h}$$

or

$$H_B = 3{,}000 \text{ lb} \rightarrow$$

The equilibrium requirement for the vertical forces is satisfied if

$$\overset{\downarrow}{P}_1 + \overset{\downarrow}{P}_{2v} - (\overset{\uparrow}{R}_A + \overset{\uparrow}{R}_B) = 0$$

or

$$R_A + R_B = 1{,}000 + 4{,}000 = 5{,}000 \text{ lb}$$

This relationship is not enough to determine R_A and R_B separately. To do this we need to use the equilibrium relationship for the moments of the forces.

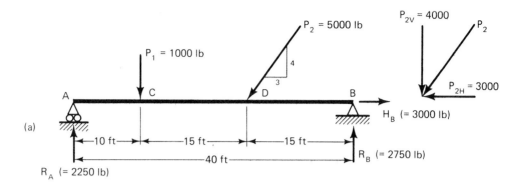
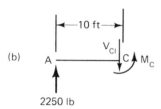
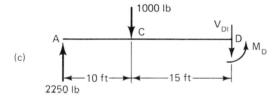
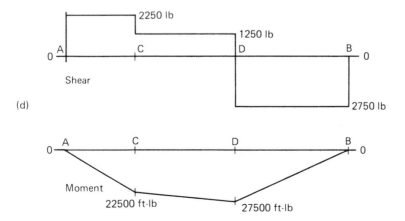

Figure 2.18

Summing the moments of all the forces about point A, for example, yields

$$10P_1 + 25P_{2v} - 40 R_B = 0$$

or

$$40 R_B = 10(1,000) + 25(4,000)$$

$$= 10,000 + 100,000 = 110,000$$

from which

$$R_B = \frac{110,000}{40} = 2750 \text{ lb} \uparrow$$

Now, using this result and

$$R_A + R_B = 5,000$$

yields

$$R_A = 2,250 \text{ lb} \uparrow$$

The forces R_A, R_B, and H_B ensure that the beam as a whole is in equilibrium. The strength of the beam is determined by computing the effects of these forces on the internal structure of the beam. In other words, how much bending is caused, and how does the beam resist the bending?

We can investigate the bending forces present throughout the beam by imagining it to be cut into two parts. At the point where the cut is made, the internal force system is exposed and may be computed. Each of the two sections must be in equilibrium, so we can use the equilibrium relationships to determine the internal forces. In Figure 2.18(b) the beam has been cut at C, just to the left of P_1. By considering the segment of the beam from A to C we observe that R_A is the only external force applied. V_{Cl} and M_C are the internal forces needed at C to maintain equilibrium. These are readily calculated by summing the vertical forces on the section A-C and summing the moments of the forces about point C. These calculations will yield

$$V_{Cl} = 2,250 \text{ lb} \downarrow$$

and

$$M_C = 22,550 \text{ ft-lb} \curvearrowright$$

By cutting a section just to the right of P_1, the only change is in the calculation of V. In this instance

$$V_{Cr} = 2,250 - 1,000 = 1,250 \text{ lb} \downarrow$$

The moment M_C is unchanged because the cut section is still only an infinitesimal distance from C.

Now, in a similar manner, the shears and moments at D [see Figure 2.18(c)] are

$$V_{Dl} = 1,250 \text{ lb} \downarrow$$

$$V_{Dr} = 2,750 \text{ lb} \uparrow$$

and

$$M_D = 27,500 \text{ ft-lb} \curvearrowleft$$

The directions indicated by the arrows are those of the forces acting on the cut section of the left part of the beam. The forces on the corresponding right parts of the beam are equal in magnitude and opposite in direction. This is necessarily so because the beam is not literally cut and the net external effect of the internal forces acting on the two parts must be zero.

The diagrams shown in Figure 2.18(d) are known as *shear and moment diagrams*. These diagrams represent the variation of the shear and moment from point to point along the beam. They are essential to the designer in that they show him graphically the strength requirements of the beam.

The principal force that characterizes the deformation and the behavior of beams is the bending moment. In an earlier section we saw that the bending of a short section of beam causes the bottom to stretch and the top to compress. If we consider each longitudinal fiber of the beam as a separate unit, we find that each fiber changes length by a different amount. It can be shown that in a beam the change in length is proportional to the distance from the neutral or unstressed fiber. The force developed in each fiber depends on the amount the fiber is stretched and on the type of material being used. For many materials the force is directly proportional to the unit change in length, that is, the change in length divided by the original length. If the force is acting on the single fiber, this is a *force per unit area*, commonly called *stress* (psi). Figure 2.19 shows a short section of beam with the stress diagram.

The beam theory needed to compute stresses is too complex to be presented in detail here. If the beam cross section is rectangular (d = depth and b = width), the maximum stresses can be calculated by the formula $\sigma = 6M/bd^2$, where σ is stress in psi, b and d are in inches, and M is the bending moment in in.-lb.

For many beams the maximum tensile and compressive stresses are nearly equal. The limiting factor in beam design therefore is to select beam dimensions b and d that limit the maximum stress with a factor of safety against failure. By maintaining the maximum stress to less than half the stress necessary to fail the material in tension or compression, a factor of safety of at least 2 can be assured.

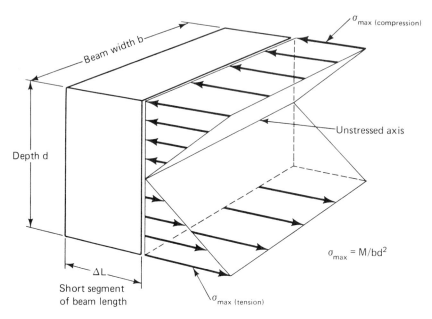

Figure 2.19
Bending stresses.

One of the limitations of beams now becomes obvious. If we select a material such as concrete with which to construct a beam, we either must modify the material by adding reinforcing steel or be handicapped by a very large beam size, because concrete has a very low tensile strength. The large compressive strength will be pretty much wasted. By adding reinforcing steel in the bottom half of the beam we increase the tensile capacity of the beam and therefore can more nearly balance the compressive strength of concrete.

A steel I-beam does not have these problems of strength, since its compressive and tensile strengths are nearly equal. However, a steel I-beam has a weakness on the compression side. This weakness is one of instability. Any member or part of a member (i.e., the top flange of the I-beam) that is slender and under compression may have a tendency to buckle before its compressive strength is reached.

2.7 STATICS OF SUSPENSION CABLES

To support a beam adequately it is necessary to provide vertical reactions at two points and a horizontal reaction at one point only. If the applied loads are all vertical, the horizontal reaction will be zero, and the horizontal equilibrium equation will be satisfied automatically. Arch and suspension bridges require a more complicated support system. This system may be examined by considering the simple suspension cable in Figure 2.20(a). The suspended cable with loads attached at two points is supported at A and at B. By considering the vertical equilibrium relationship and the moment relationship

2.7 statics of suspension cables

$$F_V = 0 \quad \text{and} \quad M_A = 0$$

we can compute $V_A = 25$ lb $\uparrow$ and $V_B = 20$ lb $\uparrow$. The relationship between the slope of the cable segment and the components of the force in the segment is based on the geometry of similar right triangles. For example, T_1 is the tension force in segment A-C and is directed along A-C, which has a slope

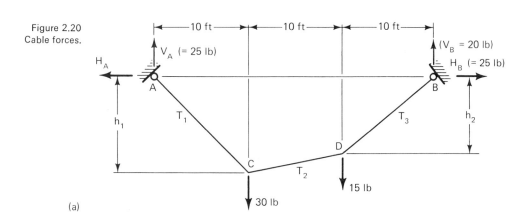

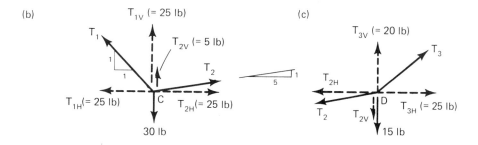

Figure 2.20 Cable forces.

of $h_1/10$. The forces at A are the vertical and horizontal components of T_1 and therefore must have a ratio equal to $h_1/10$. It follows that

$$\frac{h_1}{10} = \frac{V_A}{H_A}$$

Similarly, for segment C-D

$$\frac{h_1 - h_2}{10} = \frac{T_{2,v}}{T_{2,h}}$$

and for segment D-B

$$\frac{h_2}{10} = \frac{V_B}{H_B}$$

In each of these ratios the values of H_A, H_B, and $T_{2,h}$ must be equal. If we set $h_1 = 10$ ft, then $H_A = V_A = 25$ lb. We can next determine that

$$h_2 = 10 \left(\frac{V_B}{H_B}\right) = 10 \left(\frac{20}{25}\right) = 8 \text{ ft}$$

If we had selected $h_1 = 25$ ft, H_A would have been reduced to 10 lb and h_2 would be 20 ft.

Figure 2.20(b) shows joint C with a 30-lb load and the cable forces T_1 and T_2. Since the sag $h_1 = 10$ ft, the slope of the cable AC is 45°, and T_1 has equal vertical and horizontal components. These components are felt at the support point A, requiring that the support provide a horizontal force to the left as well as the vertical force upward. The magnitude of this force is 25 lb.

Returning our attention to joint C, we observe that $T_{1,h} = T_{2,h} = 25$ lb. Vertical equilibrium will require $T_{2,V} = 5$ lb ↑.

T_2 also acts on joint D as shown in Figure 2.20(c). If joint D is considered for equilibrium, we shall again find that $T_{3,h} = T_{2,h} = T_{1,h} = 25$ lb ↑. At this joint the cable force T_2 pulls down and to the left on point D.

The cable T_3 transfers the tension force from D to B, and the horizontal component of T_3 is exposed as $H_B = 25$ lb→. This result should be expected if the entire structure and horizontal equilibrium are considered.

The actual cable forces are

$$T_1 = 25\sqrt{2} = 35.4 \text{ lb}$$

$$T_2 = \sqrt{25^2 + (5)^2} = \sqrt{650} = 25.5 \text{ lb}$$

$$T_3 = \sqrt{(25)^2 + (20)^2} = \sqrt{1,025} = 32.0 \text{ lb}$$

It is particularly important to note that the two horizontal reactions, H_A and H_B, are absolutely necessary to the support of the cable. Both support points must resist horizontal forces, whereas only one is required for the beam.

The cable segments are in a state of pure tension because the cable itself is incapable of resisting any bending moment. This means that the cable will adjust its geometry to maintain pure tension for different loads. This ability of the cable to adjust is not very desirable if you are transporting major vehicular traffic across a bridge supported by the cable. To overcome this difficulty most modern suspension bridges are built with a stiff beam supporting the roadway loads. The beam is suspended from the main cable at a number of closely spaced points along the cable. In this way a concentrated load is distributed from one point on the roadway to all the suspenders in a uniform way.

Figure 2.21(a) shows a suspension system with a stiffening beam suspended from the cable. The suspenders each carry one-tenth of the total load on the roadway. The cable has a symmetrical configuration, with the maximum sag h at the centerline. Figure 2.21(b) shows the left portion of the cable with the internal tension at the low point C on the curve. Since the curve is horizontal at this point, the tension force is horizontal and equal to the reaction H. This satisfies the horizontal equilibrium requirement.

It is possible to compute H by summing moments of the forces on the left half of the cable about point C as follows:

$$H \times h + P_x\left(\frac{L}{10} + \frac{2L}{10} + \frac{3L}{10} + \frac{4L}{10}\right) + \frac{P}{2} \times \frac{L}{2} - 5P \times \frac{L}{2} = 0$$

$$H \times h = PL\left(\frac{S}{2} - 1 - \frac{1}{4}\right) = \frac{5PL}{4}$$

or

$$H = \frac{5}{4} \times \frac{PL}{h}$$

If we remember that $10P$ is the total load supported by the cable, then H may be written as

$$H = \frac{W_T}{8}\frac{L}{h}$$

where $W_T = 10P$. This is a well-known relationship for uniformly loaded cables. It expresses the important fact that the horizontal force at the supports, and hence the total force, decreases as the sag h increases.

It is interesting to know how to calculate the cable geometry under the action of a series of uniform loads. We can do this problem graphically if we remember that the force in the cable is directed along the direction of the cable between load points. The cable will obviously change direction at each load point.

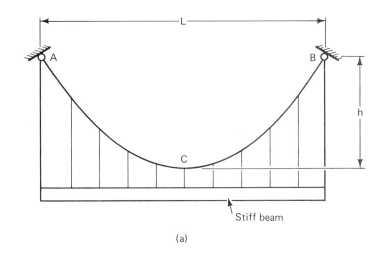

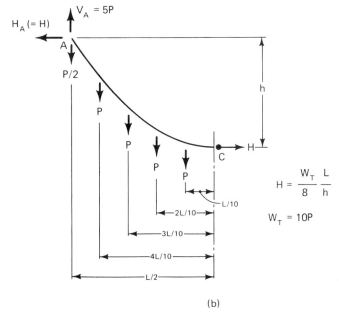

Figure 2.21
Stiffened cable forces.

Figure 2.22a shows this solution for a cable with a 48-ft span and a sag of 12 ft. In this example

$$H = \frac{W_T}{8} \frac{(48)}{12} = \frac{W_T}{2}$$

$$V = \frac{W_T}{2}$$

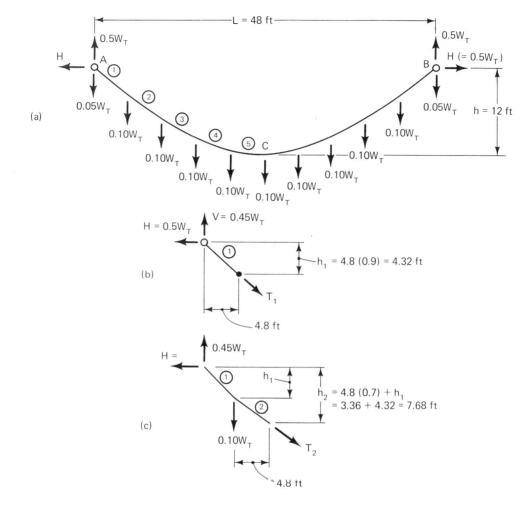

Figure 2.22
Computations for cable geometry.

and

$$P = \frac{W_T}{10}$$

The spacing between the hangers is $\frac{48}{10} = 4.8$ ft. In each segment the slope of the cable will be equal to the ratio of the vertical component of the tension to the horizontal component H. H remains constant for each segment, while V reduces from the maximum value, $.45 W_T$, in the first segment by increments of $.1P$. Starting at the first segment at A, the vertical component of the cable force is $.45W_T$ and the horizontal force is $H = .5W_T$. The slope of the cable will be equal to $.45W_T/.50W_T$, or $.90$. By constructing a line with this slope we can determine that the cable sags a total of $.9(4.8) = 4.32$ ft in the first segment [see Figure 2.22(b)]. The sag of each successive segment can be determined in a similar fashion. See Figure 2.22(c) for the construction for segment 2. The following table is a convenient way of summarizing the calculations:

Segment No.	V	H	$\dfrac{V}{H}$	$\dfrac{V}{H} \times$ Spacing	Sag (ft) $\left(\dfrac{V}{H} \times \text{Spacing}\right)$
1	$.45 W_T$	$.50 W_T$	.9	4.32	4.32
2	$.35 W_T$	$.50 W_T$	.7	3.36	7.68
3	$.25 W_T$	$.50 W_T$	.5	2.40	10.08
4	$.15 W_T$	$.50 W_T$	.3	1.44	11.52
5	$.05 W_T$	$.50 W_T$	.1	.48	12.00

The slope in the segment adjacent to the centerline is not equal to zero. If we had taken an *odd* number of equally spaced segments the central segment would have had a zero slope.

Modern-day suspension bridges are designed in such a manner as to assure that the major weight carried by the cables is uniformly distributed among the suspenders. This weight is made up of the roadway and the heavy stiffening member. The member may be a heavy beam or a truss system. The roadway and stiffening system distribute the moving traffic load over many of the cables at once, thereby reducing the distortion of the cable from its normal configuration.

2.8 STATICS OF ARCHES

As we noted earlier, we can determine the shape of an arch for a given set of loads by examining a suspended cable carrying the same loads. This arch would be a series of straight line segments, just as the cable is a series of straight segments. As the loads get more numerous and closer together, the cable and the corresponding arch take on the appearance of a continuous curve. It is this smooth curve that is so pleasing to the eye.

The design of an arch span is based mainly on the dead load of the structure. This means that the shape of the arch curve will be determined, for the most part, by the distribution of the dead load. The arch will be very close to an ideal arch as it supports the dead load. The moving traffic loads will cause a small amount of bending to be present in the arch. This bending is minor and is easily managed by a properly designed arch.

It might be interesting at this point to compute the arch shape for a given set of dead loads. Figure 2.23 shows a parabolic arch with a span of 100 ft and a rise of 20 ft at the center. The arch is supporting a distributed masonry load. The height of the masonry is 10 ft above the crown of the arch. The dead loads that the arch carries change in magnitude over the span length. The maximum load per unit length will be found at the supports, and the minimum load, over the crown.

Figure 2.24 shows the actual arch form as the solid curve. The total height of the masonry can be calculated by

$$y = \frac{4h}{L^2}x^2 + 10$$

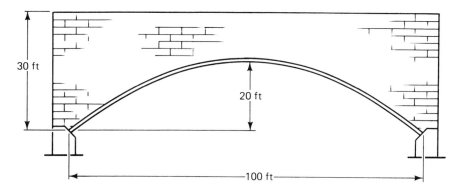

Figure 2.23
Parabolic arch.

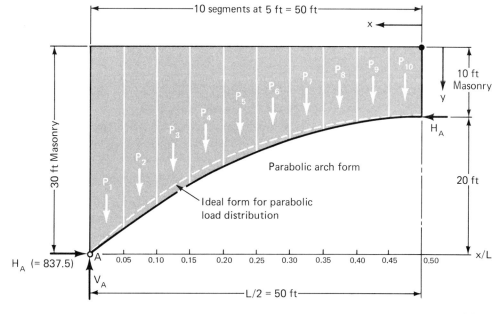

Figure 2.24

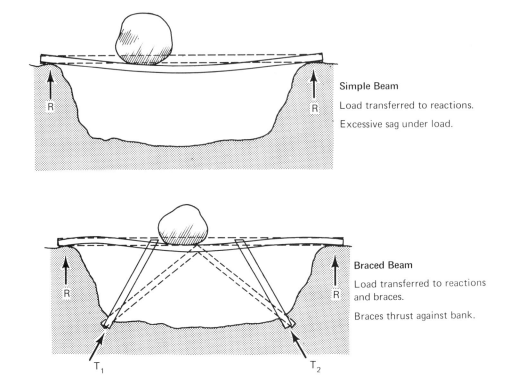

Simple Beam

Load transferred to reactions.

Excessive sag under load.

Braced Beam

Load transferred to reactions and braces.

Braces thrust against bank.

Truss ABC

Forms support structure for beam.

Lower tie AC absorbs outward thrust against bank.

Figure 2.25
Evolution of truss.

where x is measured from the crown. The forces are proportional to the height of the masonry, so each force, $P_1, P_2, P_3, \ldots, P_{10}$, can be computed. These forces are used with the equilibrium equations in exactly the same manner as was done in Section 2.7 with the cable.

The details of the calculation are omitted here, but results are plotted in Figure 2.24. The solid line is the parabola and the dashed line is the new curve. The differences in these curves are small. The parabolic load requires a shape that is slightly more circular than a uniform load.

2.9 EVOLUTION OF THE TRUSS

The truss form is the one bridge form that is solely the product of man's inventiveness. The truss does not appear in nature. This form evolved from the simple beam as man needed to extend the distance that could be spanned by the beam. The final form of the truss does not resemble the braced beam that was its beginning.

We can trace the probable steps in the development of the truss with the aid of Figure 2.25. The simple beam across the stream was too weak to support the loads over the entire distance. For some reason it was not feasible to put an intermediate pier in the center of the stream to shorten the span. The only thing that could be done was to place braces under the beam against the banks of the stream. These braces would press outward against the banks as they picked up load from the beam above. If the span was not too large, the braces could meet at the center, and the horizontal forces in the braces at that point would balance each other. This made for easier connections between the brace and the beam.

At the lower ends of the braces, the thrusts were resisted by the earth bank. There was a tendency for these braces to spread as they pushed their way into the mud of the bank. To help with this problem the two lower ends of the braces were connected together with a timber tie. This permitted the horizontal component of the thrusts at the bottom of the braces to balance each other. Now, the only reaction delivered to the banks was the vertical components.

This triangle, made up of the braces and the lower tie, becomes an elementary truss. Figure 2.26(a) shows the triangle ABC made up of three timbers pinned at the vertices. Supports are provided under the points A and B corresponding to the support of the mud banks of the stream. The load P applied at point C corresponds to the reaction of the beam on the top of the braces. This triangle is held in equilibrium by the two upward reactions. The value of these reactions can be computed by the application of equilibrium, as was done in the beam examples. In this case, because of the symmetry of the truss, each reaction is $P/2$. It should be noted that no horizontal reaction is required in this case. If the applied load were inclined, it would be necessary to resist the horizontal reaction at A or B but not at both.

Having established the external equilibrium, we can look at the internal forces in the bars.

2.10 STATICS OF TRUSSES

The bars in a truss are loaded with tension or compression forces directed along the axis of the bar. This can be demonstrated by examining the equilibrium of a typical bar in a truss. With this fact and the assumption that the

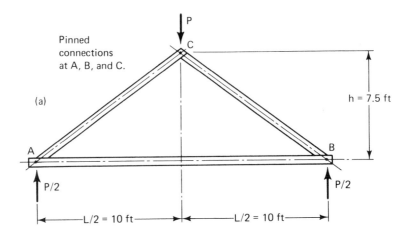

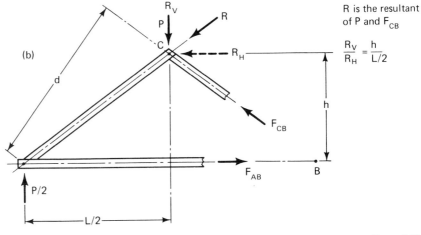

Figure 2.26
Simple truss.

members are connected at the joints with frictionless pins we are able to treat the analysis of the truss, joint by joint, as a series of *concurrent force systems*. By isolating the pin, as a body acted upon by the bar forces and other external bar forces, we create a *concurrent force system*.

As may be recalled from Section 2.2, this force system has two equilibrium requirements; therefore no more than two of the forces at a joint may be unknown. Since all the directions are known, we shall start the analysis at some joint where only two members are connected. Figure 2.27 shows joint

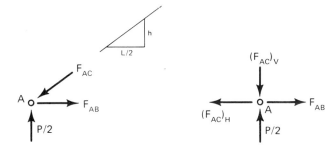

Figure 2.27
Equilibrium of truss joint.

A of the truss in Figure 2.26(a) isolated with the unknown forces F_{AC} and F_{AB} acting along with the known reaction $P/2$. The vertical equilibrium of this joint requires that

$$(F_{AC})_v = \frac{P}{2}$$

Now because F_{AC} is directed along member AC we can say that

$$(F_{AC})_h = \frac{L/2(F_{AC})_v}{h} = \frac{L}{2h}\left(\frac{P}{2}\right)$$

The horizontal equilibrium of the joint requires that

$$F_{AB} = (F_{AC})_h = \frac{PL}{4h}$$

The forces in the bars AB and AC are

$$F_{AB} = \frac{PL}{4h}$$

and

$$F_{AC} = \frac{P}{2}\left[1 + \frac{L^2}{4h^2}\right]^{1/2}$$

If $L = 20$ ft and $h = 7.5$ ft, then these values become

$$F_{AB} = \frac{2}{3}P$$

$$F_{AC} = \frac{P}{2}\left[1 + \frac{400}{4(56.25)}\right]^{1/2}$$

$$= \frac{P}{2}[1 + 1.78]^{1/2} = .83\,P$$

We still must establish whether the members are in tension or compression. Using joint A as a reference, it appears that the force F_{AC} is pushing against the pin and that the force F_{AB} is pulling on the pin. The force that pushes is the result of *compression* in member AC, and the force that pulls is the result of *tension* in member AB. This is as simple as it seems.

The previous example was particularly simple because the truss was a single triangle. The truss evolved from this simple configuration in order to extend its span. By inserting a panel between the two halves of this triangle the truss shown in Figure 2.28 is obtained. The rectangular panel $BCDE$

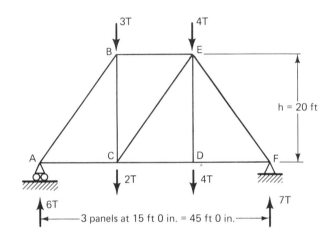

Figure 2.28
Truss analysis.

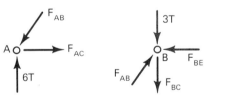

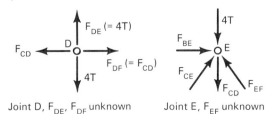

is divided into two triangles by the member *BE*. The four triangles form a rigid configuration that can support loads at any of its joints. There is no theoretical limit to how many triangles can be connected in this manner to extend the span even further. For our purposes we shall analyze this truss, since it illustrates all the important aspects of the truss.

The three-panel truss has a total span of 45 ft and a height of 20 ft. The triangles are conveniently proportioned for calculation; the sides have the ratios 3:4:5. The upper and lower panel points are loaded arbitrarily with loads totaling 13 tons. The reactions are shown on the truss at points *A* and *F*.

Solutions for the forces in the truss members are best carried out by proceeding in a systematic way. Since we shall be dealing with concurrent force systems at each joint, we must begin where no more than two unknown bar forces are present. Either support *A* or *F* would be a reasonable place to begin. We shall begin at joint *A* and proceed from there to joint *F*.

Joint A

Two unknowns: F_{AC} and F_{AB}.
Vertical equilibrium:

$$(F_{AB})_v - 6 = 0 \qquad (F_{AB})_v = 6^T$$

From similar triangles:

$$F_{AB} = \tfrac{5}{4}(F_{AB})_v = 7.5^T \qquad \text{(compression)}$$

Horizontal equilibrium:

$$\overrightarrow{F_{AC}} - (\overleftarrow{F_{AB}})_h = 0$$

$$(F_{AB})_h = \tfrac{3}{4}(F_{AB})_v = 4.5^T$$

$$F_{AC} = 4.5^T \qquad \text{(tension)}$$

Joint B

Two unknowns: F_{BE} and F_{BC}. Member *AB* known from joint *A*.
Vertical equilibrium:

$$\underset{F_{BC}}{\downarrow} \qquad (F_{AB})_v = 6^T$$

$$3^T$$

$$F_{BC} + 3 - 6 = 0 \qquad F_{BC} = 3^T \qquad \text{(tension)}$$

Horizontal equilibrium:

$$(\overrightarrow{F_{AB}})_h = 4.5^T - \overleftarrow{F_{BE}} = 0$$

$$F_{BE} = 4.5^T \qquad \text{(compression)}$$

Joint C
Two unknowns: F_{CD} and F_{CE}. Member AC known from joint A. Member BC known from joint B.
Vertical equilibrium:

$$\overset{\uparrow}{F_{BC}} = 3^T \qquad \overset{\downarrow}{(F_{CE})_v}$$

$$2^T$$

$$3^T - 2^T - (F_{CE})_v = 0 \qquad (F_{CE})_v = 1^T$$

$$F_{CE} = \tfrac{5}{4}(F_{CE})_v = 1.25^T \qquad \text{(compression)}$$

Horizontal equilibrium:

$$\overset{\rightarrow}{F_{CD}} \qquad \overset{\leftarrow}{F_{AC}} = 4.5^T$$

$$(F_{CE})_H = \tfrac{3}{4}(F_{CE})_v = .75^T$$

$$F_{CD} - 4.5 - .75 = 0 \qquad F_{CD} = 5.25^T \qquad \text{(tension)}$$

Joint D
Two unknowns: F_{DE} and F_{DF}. Member CD known from joint C.
Vertical equilibrium:

$$\overset{\downarrow}{4.0^T} - \overset{\uparrow}{F_{DE}} = 0 \qquad F_{DE} = 4.0^T \qquad \text{(tension)}$$

Horizontal equilibrium:

$$\overset{\rightarrow}{F_{DF}} \qquad \overset{\leftarrow}{F_{CD}} = 5.25^T$$

$$F_{DF} - 5.25^T = 0 \qquad F_{DF} = 5.25^T \qquad \text{(tension)}$$

Joint E
One unknown: F_{EF}. Member BE known from joint B. Member CE known from joint C. Member DE known from joint D.
Vertical equilibrium:

$$\overset{\downarrow}{F_{DE}} = 4^T \qquad \overset{\uparrow}{(F_{CE})_v} = 1^T$$

$$4^T \qquad (F_{EF})_v$$

$$8 - 1 - (F_{EF})_v = 0 \qquad (F_{EF})_v = 7.0^T$$

$$F_{EF} = \tfrac{5}{4}(7) = 8.75^T \qquad \text{(compression)}$$

Joint F
No unknowns. The results of the previous calculations may be checked for accuracy by testing the equilibrium conditions at joint F.

Vertical equilibrium:

$$(F_{EF})_v \downarrow = 7.0 \underset{\text{(from joint } E\text{)} \uparrow}{-} \text{reaction} = 7 = 0 \quad \therefore \text{ equilibrium checks}$$

Horizontal eqiuilibrium:

$$(F_{EF})_h \overrightarrow{} = \tfrac{3}{4}(7) = 5.25 - F_{DF} \overleftarrow{} = 5.25 = 0 \quad \therefore \text{ equilibrium checks}$$

2.11 HISTORICAL AND MODERN EXAMPLES

The use of the beam is perhaps the most common of all the bridge forms. The beam is used in many interesting ways that bear little resemblance to the lowly tree trunk used by our earliest ancestors. Modern beam bridges are used to solve a great variety of bridge problems, and choosing only a few examples means that we must ignore many fine specimens. Before presenting specific examples of beam bridges, it would be worthwhile to note that beams or beam elements play a very important role in practically every bridge, regardless of its type. The crossbeams and stringers associated with roadway supports are beams. The stiffening girders used to distribute loads to the suspenders on suspension bridges are really beams and function as such. The beam has an important place in most bridges.

Simple bridges constructed of beams are easily seen when you drive the new highways throughout the countryside. The bridges that provide the grade separations are multiple-beam bridges. The beams in these examples are called girders, but they have the same behavior as that described for beams. These bridges are quite uncomplicated. A number of girders are placed parallel to each other, spaced, in general, from 5 to 8 ft apart. The roadway slab is placed directly on the top of the girders. The girders are often steel beams that are standard sections made on the steel-rolling mills.

Where the spans get longer or where unusual loading requirements exist, special girders may be fabricated. The girders often have a variable depth, with the bottom of the girder resembling an arch line. Despite appearances these girders are beams in that they resist loads by bending. The shape of the girder is both functional and pleasing but does not change the beam to an arch.

Beams are somewhat limited insofar as long spans are concerned. The bending stresses (tension and compression) that the material of the beam must resist increase rapidly with increasing span length. It soon becomes inefficient and expensive to span large openings with simple beams. There are techniques by which the strength of the beam may be increased, thereby increasing the possible length of the spans. When applied to reinforced concrete beams, the technique is known as prestressing. This is a process in which axial compression is applied to the beam in addition to the bending

Figure 2.29
Rhine Bridge at Bendorf, West Germany.
(*Courtesy Morikita Publisher of Japan*)

stresses. This precompression serves to offset the tension-bending stress in the concrete and increases the efficiency of the concrete beams. The details of this technique are beyond the scope of this discussion, but the importance of the technique is clear when you consider that spans in excess of 650 ft have been erected. A bridge over the Rhine in Bendorf, West Germany, has seven continuous spans with lengths of 141, 145, 233, 672, 233, 145, and 141 ft. This bridge, pictured in Figure 2.29, is of prestressed reinforced concrete and in 1968 was the longest such bridge in the world. The Rhine Bridge is significant because of its length and uniqueness.

Another bridge that illustrates the place of the beam in modern bridge construction has a different claim to fame. The lower Chesapeake Bay Bridge-Tunnel crossing (Figure 2.30) contains, among many other interesting engineering features, 12.5 miles of bridge crossing the open water. This 12.5-mile stretch is spanned by a multiple-span beam bridge commonly referred to as a trestle structure. The individual spans are each 75 ft long and were precast on shore. They are supported on piles driven into the ocean bottom, at a height of 30 ft above mean low water. The significance of this bridge is the great economy that was realized in the repetition of the same span

Figure 2.30
Chesapeake Bay Bridge-Tunnel trestle construction.
(*Courtesy Tidewater Construction Corporation, Norfolk, Va.*)

for the full length of the bridge crossing. The unusual requirements of an oceangoing bridge, both in design (wind and wave action) and in construction, make this an exceptional application of the descendant of the rude log across the stream. Economy was not gained at the expense of beauty. There is a special appeal about this bridge when you drive over the open water. The bridge does not intrude on the experience of being over the ocean, but at the same time there is a complete sense of security.

An interesting sidelight might be added about this bridge. In the spring of 1971 a ship slipped its moorings and floated into the bridge. The damage to the bridge was restricted to two or three spans. The fact that the spans were simple beams limited the extent of the damage, and the bridge was back in full working order in a matter of weeks.

The last example of a beam-type bridge is an interesting combination of a beam with cable supports. Unlike the suspension bridge where the roadway girder is supported by suspenders from a draped cable, this form has straight cables from towers to a limited number of points along the girder. The Papineau-Leblanc Bridge spans the Riviere des Prairies near Montreal, Quebec. This bridge is known as a cable-stayed bridge. Figure 2.31 shows the arrangement of the cables that supply the equivalent of intermediate supports for the continuous girder. The single box-type girder and roadway act together to make a single longitudinal beam element. The locations of the cable connections are stiffened by a transverse diaphragm to distribute the support uniformly across the roadway. This bridge combines the elegance of slender cable supports with the simple line of the longitudinal girder to make a truly modern and exciting contribution to long span bridges.

Many of the early bridge builders used the arch form. These craftsmen did not have any theoretical knowledge of how arches work, but they had a great deal of experience. When you see the results of their work it becomes obvious that they were successful. The Romans were perhaps the greatest builders of stone arches. A number of their bridges are still in use today. The ruins of some of their structures, in particular the Pont du Gard built around 100 A.D., testify to both their amazing ability as builders and craftsmen and to their sense of beauty. The Romans constructed their arches from carefully dressed stone blocks. These blocks were placed together on falsework (temporary timber supports) with such precision that as soon as the last block, or keystone, was in place the falsework could be removed and the arch suffered no measurable deflection. The Romans did not use mortar to hold the stones together. They relied on the exactness with which the stones were cut and the shape of the arch itself to keep the cut faces in close, tight compression.

Following the Romans were the bridge builders of the Renaissance. They were even more daring than the Romans in the design of arch bridges. The Romans mainly used semicircular arches, which kept the amount of thrust against the supports to a minimum; the Renaissance builders used

Figure 2.31
Papineau-Leblanc (cable-stayed) Bridge.
(*Courtesy Civil Engineering Magazine*)

Figure 2.32
Pont D'Avignon stone arch bridge built by St. Eenezet 1177–1185.
(*Courtesy French Cultural Services*)

much shallower arches. Segments of circles as well as curves having other geometric properties were used. These arches had the advantage of much less weight, because the stone fill from the arch to the roadway level was shallower, and they were also more attractive than their Roman predecessors. They suffered by comparison with the Roman arches only in that they required much more in the way of horizontal thrust than the semicircular arches. A Roman bridge of multiple-span arches could sustain the destruction of one span without serious damage to the remaining spans, because the massive supports required for these arches could provide the horizontal thrust needed for the remaining arches. Shallow arches produce much higher thrusts and have generally smaller supports; therefore they rely on each other to balance the thrusts at a common support. If one arch is destroyed, the horizontal support is removed from the adjacent arch. The result is probably serious damage to, if not total collapse of, the rest of the bridge.

Despite this extra engineering burden on the builders of bridges, the arch became a form of special beauty. The Pont d'Avignon (Figure 2.32) is a classic example of this form. Built over the Rhone river by medieval builders about 800 years ago, it displays a delicacy that was absent in the Roman arches. The arches are not extremely shallow, and 4 of the estimated original 21 spans still stand. These spans are between 101 and 115 ft long; the remaining spans make a bridge crossing of a total of 3,000 ft in length. This was the longest bridge in its day, surpassing all the permanent Roman structures save the Roman aqueducts.

Another striking example of the stone arch bridge is the Rialto Bridge over the Grand Canal of Venice (see Figure 2.33). This single-span arch was designed by the distinguished architect Antonio da Ponte in 1587. The style is suited to the highly commercial Venice of the sixteenth century. The commercial buildings had crowded close to the banks of the canal, leaving little room for the construction of abutments. The bridge dimensions are 83 ft in span with a rise of approximately 18 ft, 6 in. The roadway width is a generous 66 ft, leaving enough room for the two rows of shops designed as an integral part of the bridge.

The arch remains a bridge form with many modern uses. Arches with spans as long as 800 ft are not at all uncommon. These arches are invariably constructed of steel or reinforced concrete. The arch also serves economically

Figure 2.33
Rialto Bridge in Venice, Italy, stone arch designed by Antonio da Ponte, 1587.
(*Courtesy Italian Government Travel Office*)

for short spans under 400 ft in length. There is a variety of treatments available in the design of modern arch bridges that enable the designer to preserve the delicacy of the arch line and still convey the immense strength inherent in the arch.

2.11 historical and modern examples

The suspension bridge is also a form with great esthetic appeal. The line of the draped cable set off against the horizontal line of the roadway is very pleasing. It is possible to introduce the esthetic dimension into the design of suspension bridges in such a way as to make them truly works of art. The inherent strength of the form led the early designers of these bridges to build very light, flexible bridges. These were undoubtedly graceful but they contained a hidden flaw that caused many of these bridges to collapse.

The very earliest suspension bridges were supported by vines or ropelike elements. These bridges have been lost to recorded history, and we can only speculate as to their details. However, as soon as the iron age was upon us, these bridges began to be built using iron chains as the suspension element. In China this custom has a long history. The beginnings of the use of suspension bridges in Europe can be traced back at least to 1617, when a design for a suspension bridge was published by Faustus Verantius.

The basic behavior of the suspension bridge was widely understood and used by 1810. In 1796 the first modern suspension bridge was built in the United States. In 1855 the Niagara Bridge was open to rail traffic. This bridge had a span of 821 ft and was 245 ft above the water. Although this was not the first suspension bridge of note in the United States, it served to stamp its designer and builder, John Roebling, as a first-rate genius who would eventually leave us the remarkable Brooklyn Bridge.

Roebling's genius was that he learned from experience and could project that experience into untried areas. The flaw in many of the early suspension bridges was the lack of rigidity. The motions of these bridges when subjected to wind or the rhythmic loading of marching feet were exaggerated out of proportion to the actual forces involved. The problem was one of dynamic instability. The bridge vibrated in resonance to the wind forces until the amplitudes of these motions were so large that the bridge tore itself apart. This was the fate of many of the early flexible suspension spans.

Roebling interpreted this behavior correctly and was able to counteract it by adding additional bracing cables from the towers to the roadway. These cables give the Brooklyn Bridge its unique appearance, as illustrated in Figure 2.34. Later this problem was solved by making the roadway structure itself much heavier and stiffer to resist the motions caused by the wind. In addition, the shape of the roadway cross section was eventually designed to make it impossible for the wind to create the oscillating forces on the bridge.

As late as 1940 the behavior of flexible suspension bridges was not fully understood. Roebling's technique of using cable stays was not generally practiced. The bridges of the modern era, that is, the first half of the twentieth century, rely on a heavy, stiff roadway system to counteract the action of winds. In 1940 the Tacoma Narrows Bridge was completed and opened to traffic. Four months later it collapsed under the action of a moderate wind of 42 mph.

Figure 2.34
Brooklyn Bridge designed and built by John Roebling, completed 1883.
(*Courtesy Civil Engineering Magazine*)

The Tacoma Narrows Bridge spanned 2,800 ft. Its roadway was 39 ft wide, and the roadway deck was supported on 8-ft-deep beams. The beam depth-to-span ratio was approximately 1/350 for this bridge, whereas most suspension bridges were designed with a ratio of 1/50 to 1/100. The beams were strong enough for all the forces, including the horizontal effects of the wind. An investigating committee absolved the designer of any fault of this nature. The problem was again with the flexible roadway supports and the comparatively narrow roadway. The roadway acted as an airfoil, causing the alternating lift and fall of the roadway to be intensified. This resonance finally built up to the point where the motions were too large to be accommodated by the hangers and cable, and the bridge literally tore itself apart. Fortunately, the disaster did not claim any lives. Figure 2.35 shows the bridge undergoing violent motion.

Figure 2.35
Tacoma Narrows Bridge, "Galloping Gertie" collapsed in 1940.
(*Courtesy R. A. Heller*)

The result of this spectacular failure spurred the engineering profession to find the answers to this puzzling problem. The twentieth-century counterpart of Roebling, David B. Steinman, designed a suspension bridge for the Mackinac Straits connecting the two parts of Michigan. This bridge, shown in Figure 2.36, has a main span of 3,800 ft and two side spans, giving a total of 8,614 ft and making it the longest suspension bridge in the world at the time it was built. Construction began in 1954 and was completed in 1957. The importance of this bridge does not lie in its immense spans or in the fact that it was built under the most difficult weather conditions. The most significant aspect of the Mackinac Straits Bridge is that it would take a wind velocity of between 622 and 966 mph to cause the type of aerodynamic instability that wrecked the Tacoma Narrows Bridge. This fact gives the bridge practically an infinite factor of safety against this type of disaster. The design was accomplished with the use of wind-tunnel models, leading to a combination of beam depth, beam profile, and, in particular, openings in the roadway slab (covered by open grating) to remove the lift forces caused by wind. This was the first completely aerodynamically stable bridge ever designed.

The truss, like the beam, is often found in bridges that are primarily of a different form. The modern suspension bridge, for example, often utilizes stiff trusses to carry the roadway. These trusses serve an extremely impor-

Figure 2.36
Mackinac Straits Bridge in Michigan, designed by D. B. Steinman, completed in 1957.
(*Courtesy American Bridge Division of United States Steel*)

tant function in the overall behavior of the suspension bridge. Likewise, many arches are not solid members but are actually made in the form of curved trusses. The truss accepts the thrust from the arch action and resists it locally in a manner peculiar to a truss. The cross bracing found in many major bridges is, in reality, an elaborate truss whose function it is to resist lateral deformations, rather than to carry the load directly.

However, the truss has its own peculiar place in history and in modern-day use. The principle of the truss was first understood by a Renaissance architect named Palladio, who completed a description of this principle in 1570. The very earliest examples of timber trusses have been lost because of the susceptibility of timber to fire and rot. However, the designs of Palladio have been preserved in his four-volume book *A Treatise on Architecture*. Two of these forms are shown in Figure 2.37. The first of these was actually used to cross the Cismone, a span of 108 ft. As is often the case, a pioneering work such as Palladio's falls into obscurity only to be revived at a later date. It was 200 years later, in 1742, before Palladio's works were translated into English and began to influence the bridge builders of England.

Some of the most interesting examples of timber truss bridges are those that were built in the 1800s in the United States. These are the famous covered bridges, which may be found in 29 of the 50 states. Far from being exclusive to New England, these bridges are in existence in every section of the United States with the exception of the southwest.

These bridges were as often the products of the imagination of the local bridge builder as they were the result of an understanding of the principles of truss behavior. Some of these bridges were combinations of trusses and arches, making a highly complex interaction between these two forms. Some of the bridges were true trusses; we can analyze them by the technique described in Section 2.10. The *queen-post truss* and the *multiple king-post*

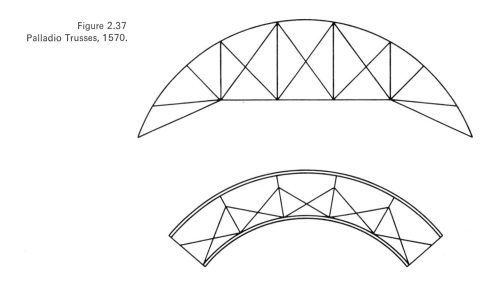

Figure 2.37
Palladio Trusses, 1570.

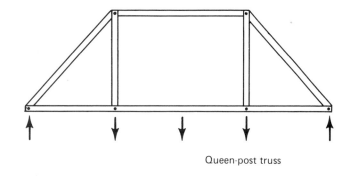

Figure 2.38

Queen-post truss

Multiple king-post truss

are typical of these simple trusses. Figure 2.38 shows these forms. Other complex forms are the *Long truss*, the *Howe truss*, and the *town lattice trusses*, named after their inventors. These are shown in Figure 2.39.

The fact that these bridges were covered is the major reason so many of them remain in use today. The covering was the most reasonable way to protect the structure against the alternate wetting and drying that soon rots the timber.

Only one example of a covered bridge will be given: The bridge shown in Figure 2.40 is a bridge over the Seymore River in Cambridge, Vermont. This little bridge is under 50 ft in span and is of the Burr truss design, which is nothing more than a multiple king-post with an arch added for extra strength. This particular bridge was taken out of service in 1955 and replaced by a modern steel and concrete girder bridge. It was moved to a farmer's field nearby, where it is used to cross farm animals and machinery over a stream. The State Highway Department maintains the bridge as one of the historical treasures of Vermont.

The introduction of iron as a construction material made possible truss bridges having much larger spans than those made of timber. At first the iron was used in place of timber for tension members in the timber trusses. Eventually it came to replace timber entirely. The advent of the railroad

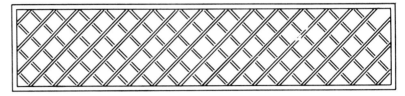
"Town lattice mode": 1820

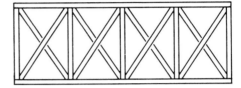
Long's multiple king-post: 1830

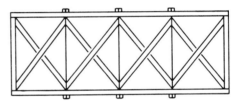

Figure 2.39   Howe's multiple king-post with iron verticals: 1830's

spurred the construction of truss bridges during the 50 years from 1840 to 1890. The bridges were invariably ugly but economical.

These iron bridges were not very reliable by modern standards. During the period from 1870 to 1880 an average of 25 bridges on the U.S. railroads failed each year. This record was not due to faulty engineering. The failures were due to the lack of uniformity in the material, cast iron, and to a lack of understanding of the behavior of large moving loads. The engineers of the time did their best with what they knew, but there was yet much to be learned. The modern era in bridge building began about 1880, when steel became available at a reasonable cost. There are many great bridges built of steel trusses; we shall discuss two of the greatest.

In Scotland, a narrow arm of the ocean that makes a deep penetration into the land is called a firth. This inlet presents a barrier to rail travel along the coastline. To avoid an expensive detour around the firth, the Firth of Forth Bridge was built in the years 1883–1890. This great bridge is a cantilever design, with the cantilevers and suspended spans formed of trusses (see Figure

Figure 2.40
Vermont covered bridge.
(*Courtesy Paul Miller*)

2.41). The cantilever principle was well known in the ancient Orient, and this bridge was one of the first modern bridges based on this system.

The principle is simple and is illustrated in Figure 2.42. Each of two spans is balanced on a single pier. The huge mass of the dead weight keeps these spans balanced with only a small amount of support needed at either end. The long inner arms of the two cantilevers reach toward each other to support a simple span between their ends. This simple span is called the *suspended span*.

The Firth of Forth Bridge has three cantilever spans with two suspended spans. The outer ends of the two outside cantilever spans were counterweighted to balance the load of the suspended span. The two center openings have a clear span of 1,700 ft. The suspended spans are each 350 ft. The trusses of the cantilevers are 350 ft deep over the piers, tapering to 50 ft deep at the suspended spans. The suspended spans are 150 ft above water. This bridge is no beauty. It has been charitably described as awkward. It does have an overpowering presence that dominates the landscape with a seemingly invincible strength. It stands as a monument to its designer, Benjamin Baker (1840–1907).

The second truss bridge to be described is part of the Chesapeake Bay crossing discussed earlier. The 17.5-mile-long bridge crosses three navigation channels. Two of these navigation channels are kept open by the use of tunnels under the channel. The third channel is spanned by a high-level bridge 325 ft long.

The bridge itself is not too unusual. It is a normal through-type truss set on piers 80 ft above normal water level. The unusual aspect of this bridge

Figure 2.41
Firth of Forth cantilever, design by Benjamin Baker, 1840–1907.
(*Courtesy Civil Engineering Magazine*)

is that it was fabricated and erected on a barge and floated into position. This barge was equipped with a temporary set of supports that held the erected truss high enough to clear the pier caps. When the barge was in place, the bridge was lowered onto the caps and it, in turn, was in place. It was only necessary to place the roadway slab on the bridge stringers before the

Figure 2.42
Cantilever system.

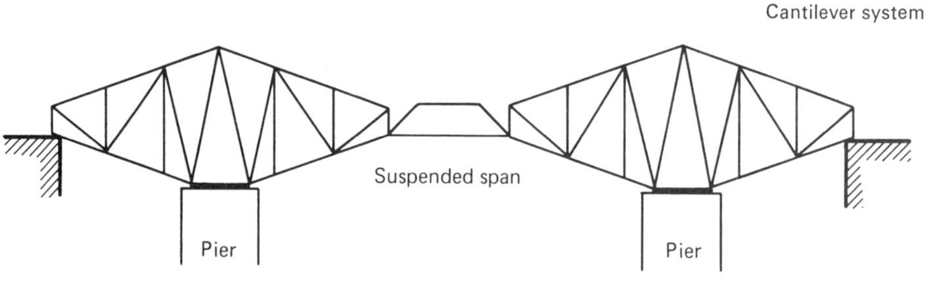

bridge was ready for use. This erection technique saved much in the way of hazardous exposure and kept the navigation channel open for ship traffic. Figure 2.43 shows the trusses being lowered onto the piers.

2.12
DESIGN PROJECT

The discussion of bridge forms in the previous sections is necessarily brief and incomplete. A thorough understanding of many related subjects such as statics, dynamics, strength of materials, and the many details of design is necessary before we can really appreciate the achievements of the designers and builders of bridges.

However, we now know enough to attempt a small experiment. Consider the following problem: We wish to span a clear opening of 30 in. The bridge form we use must provide for a roadway width of 4 in. and a maximum bridge width of 4.5 in. The bridge will be loaded by a simulated traffic load that applies four loads of equal magnitude on the bridge. The details of the spans and loading arrangement are shown in Figure 2.44.

To simplify the actual construction of the bridge model, we shall limit ourselves to balsa wood, string, and Elmer's glue for materials. The balsa

Figure 2.43
Chesapeake Bay Bridge-Tunnel, placing of the through truss span.
(*Courtesy Tidewater Construction Corporation, Norfolk, Va.*)

wood should be available in a variety of cross-sectional dimensions and in 36-in. lengths. The commonly used sections are $\frac{1}{8}$, $\frac{1}{16}$, and $\frac{1}{4}$ in. square and $\frac{1}{16} \times \frac{1}{4}$, $\frac{1}{8} \times \frac{1}{4}$, and $\frac{1}{16} \times \frac{1}{4}$ in. rectangular sections. Some $\frac{1}{16}$ in. flat stock will be needed for fabrication of connections.

The project is to design and construct a bridge to the above specifications. The bridge designs may be evaluated by a number of criteria such as appearance and workmanship. However, a more realistic evaluation might be the determination of how much load the bridge will carry before failure. To make a valid comparison between two different designs, we must account for the amount of material used in the bridge model. In general, the more material efficiently used in the bridge, the greater the failure load. To judge the efficiency of the designs we ought to compare the ratios of the load carried to the weight of material in the bridge. This rough efficiency factor is a good measure of the effectiveness of the design.

A few helpful hints are in order at this point. In making small-scale models the most difficult problem encountered is in forming the connections. The normal contact areas between the small cross-section members are too small to develop any appreciable strength in a simple glued joint. To increase this effective area we can glue each member to a common plate called a *gusset plate*. The gusset plates are made from the flat stock. Since there is a pronounced difference in the strength of the flat balsa wood sheets between the long direction and the short direction, we should prepare the gusset plates by first laminating two pieces of the flat stock at right angles to each other (see Figure 2.45). Two such gussets should be used at each

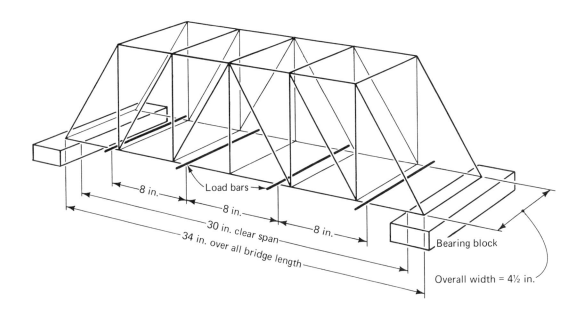

Figure 2.44
Typical bridge model and load arrangement.

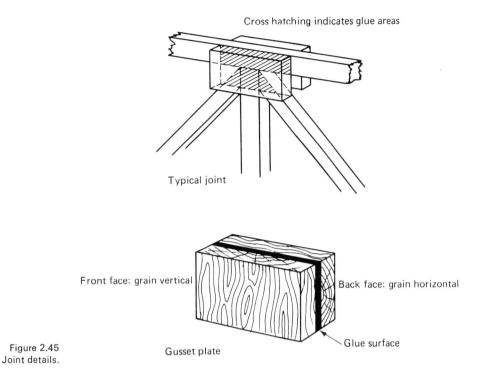

Figure 2.45
Joint details.

joint to provide additional strength and to maintain symmetry in the joint. A typical joint is shown in Figure 2.45.

Another useful technique, for curved members that are to be laminated for arch models is to preform the member by wetting the wood before bending. After the member has dried in its bent form it will take less force to maintain the proper shape. The glue used in laminating the strips together is not wasted in trying to resist the tendency to straighten out.

The last suggestion to improve the final behavior of the bridge model is the proper use of cross bracing. Since there are two main bridge forms supporting the roadway, it becomes imperative that the two be interconnected so that they will share the load as evenly as possible. A second reason for cross bracing is to provide for intermediate supports for long compression members. The top members of a truss are likely to be in compression and may buckle laterally before they fail in compression. By bracing these members to the other truss, this tendency to buckle will be reduced.

The construction and testing of these bridge models will give us an idea of the problems associated with real bridges. Seeing a model being loaded to failure and trying to predict how the failure will take place is a unique learning experience. The comparisons that can be made between different models is also an interesting experience. With a little care and attention to detail these bridge models can be made to carry over 500 pounds before failure. The author has tested three models that have carried over 800 pounds. How about yours!

BIBLIOGRAPHY

1. ALLEN, R. S., *Covered Bridges of the Northeast.* Brattleboro, Vt.: The Stephen Greene Press, 1957.
2. BECKETT, D., *Great Buildings of the World, Bridges.* New York: The Hamlyn Publishing Group, 1969.
3. DEMERS, J. G., and O. F. SIMONSON, "Montreal Boasts a Cable-Stayed Bridge," *Civil Engineering*, Aug. 1971.
4. EELES, A., *Canaletto.* New York: The Hamlyn Publishing Group, 1970.
5. GIES, J., *Bridges and Men.* Garden City, N.Y.: Doubleday & Company, Inc., 1963.
6. NARUSE, K., K. AOKI, and E. MURAKAMI, *Bridges of the World.* Tokyo: Morikita Publishing Company, Ltd., 1968.
7. SLOAN, E., *American Barns and Covered Bridges.* New York: Funk & Wagnalls, 1954.
8. SMITH, H. S., *The World's Great Bridges.* New York: Harper & Row, Publishers, 1965.
9. STEINMAN, D. B., and S. R. WATSON, *Bridges and Their Builders.* New York: Dover Publications, Inc., 1957.
10. STRAUB, H., *A History of Civil Engineering.* Cambridge, Mass.: The M.I.T. Press, 1952.
11. SVERDRUP, F., "Engineering Design, Chesapeake Bay Bridge-Tunnel Project," *Civil Engineering*, Dec. 1963.
12. TYRRELL, H. G., *Bridge Engineering.* Copyright 1911 by Henry Grattan Tyrrell, C. E.

3

mass transportation

D. E. Cromack

W. P. Goss

3.1 INTRODUCTION

This chapter is designed to familiarize the beginning engineering student with the terminology, concepts, and current status of mass transportation. Nearly all fields of engineering are encountered during or within the complexities of mass transportation.

Mechanical engineers are involved in the design and manufacture of vehicles and components. Civil engineers lay out, design, and supervise the construction of highways and guideways for transportation systems. The problems of command, control, and communications are within the realm of electrical engineering. Scheduling procedures and management techniques utilize the knowledge generally covered in an industrial engineering curriculum. Chemical engineers, generally associated with the petroleum industry, are important in the field of mass transportation, since it is the largest consumer of petroleum products. Furthermore, a broad spectrum of engineering expertise is needed in the planning, implementation, and operation of transportation systems.

A wide range of job opportunities is available in transportation-related fields as indicated above by the involvement of all engineering disciplines. These opportunities are expected to become more prevalent, without any regional restrictions, as more and more emphasis is placed on the improvement of mass transportation.

Mass transportation, as defined here, encompasses *all modes* of transportation so far devised by man to move people from one place to another. The transportation of goods will not be considered, although much of the material in this chapter applies equally well to the movement of goods. Throughout history the trend in mass transportation has been to increase speed and to improve economy. Transportation systems have developed according to the needs of society and to the suitability to the region, terrain, or times. Yet we find ourselves in the dilemma of inadequate transportation for most of the disadvantaged, the handicapped, the elderly, and the young in our society. Questions that we shall look into in this chapter are the following: How have we gotten to the state of mass transportation as it now exists, how can we plan for better mass transportation, how can we select the best of a wide variety of possible mass transportation systems, and how can we implement that system or systems with minimal impact on man and his environment?

The overall national transportation picture now consists of a massive Interstate Highway System for automobiles, trucks, and buses, an attempt at regrowth of passenger trains to operate at higher speeds with improved service and comfort, along with an airway system that is rapidly reaching saturation conditions insofar as many airport facilities are concerned. Associated with these complications for long-distance travel are the problems of congested urban mass transportation. Cities are rapidly becoming, or are already, clogged with automobiles. New streets, freeways, and parking lots are using up available land and are still inadequate to accommodate the ever-increasing numbers of automobiles. It is estimated that approximately 40% of the land of the City of New York is devoted to the motor vehicle. To indicate the almost total dependence that we have on the automobile in the United States, consider the data given in Tables 3.1 and 3.2. Table 3.1 shows that overall 86% of the people who use transportation for any reason use the automobile. This is not simply an indicator of short trips under 50 miles where 95% of the people use the automobile; it also applies

Table 3.1
Use of Transportation Modes

Mode of Transportation	Travelers (% Use)
Automobile	86%
Bus	3
Commercial air	8
Train	1
Other	2

Source: 1971 Automobile Facts and Figures, Automobile Manufacturers Association, Inc.

to long trips over 1,000 miles where 56% of the people use the automobile. From Table 3.2 it can be seen that many low-income households do not own an automobile and must therefore rely on other modes of transportation. Equally interesting is the fact that although the total number of registered motor vehicles has risen from 60 to 90 million from 1960 to 1970, the total number of households without automobiles has remained constant at about 13 million. The additional automobiles have largely been absorbed by the increase in multicar households. Thus, we are in a situation where we have almost total dependence on the automobile for mass transportation, while inadequate transportation exists for a large portion of the low-income households in our society. A similar situation exists for the elderly, where 45% of the household owners over age 65 do not own an automobile.

Table 3.2
Households Owning Automobiles

Household Income ($)	No Car (%)	One Car (%)	Two Cars (%)	Three or More Cars (%)
Under 3,000	57.5	38.0	3.8	.7
3,000–4,999	30.8	55.9	11.8	1.5
5,000–7,499	13.6	64.3	19.5	2.6
7,500–9,999	8.4	56.9	29.9	4.8
10,000–14,999	4.1	47.5	40.2	8.2
15,000 and over	3.8	33.1	50.4	12.7
Total all households	20.4	50.3	24.6	4.7

Source: 1971 Automobile Facts and Figures, Automobile Manufacturers Association, Inc.

This picture does not look pleasing, but it does show the need for extensive efforts to improve many aspects of our mass transportation systems. In this chapter we shall be discussing some of the specific problems, looking at some of the possible solutions, and, hopefully, generating some interest for future work in the area of transportation.

3.2 HISTORY OF MASS TRANSPORTATION

3.2.1 Historical Background

Rather than go through a dull presentation of the history of mass transportation in the United States, we shall draw upon the knowledge of the reader. Most of you are familiar with important dates in United States history, such as

- 1620s: Colonization
- 1776: Independence
- 1860s: Civil War
- 1910s: World War I
- 1940s: World War II

So that everyone will have a common reference point, we shall add the following dates that are important to transportation:

- 1700s: Steam engine invented
- 1700s: Canals started
- 1730s: First stagecoach line

1800s: First turnpikes
1800s: Steam ships
1830s: First railroad
1900s: First automobiles
1900s: Wright brothers' first flight
1950s: Jet aircraft
1950s: Start of the Interstate Highway System
1960s: Establishment of the U.S. Department of Transportation (D.O.T.)

With this information, we shall ask you to imagine yourself in one city desiring to travel to another city several hundred miles away at the following points in time:

1650: 30 years after colonization
1750: 25 years before the Revolutionary War
1850: 10 years before the Civil War
1950: 5 years after World War II
2050: The future

Next, consider the various means of transportation, from walking to flying, that are possible at each of these times. For each of these dates, establish lines that are proportional in length to the credibility of traveling by way of the various modes you are considering. By credibility, we mean the probable outlook of middle-class citizens of each particular time, as they consider the means of transportation available to them for a particular journey. The citizen would probably consider to some degree the costs, convenience, availability, trip times, and relative safety associated with the trip. For example, let's consider the possibility of traveling from Boston to New York City in 1900. The automobile is just being developed, the Wright brothers are several years from their first aircraft flight at Kitty Hawk, and travel by boat is still a valid possibility. Figure 3.1 shows an approximate listing of the

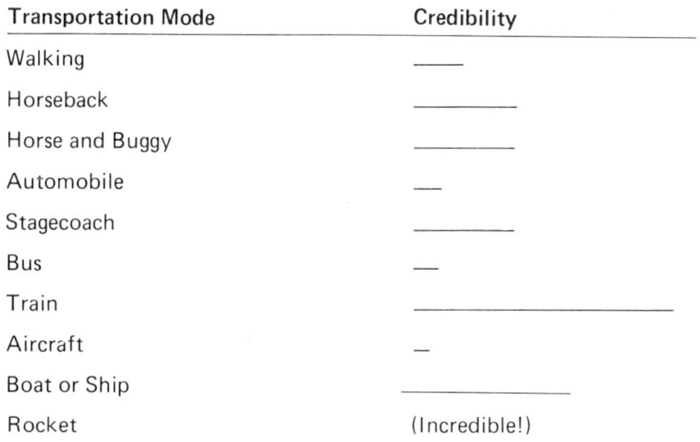

Figure 3.1
Credibility of various means of travel from Boston to New York City in 1900.

Transportation Mode	Credibility
Walking	____
Horseback	_____
Horse and Buggy	_____
Automobile	__
Stagecoach	_____
Bus	__
Train	_____
Aircraft	_
Boat or Ship	_____
Rocket	(Incredible!)

credibility of the different modes that a citizen of Boston in 1900 might consider. Note that although the length of the credibility lines may not be exactly proportional to the actual usage of a particular transportation mode, the lines can be used to produce a reasonable estimate of "relative usage." Once you have determined the credibility for the various dates suggested, these dates may be plotted on a graph of credibility versus time, as shown in Figure 3.2. Only one mode of transportation, the train (or rail), is plotted. Note that the drop in relative usage during the depression of the 1930s, after the stock market crash, may not occur to each of you; however, some of you might surmise this occurrence. Even if this were not recognized, you should arrive at the general trend of increase in rail travel in the upper portion of the Northeast Corridor (Boston to New York City) until the end of World War II, at which time automobile travel started to expand rapidly. The most interesting portion of this exercise is in trying to predict the future of the various modes of transportation. Will rail travel significantly increase as shown on Figure 3.2 because of the creation of the quasi-public rail corporation AMTRAK (described in Section 3.3), or will it be pushed to extinction by the automobile and aircraft? Questions of this type are of prime interest in this chapter and to answer them requires insight into the various types of transportation systems now being used or developed for future use, the impact of these systems on man and his environment, the way people select transportation systems, and the need to plan adequately for implementation of selected systems. The establishment of a rough credibility (or relative usage) versus time plot for all possible modes helps give some insight into answering these problems. (This development is left as an exercise; see Exercises 1 and 2.) In general, the mass transportation trends will show the demise

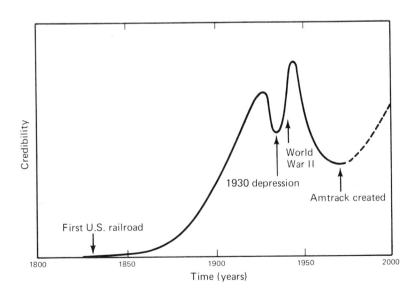

Figure 3.2

of horse-drawn land vehicles and water vehicles as railroads developed, and then the demise of railroads as the use of the automobile and the airplane rapidly expanded.

3.2.2 Federal Developments

The trends previously discussed were recognized in the early 1960s on both a national and regional basis. The Northeast Corridor between Boston and Washington is a prime example where rapid changes in transportation usage have created the need for regional studies and planning to meet the need for personal mobility. Taking the initiative, the federal government created, in 1966, the U.S. Department of Transportation (D.O.T.), with the Secretary of Transportation as a member of the President's Cabinet. The prime objective of the D.O.T. is to establish a more comprehensive program in federal transportation policy to allow the development of balanced transportation systems. The basic organization of the D.O.T. is shown in Figure 3.3. The principal components and their functions are

Urban Mass Transportation Administration (UMTA): Administers capital grant, research, development and demonstration, and technical study programs in the area of urban mass transportation

Federal Railroad Administration (FRA): Deals with railroad safety, high-speed ground transportation research, and operation of the Alaska railroad

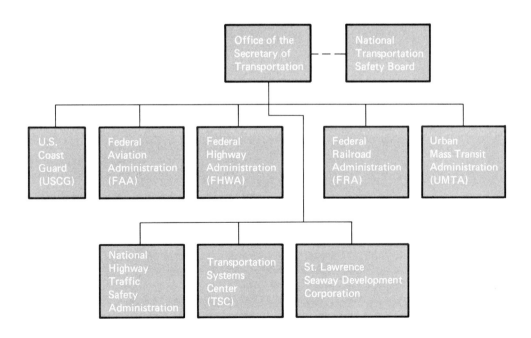

Figure 3.3
Organization chart, U.S. Department of Transportation.

Federal Highway Administration (FHWA): Administers the federal highway construction programs and highway safety and improvement programs

Federal Aviation Administration (FAA): Administers federal airport construction programs and aircraft safety programs and provides aircraft traffic and control facilities in the United States

U.S. Coast Guard (USCG): Administers coastal and inland waterway regulation programs and provides water safety programs

National Highway Traffic Safety Administration: Provides basic information and carries out research on highway safety

In addition to these major administrations there are also the following organizations:

Transportation Systems Center: Provides technical transportation expertise to the other administrations and to the Secretary of Transportation

St. Lawrence Seaway Development Corporation: Operates the St. Lawrence seaway facilities

National Transport Safety Board: A semiautonomous organization that has major responsibility for transportation safety

Although not shown in Figure 3.3, the Secretary of Transportation has a number of assistant secretaries and special offices and boards that perform vital functions and report directly to the Secretary.

From these various components of D.O.T., several new programs have developed. One was the Northeast Corridor (NEC) transportation project, which carried out a major study of all modes of mass transportation in the Northeast Corridor region. The results indicated that the region has major

Figure 3.4
V/STOL aircraft.
(*Courtesy
LTV Aerospace Corporation*)

transportation problems that might be alleviated by the use of high-speed rail systems, a second North–South interstate highway made up from modifications of the existing highway system, and the introduction of vertical and/or short takeoff and landing aircraft systems (see Figure 3.4). In addition, several rail demonstration programs, which are described in Section 3.5.3, have been carried out by the FRA in the Northeast Corridor between Washington and New York City and Boston. The Office of High Speed Ground Transportation (OHSGT), formerly within the FRA, administered the Northeast Corridor studies and also carried out rail demonstration programs and intercity high-speed ground transportation research and development studies. The functions of this office are now divided between the FRA and the Office of the Secretary to Transportation.

Programs in tracked air cushion vehicle (TACV) development and demonstration by both the FRA and UMTA are currently underway, as is tube vehicle system (TVS) research by the FRA (see Figure 3.5). The FAA has a continuing development and demonstration program on vertical and/or short takeoff and landing (V/STOL) systems and is also working on advanced aircraft control systems. The UMTA has been and is involved with a large number of major programs to improve urban mass transportation. These programs are the direct result of Congress passing the Urban Mass Transportation Assistance Act of 1970, which authorized $10 billion in federal assistance to solve local and regional urban transportation problems. Some of these programs are rail rapid transit development and innovative bus transit demonstrations. Most of the systems associated with these programs are described in Section 3.5.

A major offspring of the D.O.T. has been the National Rail Transit Corporation, called AMTRAK, which went into business in May 1971. AMTRAK, which is funded by Congress, operates as a semipublic corpora-

Figure 3.5
TACV concept.
(*Courtesy
LTV Aerospace Corporation*)

tion whose main function is to operate the feasible portions of the U.S. passenger railway services that private railroads are not interested in operating.

In addition to the above responsibilities and programs that come under the jurisdiction of the D.O.T., there are several other federal organizations outside the D.O.T. that are involved with transportation. These include the Interstate Commerce Commission (ICC), which regulates most rail, truck, and water transportation rates; the Civil Aeronautics Board (CAB), which regulates most aircraft transportation rates and investigates aircraft disasters; and the Environmental Protection Agency (EPA), which, among its many activities, is involved with transportation system air pollution regulation and the development of clean transportation propulsion systems.

3.2.3
State and Regional Developments

Federal initiative has spurred the development of local, state, and regional endeavors such as the establishment of departments of transportation in many states. It has also strengthened the programs of many local and regional transportation authorities such as the Port of New York Authority and the Metropolitan Transportation Administration (M.T.A.) in the metropolitan New York area, the Massachusetts Bay Transportation Authority (M.B.T.A.) in the Greater Boston area, the Chicago Transit Authority (C.T.A.), the Southeastern Pennsylvania Transportation Authority (S.E.P.T.A.), the Port Authority Transit Corporation (P.A.T.CO.) in Greater Philadelphia, the Bay Area Rapid Transit (B.A.R.T.) District in Greater San Francisco, and the Cleveland Transit System. Several new authorities have also been created, such as the Washington Metropolitan Area Transit Authority (W.M.A.T.A.), which is constructing a new rapid transit system (Metro) for the Washington, D.C., area, and the Metropolitan Atlanta Rapid Transit Authority (M.A.-R.T.A.), which is developing a rapid transit system in Greater Atlanta.

These activities indicate that there is a growing interest in mass transportation on both state and regional levels that needs proper coordination in the planning of a balanced national transportation system. (See Figure 3.6.)

3.3 DEMAND AND PLANNING

Before looking at specific transportation system concepts, it is essential to be aware of the problems of demand, and the need and importance of transportation planning.

3.3.1
Demand

Demand is simply the need for transportation. The measurement and/or prediction of this need, however, is not a simple task. Many times we find a latent demand for a particular mode of transportation that does not show up in any surveys but occurs after the transportation mode is introduced. This has obviously been the situation for many urban highways. In other words,

(a)

(b)

Figure 3.6
Balanced transportation concepts. (*Courtesy General Electric Company*)

many times the introduction of a particular transportation system creates a demand that had not previously existed. With this in mind, we use the definition that demand is simply the need for transportation.

Transportation planning is the process of looking at all the facets that influence or are affected by transportation. This planning must be carried out from the very beginning through the final implementation stage. This has not been the case in the past when adequate planning has not been considered to be of suffi-

3.3.2
Planning

cient importance. Because of this past attitude, many of today's transportation problems are due to insufficient short-and long-range planning.

The systematic transportation planning process is described in reference 2. It should be emphasized that the function of transportation planning, while only briefly covered here, is perhaps one of the most important, the most difficult, and the most encompassing aspect of the entire transportation process.

3.4 CRITERIA

Some set of criteria must be established as part of the overall transportation planning process whether the transportation system being considered is a new section of road or a balanced transit system. The problem in the past has been that we have not considered the total impact of a particular transportation mode during any phase of the planning process. Certainly, the planners of the interstate highway system did not consider the impact of air pollution from automobiles, nor did the planners of the railroads or urban rapid transit systems seriously consider the effect of noise. We have therefore divided the transportation selection criteria into two parts: classical criteria that have been used in the past and will still be used in the future, and current criteria, which planners are now starting to consider.

3.4.1 Classical Criteria

The selection of the best highway location or best transportation system has for many years been based on certain classical criteria such as cost and speed. These criteria are no longer entirely sufficient at present, but they are still of prime consideration. The following items of classical selection criteria are listed in the approximate order of society's preference. They are cost, convenience, speed, reliability, and safety.

(a) Cost

A most important consideration in the selection of modes of transportation is cost. However, most people do not realize what the true cost of a particular mode of transportation is. For example, consider the following two situations:

1. The 5-days-per-week trip to and from one's place of employment.
2. A vacation trip to a distant city.

Suppose that in the first situation the employee has the choice of purchasing a second automobile or using a public transit (bus, subway, etc.) system. What costs should he consider? Certainly he will consider the purchase cost of the second automobile, say $1,000. He may also consider the cost of fuel and insurance, but he will probably not consider the cost of repairs, tires, batteries, property tax, etc. Certainly, the impact of noise, air pollution, etc., on the environment will probably be of little concern to him. Thus, unless he cannot afford the initial $1,000 or unless traffic congestion is so bad and the public transit service is quite convenient, the employee will probably purchase a second automobile. But he has not based his decision on all the factors involved.

Suppose the same employee, still with a single automobile, has to decide which mode of transportation to use in the second situation. He may consider the cost of traveling by some type of common carrier such as aircraft, bus, or rail. He may also consider the problem of transportation when he gets to the distant city. As for using his automobile, he will estimate the number of times that he has to fill up the gas tank and the cost of tolls. Very little consideration will be made on the basis of typical automobile costs per mile (from 8 to 12 cents/mile) since it is usually justified that some of the fixed costs such as insurance, property tax, etc., would be incurred whether the automobile is used or not. Thus, unless time considerations were of major importance, a decision would probably be made in favor of the automobile (remember the statistic of 56% for trips over 1,000 miles). On the other hand, if travel time is important, the use of commercial aircraft may be the choice. Again, very little consideration will be devoted to the other impacts of transportation mentioned previously.

Thus, many decisions about methods of transportation are not based on the actual costs incurred by the user, and seldom are the costs incurred by others considered. Ideally, we would like to have the user consider all the costs of transportation, and this requires two processes which are simple in theory, yet difficult to achieve.

One is the internalization of external costs of transportation. For example, we are starting to pay for the external costs of automobile air pollution by having to pay for air pollution control devices on new automobiles. Thus, a city dweller may have fewer medical costs because the car owner is paying for air pollution control equipment on his automobile. The medical costs which were originally external to the automobile purchase costs have now been internalized so that the auto owner is paying for some of them when he buys his automobile. The second process which must occur is the education of the transportation user to be aware of all the various costs incurred in the use of various transportation modes. This can best be achieved by internalizing as many external costs as feasible (i.e., make the user pay the true costs of transportation) and by providing cost comparison of various modes of transportation through the mass media (newspaper, magazines, radio, and television). Only then will the transportation consumer have the opportunity to make a truly rational choice.

(b) Convenience

The individual automobile is not always the cheapest mode of transportation available, but the convenience factor has fostered its wide use. Other modes of transportation have experienced considerable difficulty in competing with the automobile because of its convenience. As highways become more congested, tolls increase, and parking becomes more difficult and expensive, other mass transit systems may become more acceptable. These systems will not become acceptable, however, until they also become more convenient or until their costs are so much lower than the automobile that they offset the inconvenience. We should also point out that in many applications, such as rural transportation, the automobile is by far the cheapest and most convenient mode of transportation. Thus, in this case, we should not seriously consider replacing it with other modes.

(c) Speed

The attractiveness of the Interstate Highway System and the success of the airplane are not primarily due to convenience or cost but are due to the speed factor. Maximum speed is not necessarily the most important criterion, however. Total trip time, the time required to go from origin to destination, is quite important and is generally used when comparing types of transportation systems. For air travel, the time to get to and from the airport has continued to increase. Airport access time has increased to such a degree that it now represents nearly half of the total trip time from Boston to a New York city.

(d) Reliability

For some, reliability determines which make of automobile to purchase. Consumer Reports uses "rate of repair" as a measure of reliability for automobiles.

Extensive preventive maintenance and maintenance check procedures make the airplane very reliable mechanically. Weather, however, greatly reduces the reliability of the airplane as a mode of transportation. What, for example, is the probability that a particular flight will leave on schedule from an airport in the northeast during the winter months?

Public transit systems, except in certain situations, have not been kept up to date and in good repair. These systems have also become less reliable in their schedules, which only adds to the decline in their use.

(e) Safety

Safety, in most modes of transportation, has only recently received much attention. Automobile manufacturers, for example, are now under considerable pressure to improve the safety of the vehicles sold. In addition, highway design, lighting, and traffic signs are all being modernized to improve automotive safety. We are seeing the internalization of automobile accident costs by the addition of seat belts, padded dashboards, adequate bumpers, safer highways, and higher insurance rates.

On the other hand, the commercial aircraft industry and the federal government have put a lot of emphasis on commercial aircraft safety through excellent preventive maintenance programs, improved air traffic control systems, better aircraft design, and safer airport layouts. The user, of course, pays for most of this through the ticket purchase price. To illustrate the relative safety of various modes of mass transportation, Table 3.3 compares the deaths per 100 million passenger miles for automobiles, buses, trains, and aircraft. When one considers the relative dangers of commercial aircraft and automobiles, it is clear that commercial aviation has done a commendable job. Of course, the comparison between private (general aviation) and commercial aviation shows that the relative safety of commerical aviation

Table 3.3
Transportation Death Rates per 100 Million Passenger Miles

Transportation Mode	Deaths (per 100 million passenger miles)
Automobiles (drivers included)	2.30
Automobiles on turnpikes	1.20
Buses	.22
Railroad passenger trains	.07
Scheduled airlines	.13
General aviation (including pilots)	18.00

Source: National Safety Council, Transportation Accident Death Rates, 1969.

is partly due to its ability to select and regulate the individuals who operate the aircraft. This is one of the points the automobile industry likes to make ("We can control the quality of all of the nuts, bolts, etc., that make up the automobile, but we can't control the nut behind the wheel"). However, it is now being generally accepted that since this is the case, we should make every effort to protect that "nut" and the innocent people he can hurt by making the basic automobile much safer.

In addition to the classical criteria for selecting mass transportation systems, we are now also considering other criteria such as air pollution, land utilization, energy utilization, water pollution, noise pollution, natural resource depletion, population movement, and various social issues. Ultimately, we would like to internalize any external costs incurred by these criteria, whether it be the direct noise pollution from jet aircraft or the indirect central power-plant pollution caused by electric rapid transit systems.

3.4.2 Modern Criteria

(a) Air Pollution

Air pollution has become a major issue during the past decade because of the serious problems with smog and the increase of respiratory ailments such as emphysema in urban areas. The transportation system which has taken the brunt of the air pollution regulations (the major one being the 1970 Clean Air Act by Congress) has been the automobile. This is not altogether unexpected when we consider the data from Table 3.4, which show

Table 3.4
Amount of Air Pollution by Source (1969)

SOURCE	POLLUTION TYPE PER MILLIONS OF TONS				
	HC^*	CO^*	NO_x^*	Particulates	SO_x^*
Transportation	19.8	111.5	11.2	.8	1.1
(automobile)	(16.9)	(96.8)	(7.6)	(.2)	(.2)
Stationary power sources	.9	1.8	10.0	7.2	24.4
Industry processes	3.5	12.0	.2	14.4	7.5
Solid waste disposal	2.0	7.9	.4	1.4	.2
Miscellaneous	9.2	18.2	2.0	11.4	.2
Total emissions	35.4	151.5	23.8	35.2	33.4

*HC, unburned hydrocarbons; CO, carbon monoxide; NO_x, nitrogen oxides; SO_x, sulfur oxides.

Source: *Air Pollution Control Office, Environmental Protection Agency, 1971.*

that transportation (primarily automobiles) produces the majority of the volume of air pollution. However, if one considers the relative toxicity of the major air pollutants, the systems which emit more sulfur dioxide become more important polluters. However, this does not mean that we should not regulate automobile emissions; it just means that we have to develop realistic and health-conscious air pollution regulations for all sources. Even then, the automobile is still a major contributor, since about 15 to 20% of U.S. industrial production, such as steel and petroleum products, is related to the automobile.

One of the problems with introducing any pollution regulation is that most of the systems it affects were not developed with pollution regulations in mind. Certainly this is the case with the automobile, where the internal combustion engine has been extensively developed mainly from a performance viewpoint. The introduction of automobile air pollution regulations (see Table 3.5) has severely reduced their performance so that in some cases it presents a safety hazard: Many new automobiles may stall at inopportune times. It is quite possible that by the 1980s a new type of automobile propulsion system may be developed, possibly rotary internal combustion (Wankel), external combustion (Rankine, or better known as the steam engine), elec-

Table 3.5
Federal Automobile Air Pollution Regulations

Year	Exhaust Emissions (grams/mile)			
	HC*	CO*	NO_x*	Particulates
Uncontrolled car	17	125	6	.3
1971	4.6	47	NR*	NR*
1972	3.4	39	NR*	NR*
1973	3.4	39	30	NR*
1975	.41	3.4	3.0	.1
1976	.41	4.4	.4	.1

*HC, unburned hydrocarbons; CO, carbon monoxide; NO_x, nitrogen oxides; NR, no requirements.

tric, gas turbine (Brayton), or some combination (hybrid) of the above systems. We may also see changes and/or modifications in the propulsion systems used for other mass transit systems.

(b) Land Utilization

At the present time, the concept of land usage is becoming an important criterion in evaluating transportation alternatives, especially in urban areas where land costs have spiraled. Two transportation systems that have had a major impact on urban development have been the elevator and the automobile. The development of the high-speed elevator created the possibility of building tall buildings in urban areas, which yield a much more economic utilization of available space. On the other hand, the automobile has led to an inefficient use of urban land. New York City, with 300 square miles of land, has approximately 4,500 miles of paved streets. If we assume an average street width of 75 ft, then 64 square miles (or 21%) of the total area is tied up in a land use that provides no property tax revenue to the city. It does, however, require surface maintenance, snow removal, traffic controls, street signs, markers, and police patrol, all of which represent a drain on the city's budget. Similarly, in the city of Boston, approximately 15% of the land is used for highways and streets, and in the downtown area 30 to 35% of the land area is devoted to the automobile. When viewed in this way, the arguments for auto-free zones in urban areas makes a great deal of sense.

Another problem in land usage that the automobile is creating is due to the building of expressways (or highways) through urban areas. Many urban expressways have, only recently, been put off or abandoned because of citizens protesting the breaking up of residential neighborhoods, the noise and air pollution, and the isolation of urban sections associated with urban expressway use. Some individuals argue that it makes much more sense to have urban perimeter beltways and parking with rapid rail or bus transit into the central city. This argument does have some validity, although it may also accelerate the movement of business and industry, along with middle class employees, to the beltway regions, thus leaving fewer employment opportunities to the less mobile inner-city residents. Consequently, the question of proper land use and transportation planning are closely interrelated. (See Figure 3.7)

(c) Energy Utilization

At the present time, transportation consumes annually from 50 to 60% of all the petroleum used in the United States and about 25% of all the energy used per year. This is extremely critical since it now appears that the world's

Figure 3.7
Land use, Los Angeles.
(*Courtesy General Electric Company*)

known and possible petroleum reserves will last only about 100 to 200 years. A criterion based on energy considerations for evaluating the transportation/energy efficiency of various transportation systems is detailed in Section 3.6.2.

(d) Water Pollution

While most transportation systems do not directly cause water pollution, there are many situations where there are indirect effects. For example, in the construction of transportation vehicles, large amounts of steel, plastic, glass, and electric energy are consumed. Many of the by-products of the processing of these materials find their way into our streams, rivers, lakes, and oceans, thereby contributing to water pollution. Furthermore, the thermal pollution caused by electric power plants must be considered as an external cost of any electrical energy utilized in the construction or operation of transportation vehicles. All of this is especially relevant to the automobile where 8 to 10 million units are built each year and where about 20% of the gross national product (G.N.P.) is related to automobile use (automotive, petroleum, steel, etc., industries). In addition, the winter salting of our highways for automobiles, buses, and trucks is now starting to affect water supplies significantly in the northern states.

(e) Noise Pollution

Transportation systems create many different noise problems such as aircraft engine noise, city traffic noise, and noise from large intercity trucks. A major technical problem with all aircraft is the reduction of noise. Noise is a major factor holding back the rapid introduction of short and vertical takeoff and landing (V/STOL) aircraft for short-haul intercity applications. The quieting of engines on most transportation systems is now undergoing extensive study and will ultimately add to the costs of transportation.

(f) Natural Resource Depletion

As mentioned in Section (c), transportation is rapidly depleting our petroleum supplies. In addition, large amounts of steel, aluminum, and other materials are being used up for transportation purposes. Major efforts to recycle used transportation systems, especially automobiles, need to be undertaken. If transportation vehicles were judged solely from a natural-resource viewpoint, the policy of planned obsolescence in automobiles would certainly be viewed as absurd. (Note that the planning is not in creating an automobile that will fall apart after a certain time but represents a lack of planning in design for long-lasting, durable vehicles that can be recycled after they have finally outlived their usefulness.)

(g) Demographic Issues

Demography is the statistical study of human population, particularly with respect to size, density, and distribution. Certainly, the history of the urbanization of the United States is related to the introduction of different transportation systems. Centers of cities were normally quite small until the introduction of public transit. Furthermore, the phenomenon of the suburb was certainly helped along by the rapid increase in the use of the automobile. The automobile has made it possible for business and industry to locate in the suburbs and for suburban shopping centers to supplant much of the downtown shopping business. This has created the situation where a large portion of the trips in urban areas are now crosstown or within a suburban area. Furthermore, this makes it more difficult to introduce bus or rail service since the transportation needs of many individuals are so diverse that only a personal transportation system can satisfy these needs. In the meantime, the individuals who do not have access to an automobile, such as the elderly, the poor, and the young, have to rely on generally inadequate public transportation to meet their needs.

(h) Social Issues

Many social problems are created through improper or inadequate transportation planning. Some of these, such as the isolation of inner-city residents, the stranding of the elderly, and the creation of poor living conditions through the effects of pollution, have already been mentioned. In addition, other social effects related to transportation have to be considered when selecting a particular transportation plan. For example, suppose that automobiles were banned from most metropolitan areas. This could have a great effect on the many automobile workers, who may lose their jobs because of a decrease in production; also consider the concomitant increase in welfare costs, the increased crime rates, and many other social issues.

3.4.3 Summary

The selection of a "best" transportation system or network depends on many criteria, some (classical) which have been considered in the past and others (modern) that are just now being considered. Ultimately, we would like to have the actual cost of a particular transportation mode mirror its true costs or impact on society. Thus, we would like to internalize some of the external costs that are described under Modern Criteria.

3.5
SYSTEMS CONCEPTS

Mass transportation systems can be classified in many different ways, and no matter what scheme is used there will be considerable overlap between these classifications. Furthermore, nearly all types of transportation systems that move people also move materials. The systems described here are concerned only with the transportation of people.

3.5.1 Classification of Systems

Transportation systems will be divided into the two major classifications of intraurban and interurban. Intraurban systems are those systems providing transportation within a specified urban area. Interurban systems are transportation systems between urban centers. It immediately becomes apparent that some systems belong in both categories, as will be discussed later.

Further classification will be according to demand, or user-initiated, systems as opposed to scheduled, or operator-initiated, systems. As an example, an elevator is a demand-type system, as is a taxi. The user sets the system in motion on demand to meet his needs for transportation. An escalator is usually a scheduled-type system that is scheduled to operate continuously, whereas airlines are also scheduled but with prescribed times of departure.

Demand systems generally fall within the intraurban classification, while scheduled-type systems may be either intraurban or interurban systems.

3.5.2 Intraurban Systems

Intraurban transportation systems are generally characterized as low speed and normally operate over distances of under 30 miles. It is important for these systems to be efficient, reliable, and convenient. The primary objective of most intraurban systems is the decrease in highway and parking congestion achieved by providing a convenient, pleasant alternative to the private automobile. Speed is not the main criterion for most intraurban systems. Average speeds (called block speed) on the order of 30 mph for downtown areas are quite sufficient, while block speeds of 50 to 70 mph for suburban to urban travel are reasonable. Other problems of concern for intraurban transportation have to do with land utilization, decreasing emissions, and reducing noise levels. (See Figures 3.8 and 3.9)

Figure 3.8
Traffic congestion before BART, San Francisco.
(*Courtesy General Electric Company*)

Figure 3.9
Traffic flow with CTA,
Chicago.
(*Courtesy General
Electric Company*)

For many urban areas, a pressing and typical transportation problem is that of airport access. For example, the time required to get to and from the airports is quite often greater than the flight time from Boston to New York or from New York to Washington.

(a) Demand Systems

Demand systems are all those transportation systems that are initiated or set in motion in response to the user or on demand. These systems are generally very flexible at least with respect to the time used, but they may also be flexible with respect to route. They are also of relatively low density, (i.e., they carry only a few persons at a time but may, in the course of an hour, move a large number of people). Examples of demand-activated systems are

1. *Elevator.* The elevator is a demand type of system that is flexible in time of operation and is user-initiated but operates within a fixed right-of-way.
2. *Taxi.* The taxi is another conventional demand system. Like the elevator, the taxi is user-initiated and is flexible in terms of time and is also flexible as to its route of operation.
3. *Automobile.* The private automobile is generally considered to be the most convenient means of personal transportation for the vast majority of adults. It is convenient both in terms of availability and flexibility. This convenience factor decreases, however, with the ever-increasing highway congestion and parking problems. Automated highways have been proposed as a means of increasing safety and at the same time

increasing the capacity of super highways. On the automated highway, each vehicle would be fully controlled by the system both in speed and direction. This degree of control would permit a reduction in headway (the distance between vehicles) required for safety and an increase in speed. Fully automated highway systems, although already proposed and under investigation, may be as much as 20 years away.

4. *Dial-a-ride*. This type of system, which goes by many names, consists basically of a fleet of small buslike vehicles that are routed according to user demands. The buses may be of the type described below or they may be of the van type: bodies with additional seats such as those produced by Dodge, Chevrolet, and Ford. Using vehicle/dispatcher communications en route, vehicles are rerouted as efficiently as possible to provide door-to-door service. Unlike the conventional taxi, which generally picks up and delivers a single person at a time, the dial-a-ride system would provide transportation for several people with similar origins and destinations. A dial-a-ride demonstration project, currently being conducted in Haddonfield, New Jersey, is financed by the U.S. Department of Transportation and the New Jersey Department of Transportation. (See Figure 3.10.)

5. *Personal rapid transit* (PRT). A number of variations of the so-called personal rapid transit system have been proposed (see Figure 3.11) A PRT system is any system of small vehicles operating in a network of guideways on exclusive rights-of-way that take passengers from origin to destination. Some proposed systems are truly personal, with room for only two people, while other types are larger and may carry as many as 40 passengers.

(b) Scheduled Systems

Most transit systems in operation today are scheduled systems. Schedules are fixed both with respect to published timetables and predetermined

Figure 3.10
Dial-a-ride concept using mini-buses.
(*Courtesy Twin Coach/Highway Products, Inc.*)

Figure 3.11
Personal rapid transit (PRT).
(*Courtesy General Motors Corporation*)

routes. Both routes and timetables are set according to forecasted demands, and individuals must then travel according to these schedules. Some degree of convenience is obtained through increased frequency of operation.

Scheduled systems are operator-initiated, are generally of high capacity, and are scheduled to operate between or through highly populated areas. Examples of scheduled systems are

1. *Escalator.* The escalator, although generally thought of as a moving stairway, may also be a moving sidewalk. At the New York World's Fair and, more recently, at the San Francisco International Airport, moving sidewalks were used. Moving sidewalks, of necessity, must be slow to allow access and egress but are capable of moving a large number of people over relatively short distances.
2. *Buses.* Buses have long been operated as scheduled transit systems. In recent years, the size of the buses has been varied so as to economize on operating expenses and to provide greater flexibility by operating into less densely populated areas. Most intraurban buses (sometimes

called transit coaches) are designed for low-speed (under 50), stop-and-go operation. Diesel engines are the most common propulsion systems, and it is not unreasonable for a properly maintained transit coach to last for 500,000 to 1 million miles (8 to 15 years). Recently, smaller transit buses (20 to 35 seats as compared to 40 to 60 for regular-sized transit coaches) called mini-buses have shown an applicability to small urban operation as well as to special large urban operation, thus providing greater flexibility through short headways. (See Figure 3.12)

Two recent innovations in bus operation have provided an upsurge in their use and acceptance as a viable alternative to the private auto. One innovation has been the use of an exclusive bus lane on super highways leading into and out of city centers such as New York, Washington, and Boston. These exclusive bus lanes permit the buses

Figure 3.12
Future bus concept.
(*Courtesy General Motors Corporation*)

(a)

(b)

to move at higher speeds and thus maintain better schedules since they avoid the usual delays of automobile congestion. Large outlying parking areas are provided for these bus transit users, who are now finding this mode of transportation into the central business district (CBD) far more convenient and pleasant. The second innovation is the so-called mixed or multimode system. Buses that operate in a conventional mode over the outlying streets may, when full, become an express bus and travel on a separate right-of-way into the CBD. This right-of-way may be the exclusive bus lane previously mentioned or it may be a sophisticated automatic guideway.

3. *Trains.* Trains are strings of vehicles that operate on a separate right-of-way and are guided by rails. This definition encompasses more than the conventional train. A train may operate above or below ground, whichever is best suited to the local terrain and utilization of land.

Under the classification of trains are the so-called monorail systems, which hang from or are supported on top of a single rail, as well as the subway trains that operate below ground. In transportation jargon, urban trains such as New York City's subway system are called rapid transit systems, while the Long Island railroad trains are called commuter trains. The vehicles themselves are different in that rapid transit vehicles tend to be smaller (50–70 ft long) than commuter trains (70–85 ft long) and they normally operate from a third-rail electric power pickup, while the commuter trains are either diesel driven or have overhead electrical power pickup. Usually, rapid transit systems and commuter trains are operated by independent agencies, and no provision is made to allow the vehicles to operate on the other system's rights-of-way. (See Figure 3.13.)

3.5.3 Interurban Systems

Most interurban systems, other than the automobile, are scheduled. The automobile was discussed previously so we shall only consider scheduled interurban systems here. Interurban systems can be classified as medium-distance systems operating from approximately 30 to 300 miles and long-distance systems operating over 300 miles.

(a) Medium Distance

In this range, several of the systems previously described under intraurban systems, such as commuter trains, may be applicable for the shorter distances within the medium-distance range. Likewise, small jet aircraft, which are described under long-distance interurban systems in Section (b), are applicable to the longer distances within this range. Thus, we shall discuss only those systems that are primarily applicable to medium-distance travel where speed is now becoming a more important factor.

1. *Highway buses.* These are single-door buses that have reclining seats and rest rooms and are designed for highway use. The vehicles are designed for high speed (60–75) and can carry from 40 to 60 passengers. The cost of one of these intercity buses is about $60,000, but they are

Figure 3.13
Bay Area rapid transit system.
(*Courtesy General Electric Company*)

capable of operating for 20 to 25 years with from 1 to 2 million miles of travel. Intercity buses in the United States are run mostly by private operators with the two largest being Greyhound and Continental Trailways. Most of these buses are diesel-powered, although it is quite possible that future intercity buses will be gas-turbine-powered.

2. *Rail systems.* Rail systems are generally considered to be vehicles that operate at ground level on fixed-rail guideways. Support of the vehicle is by wheels in direct contact with the rails. Anticipated speeds are up to 200, which is thought to be the limit of adhesion of power-driven steel wheels on steel rails. This might be extended by the use of other wheel/rail combinations or by having the wheel just for support and using some other means of propulsion. A recent development in U.S. rail systems has been the operation of new equipment on present railroad right-of-ways. The Metroliner is a demonstration project which, depending on the number of stops, completes the 220-mile New York City/Washington, D.C. run in from 3 to 3 1/2 hr. This project has been fairly successful. A second demonstration project, the United Aircraft Turbotrain, operates on the 210-mile shoreline route between Boston and New York City in approximately $3\frac{3}{4}$ hr. The project has been plagued with many problems, including the financial problems of first the New Haven and then the Penn Central railroads. Turbotrains have been operated successfully on the Montreal/Toronto run in Canada and are now operating between Parkersburg, West Virginia and Washington, D.C. (See Figure 3.14.) A third more extensive project started when Congress created the new semipublic corporation named AMTRAK, which was described previously.

Foreign rail systems, such as the Japanese and European systems, will not be discussed, since the demographic and social conditions that they operate in are different from those in the United States. However, some of the experience gained in new equipment and operating methods

113

of foreign systems should be applicable to U.S. rail development. This is especially true for the new Advanced Passenger Train (APT) now being developed in Great Britain.

In general, rail systems have an advantage in that they can be developed at the present time on current rights-of-way using current technological capabilities. Thus, the principal costs are for new equipment, although, as rail systems develop, new rights-of-way will certainly add to these costs. The main disadvantages are those that have caused the demise of rail travel in the United States: the problem of operating passenger trains on the same right-of-way as freight trains as well as high labor costs. The present successful railroad organizations make their profits on freight, and it follows that more profitable freight moving at slower speeds would have an adverse effect on higher-speed passenger trains unless sophisticated control systems were used. The problem of high labor costs is one that makes it difficult to allocate funding for new equipment, to develop new operational techniques and, in general, to carry out the research and development programs necessary to maintain a vigorous passenger rail system. As an example of current high labor costs related to urban mass transportation, the Massachusetts Bay Transportation Authority, with an annual budget of $150 million, expends approximately two-thirds, or $90 million, for labor. Another problem with rail transportation has been overregulation by the state and federal governments. The possibility of deregulation of interstate transportation rates is one area that might help the present rail situation.

Figure 3.14
Turbotrain.
(*Courtesy United Aircraft Corporation*)

The primary operational disadvantages of rail systems are those of safety, noise, and rail gage. The safety aspect is due to the use of both high-speed passenger trains and slow-moving freight trains on rights-of-way that do not have adequate control systems and are not isolated enough from external effects such as automobile traffic at grade crossings. The noise problem is due to the on-board propulsion such as gas turbines and diesels and steel wheel/rail support systems. Electric propulsion can help alleviate the noise problems. However, as higher speeds are attained, the noise and safety problems will become even more prevalent. The problem of rail gage is as follows: It is quite difficult to design adequate suspension systems for an 8-ft-wide by 10-ft-high vehicle that must ride on a 4-ft, 8-in. gage track. This disadvantage of rail systems and several others could be alleviated by operating on new, separate rights-of-way with larger-gage tracks.

3. *Tracked air cushion vehicles.* Tracked air cushion vehicles (TACVs) are systems where the vehicle is suspended above the guideway by pressurized air that is forced between the vehicle and the track surfaces. The track is used to provide a suitable support surface and to prohibit excess lateral motion of the vehicle. TACVs currently under development are expected to be operational by 1975–1980 and will move at speeds in excess of 150 mph and ultimately at speeds of 350.

TACVs have been under development in France, Great Britain, and the United States. The French have developed a tracked air cushion vehicle, called the Aerotrain, that will go into the demonstration phase in the near future. Two developmental vehicles using aircraft-type propulsion systems have been operated on an inverted "T" guideway for the past several years. The British tracked air cushion vehicle, called the Hovercraft, is propelled by a linear induction motor. A 6-ft-long operating model using peripheral jet air cushion pads for support has been tested since 1966. The U.S. Department of Transportation recently completed construction of a prototype tracked air cushion vehicle that will be extensively tested and developed at the Pueblo, Colorado test facility over the next several years. The U.S. vehicle was designed and built by Grumman Aircraft and will be tested in conjunction with a linear induction motor propulsion system, designed and built by the Garrett Corporation. A second tracked air cushion vehicle for intra-urban application is now being built by Rohr Industries, Inc. for testing at the same test facility.

The main advantages of TACVs are their expected superior-ride quality at high speeds and that they will travel on right-of-ways that are separate from other traffic. The principal disadvantages are the high initial costs for the vehicles, for the acquisition of the rights-of-way, and for the guideway construction. Additional disadvantages are the possibility of excess noise (which would by greatly reduced by the use of a linear induction motor propulsion system) and the problems of safety. At 350 mph, it could be disastrous if the vehicle struck debris on its right-of-way or if it were struck by an airborne object. The last two problems of noise and safety could be diminished by having TACVs travel in some type of enclosure.

4. *Tube vehicle systems.* Tube vehicle systems (TVSs) (See Figure 3.15) are defined here as any type of vehicle that travels in an enclosure at

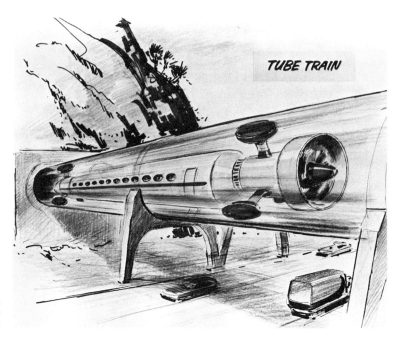

Figure 3.15
Tube vehicle concept.
(*Courtesy General Electric Company*)

speeds sufficiently high so that the motion of the vehicle is affected by the proximity of the enclosure. If the enclosure is circular in cross section and the vehicle surfaces are relatively close to the tube wall surfaces, the vehicle tends to act like a loose-fitting piston traveling through the tube. It is this "piston effect" that separates TVSs from other types of high-speed ground transportation systems. TVSs are expected to operate at speeds in the range of 150 to 500 mph. The earliest expected implementation of TVSs is 1980 to 1985, depending on future research and development efforts. A TVS research program is currently underway by the U.S. Department of Transportation. However, plans for the design of a prototype system have not been delineated.

The advantages of TVSs are, in addition to their having their own right-of-way, their relative safety from debris, their all-weather capability, and their relatively small impact on the environment. Environmental effects will be discussed in more detail later. The primary disadvantage of TVSs is the large initial costs for construction of the tubes and tunneling if located underground. However, it is quite possible that, as urban and suburban ground-level right-of-way costs continue to increase, it will become more economical to construct tunnels and tubes underground. A second disadvantage of TVSs is the technological development necessary in the areas of stability and control, aerodynamics, suspension dynamics, and tunneling.

TVSs can be classified further as pneumatic and nonpneumatic. A pneumatic system is defined as one in which there is a pressure difference across the vehicle and no air flow is allowed past the (high-blockage-ratio) vehicle. Pneumatic propulsion results from the pressure differential across the vehicle. In a nonpneumatic system, the vehicle does not completely fill the tube cross section, thus permitting air flow past the (low-blockage-ratio) vehicle. Owing to this flow past the vehicle, additional aerodynamic drag is encountered.

It should be noted that under these definitions, partial evacuation of the tube is possible for both types of systems, as is the possibility of gravity augmentation. Gravity augmentation is achieved by having the tube located at great depths below ground level so that a vehicle is initially accelerated by gravity as it starts on a trip and is decelerated by gravity as it completes the trip. In the following, several current and proposed TVS concepts will be discussed.

(a) *Tube flight.* The tube flight concept, as proposed by Foa at Rensselaer Polytechnic Institute, consists of a ground transportation scheme in which vehicles, aerodynamically supported and propelled, travel at speeds in excess of 300 mph through nonevacuated tubes. Large clearance fluid support is provided by peripheral jets located around the vehicle. Propulsion may be accomplished through the use of a turbojet engine or turbine or electrically driven conventional or so-called bladeless propellers. Significant aspects of the tube flight concept are the large clearance fluid support and a "matched internal" mode of propulsion. Matched internal propulsion means that the fore-to-aft transfer of air is accomplished in such a manner and at such a rate that the flow disturbances throughout the tube are kept at a minimum.

The high drag caused by operation in an atmospheric tube is significantly reduced by the internal matched propulsion mode and by the use of either a low-blockage-ratio vehicle or by the use of tube wall perforations.

(b) *Linear induction motor/wheeled system.* This system consists of wheel supported vehicles propelled by a linear induction motor (LIM) and operating in an evacuated tube. The partially evacuated environment provides for a reduction in drag, while the LIM propulsion mode and wheel suspension permits a reasonable clearance between the vehicle and tube. It is proposed to use gravity augmentation with this system. Extensive studies on this type of system have been carried out and it appears to be one of the most feasible candidates for future development.

(c) *Gravity vacuum transit.* The GVT system consists of a high-blockage-ratio vehicle, supported by steel wheels on steel rails and propelled by gravity that is augmented by a pneumatic propulsion scheme to overcome frictional losses. Gravitational acceleration is accomplished by constructing the tunnels to depths on the order of several thousand feet. This system, along with the LIM/wheeled system, appears to be feasible enough to consider for future development.

(d) *Fluid entrained vehicle concept.* Another type of tube vehicle system is the fluid entrained vehicle concept. In the proposed tube vehicle system, a passive vehicle is propelled and suspended by a high-velocity air stream. Nozzles, encircling the tubular guideway and spaced along its length, inject a flow of air that entrains the vehicle. In addition to providing support, this injected flow of air propels the vehicle along at a speed somewhat less than the flow velocity.

(e) *Uniflow.* The uniflow personal rapid transit system consists of small (eight-passenger) passive vehicles enclosed in a guideway and supported by jets of air stationary in the guideway. These same jets provide propulsion through a linear turbine mounted on the underside of the vehicles.

(f) *Magnetic suspended system.* The D.O.T. is funding studies of a new TVS concept proposed by Stanford Research Institute, Menlo Park, California. This concept envisions linear induction motor-propelled vehicles traveling through partially evacuated tubes. Suspension is to be provided by cryogenic superconducting magnets.

5. *V/STOL aircraft.* Over the past several years, the increased convenience, economy, and public acceptance of air travel has resulted in "commuter" air traffic between urban centers. This demand for air travel has greatly increased the air traffic congestion at the airports. Equally congested is the surface transportation to and from the airports.

One solution to the airport access problem has already been discussed, that of improved ground transportation systems. Another solution is the use of V/STOL aircraft. The term V/STOL, as used in this chapter, means a transportation system consisting of aircraft and landing facilities. These aircraft may operate between conventional outlying airports and downtown metropolitan centers (interurban) or they may operate over greater distances between different urban centers.

The terms VTOL (vertical takeoff and landing) and STOL (short takeoff and landing) are generally considered as the type of aircraft classified according to their operational characteristics. The classification of VTOLs is typified by the helicopter and variations of the helicopter principle. STOL aircraft are, in general, modifications of conventional aircraft in which the short-takeoff and landing capabilities are obtained through the use of various high-lift devices.

Because of their shortened landing and takeoff capabilities, the requirements for the airport facilities are considerably different from those of conventional airports. For instance, heliports have been located on tops of buildings in several cities throughout the country. Because of the short runway requirements, heliports and V/STOL ports can be located near or within metropolitan areas. This urban location, however, presents some unique problems and constraints both on the aircraft and on the landing facilities. Perhaps the most important problem areas for V/STOL systems are safety, noise, cost, and public acceptance.

For a safety standpoint, any V/STOL system must possess all-weather capability and pinpoint controllability. This means the navigational equipment and aircraft-handling qualities must meet or exceed the most advanced systems that are currently available. Proof of these capabilities will go a long way toward speeding up public acceptance.

Noise, both internal and external to the aircraft, has recently received considerably more attention because of larger aircraft and the increased operation near densely populated areas. The noise problem will demand a great deal of technical effort in the near future.

Costs for V/STOL systems may be viewed in the four general categories of V/STOL ports or facilities, further development of aircraft, cost of aircraft, and direct operating costs.

To properly implement V/STOL service, i.e., to take full advantage of the capabilities, nearly all areas will require new landing facilities in or near the high-density activity centers and separate facilities designed for the V/STOL aircraft adjacent to the conventional airports.

Currently there are very few STOL aircraft, and essentially the the helicopter is the only VTOL aircraft that is commercially operational. This means that considerable development work, hence money, is needed before more than a moderate V/STOL system can be implemented.

Estimated prices of V/STOL aircraft with the desired capacities of 30 to 120 passengers are between $1.2 and $8.2 million. In addition, the direct operating costs of STOL versions range from 2.1 cents to 2.5 cents per available seat mile. Because of the greater direct-lift capability, VTOL designs will have somewhat greater direct operating costs under present technology.

Public acceptance and usage of V/STOL systems depends directly on these costs in addition to such factors as increased safety and the reduction of noise. V/STOL aircraft would be in operation today to a much larger extent except for the reluctance on the part of the public to permit the construction of facilities in some of the densely populated areas. Public acceptance will continue to be a major stumbling block to operational V/STOL systems in some urban centers.

(b) Long Distance

Rail systems, tracked air cushion vehicles, and tube vehicles can all be used for long-distance travel as well as the automobile. In addition, conventional takeoff and landing aircraft (CTOL) can be utilized.

Conventional aircraft. Conventional aircraft for commercial use have grown steadily from the 1928 Ford Tri-Motor, which carried 14 passengers, to the Boeing 747 jumbo jet (see Figure 3.16), which can carry up to 490 passengers. Speeds of commercial aircraft have increased from 120 to 650 mph and, in the foreseeable future, will increase to the supersonic speed of 1400 mph for the British French Concorde.

Major problems exist with the present conventional aircraft systems. Some of these problems are noise, air pollution, air traffic control, and airport congestion.

Problems of noise and air pollution are essentially the same as for V/STOL aircraft. The engines, on landing and takeoff, are the prime contributors to the noise. Major advances have been made recently in reducing both the noise and air pollution. On the Boeing 747, the engines are only half as loud as the engines of the 707 and are virtually smoke-free as a result of more thorough burning of the fuel.

Air traffic control is a problem directly related to the air traffic congestion. Scheduled flights are often delayed at most major airports because of the large numbers of aircraft attempting to land or take off. Improved air traffic control equipment and techniques are currently being developed to help alleviate the congestion and improve the flight safety, particularly near the large municipal airports. Operation of V/STOL and general aviation

(a)

(b)

Figure 3.16
Jumbo jet.
(*Courtesy Lockheed Aircraft Corporation*)

aircraft at separate landing facilities will go a long way toward easing the problem of air traffic congestion and control.

Problems of airport congestion, i.e., baggage handling, passenger transit around the airports, and airport access, are all gradually being reduced. Mass transit systems to and from airports reduce the auto traffic and parking, while smaller transit facilities around the airports reduce the passenger travel time. Automated and containerized baggage handling aids in reducing the ground time for scheduled airlines.

Attacking the air transport problems from this systems standpoint should provide many improvements in the overall transportation picture in the near future.

3.6 ENVIRONMENTAL CONSIDERATIONS

This section includes many of the previously mentioned current criteria related to environmental concerns. These are repeated here because we feel that these criteria will be the major driving forces behind the development of our future transportation systems. In the following sections we shall consider three of the most important environmental problems associated with transportation systems: air pollution, noise pollution, and energy considerations.

3.6.1 Air Pollution

At present, transportation causes a major portion of the air pollution in the United States. Automobiles are the primary transportation source of air pollution and, in some areas such as the Los Angeles Basin, are primarily responsible for the phenomenon of photochemical smog. Smog is formed from sunlight causing a chemical reaction between unburned hydrocarbons and nitric oxides. At times when the weather conditions are just right (or wrong?) the smog becomes so obnoxious that school children are not allowed outside for recess. The occurrence of this smog is one reason California has been a leader in the area of automotive air pollution regulations. However, does this mean that the same regulations, through federal laws, should apply to automobiles in Montana or elsewhere? Furthermore, how do we make sure that the automobiles will still meet these regulations as they get older? Massive inspection is possible but extremely expensive.

What about the type of engines that will be necessary to meet the more restrictive (because the number of automobiles is increasing) regulations? Will it be the present internal combustion engine with additional emission control equipment? Will it be necessary to eliminate the lead in gasoline because it fouls the operation of catalytic mufflers, which are supposed to help the exhaust gas burn to completion? How much will all this equipment cost? Will we have to abandon the internal combustion engine in favor of a slightly cleaner but more inefficient (from a fuel or energy standpoint) propulsion system?

All the above questions are presently under consideration, and the ultimate answers or solutions to them may have a significant impact upon our future transportation systems.

3.6.2 Noise Pollution

The problem of noise pollution, as mentioned in the discussion of the V/STOL system, is a major problem facing transportation today. Noisy city and highway traffic is a problem that must be studied and solved. One problem under major study now is that of truck tire noise. For example, it has been shown that variations in tread design can have a significant effect upon the emitted noise levels.

Considerable research and development is now underway in an attempt to reduce propulsion system noise in aircraft, buses, trucks, and automobiles. As noise regulations are introduced, we may find that electric propulsion systems such as electric rotary motors and linear induction motors will be the best solution for ground transportation systems. We may also find that enclosing the ground transportation systems in tubes will be necessary to meet future noise standards.

Again, this indicates how environmental considerations may be of major importance to the design and selection of our future transportation systems.

3.6.3 Energy Expenditures

In trying to compare different transportation systems, we should first ask the question, What is the prime function of transportation? The answer is that the function of transportation is to transport people and/or goods safely, conveniently, economically, and efficiently from one point to another. The use of the word *efficiently* means that the travel time should be reasonable. For example, consider the travel time necessary for a 10-mile suburban-urban stage length (surface distance between stations or terminals). At a block velocity (defined as the stage length divided by the travel time) of 100 mph, the travel time is 6 min, while a block velocity of 200 mph yields a travel time of 3 min. Obviously, a block velocity of 100 mph, with its attendant lower energy requirements, may be considered relatively efficient for a 10-mile stage length.

Returning to the comparison of different transportation systems, once adequate designs for safety and convenience and efficiency are achieved, the principal comparison criterion used is economics. One well-used parameter for mass transportation systems is the cost per seat mile. However, this type of comparison parameter has several failings. One is that since it is based on both the prorated initial and average operating costs, the relative size of the guideway or highway component of the initial costs tends to obscure the significance of the other costs. Thus, it might be more appropriate to compare the initial and operating costs separately. Certainly, when a municipality, transportation authority, or business organization looks at various transportation alternatives it has to consider three factors: the initial costs (related to loans, bonds, etc.), the operating costs (yearly expenditures), and, finally, the total costs (to determine the overall profitability).

When looking at some of the external costs of transportation systems, one of the prime environmental effects is created in the production of energy to propel the vehicle(s). From the point of view of a natural resource, transportation in the United States utilizes 50 to 60% of the nation's total energy consumption. From a pollution point of view, transportation produces a large portion of urban/suburban pollution directly through the burning of fuel for propulsion. In addition, there is an indirect effect through the generation of electricity used in the production of steel, glass, etc., necessary to manufacture buses, trains, aircraft, and 9 million automobiles a year. This leads to one possible method of comparing operating costs, the use of a parameter defined as the transportation/energy efficiency, which is the number of seat miles per gallon (where a gallon is defined as a unit of energy equal to 136,000 Btu or 40 kWh).

Use of this parameter as a basis for comparison of transportation systems allows one to have some measure of the external effects and also allows a reasonable estimate of operating costs once the cost of a gallon of fuel or a kilowatt-hour is known. Table 3.6 gives some representative values of the transportation/energy efficiency and estimated block velocities for various transportation systems.

Table 3.6
Propulsion Efficiency of Various Transportation Systems

Transportation System	NPE (seat miles/gal)	Block Velocity (mph)
Automobile	64	30
Bus	220	30
Commuter train	200	50
Metroliner	75	70
Hovertrain	40	150
Gravity vacuum tube	60	150
STOL	33	200
Small jet	34	300
Jumbo jet	40	400
SST	22	1,500

An examination of the table indicates that (1) the lower-speed vehicles with a large number of seats relative to the size of their propulsion system have high transportation/energy efficiency, and (2) the lifting of a vehicle off the ground requires additional energy and therefore a lower propulsion efficiency.

Although we have not considered this here, a more valid comparison of these transportation systems is to use the transportation/energy efficiency in passenger miles per gallon. The first may be thought of as the potential (seats) for transportation, while the second form is the actual utilization (passengers) of the transportation system. They are related to each other through the demand for a particular transportation system.

3.7 SUMMARY

In this chapter we have looked into the many aspects of mass transportation. There are many interesting and exciting engineering problems that mass transportation possesses, yet we must remember that the engineering is only one portion of the solution. We are interested in moving people, and the methods of movement may also create environmental and social problems. Solutions to the problems of mass transportation are therefore not simple technological ones but part of the solution to a complex social-technological-political problem.

Therefore, we cannot simply say "Eliminate the automobile," because to do so would create a much greater problem. We can, however, say "Let's modify, restrict, and augment the use of the automobile when the situation warrants it." Only by this way of thinking can we develop an integrated and balanced mass transportation system that meets the needs of all segments of our society.

REFERENCES

Books

1. BUEL R. A., *Dead End—The Automobile in Mass Transportation*. Englewood Cliffs, N.J.: Prentice-Hall, Inc., 1972.
2. CREIGHTON, R. L., *Urban Transportation Planning*. Urbana: University of Illinois Press, 1970.
3. HELLMAN, H., *Transportation in the World of the Future*. New York: M. Evans and Company, Inc., 1968.
4. LEAVITT, H., *Superhighway—Superhoax*. New York: Ballentine Books, Inc., 1970.
5. MUMFORD, L., *The Highway and the City*. New York: Harcourt Brace Jovanovich, Inc., 1963.
6. STONE, T. R., *Beyond the Automobile*. Englewood Cliffs, N.J.: Prentice-Hall, Inc., 1971.
7. WILLIAMS, E. W., Jr., *The Future of American Transportation*. Englewood Cliffs, N.J.: Prentice-Hall, Inc., 1971.

Magazines and Journals

1. *Astronautics and Aeronautics* (The American Institute of Aeronautics and Astronautics).
2. *Automotive Engineering* (The Society of Automotive Engineers).
3. *Defense Transportation Journal* (The National Defense Transportation Association).
4. *Going Places* (General Electric Co.).
5. *High Speed Ground Transportation Journal*.
6. *Highway Research Record* (National Research Council/Highway Research Board).
7. *IEEE Transactions on Vehicular Technology* (Institute of Electrical and Electronics Engineers).
8. *Institute for Rapid Transit Digest*.
9. *Mechanical Engineering* (The American Society of Mechanical Engineers).
10. *Metropolitan* (Bobit Publishing Co.).

11. *Passenger Transport*, weekly (American Transit Association).
12. *Traffic Engineering* (Institute of Traffic Engineers).
13. *Traffic World* (The Traffic Service Corp.).
14. *Transitrends—Urban Transportation's Weekly Newsletter* (Bobit Publishing Co.).
15. *Transport Economics* (Bureau of Economics, Interstate Commerce Commission).
16. *Transportation* (American Elsevier Publishing Co.).
17. *Transportation Engineering Journal* (American Society of Civil Engineers).
18. *Transportation Journal* (American Society for Traffic and Transportation).
19. *Transportation Research—An International Journal* (Pergamon Press).
20. *Transportation Science* (Operations Research Society of America).
21. *Transport Topics, National Newspaper of the Motor Freight Carriers.*
22. *What's Happening in Transportation* (Transportation Association of America).

Exercises

1. Establish rough credibility (or transportation "usage") versus time curves for the dates 1650, 1750, 1850, 1950, 2050.
2. Establish rough credibility (or transportation "usage") versus time curves for the dates 1600, 1700, 1800, 1900, 2000.
3. Using the selection criteria presented in Section 3.4, compare the various possible methods of transportation for one of the following situations:
   a. Your school and your home.
   b. Your home and the nearest city.
   c. Your home and the closest commercially served airport.
   d. Your school and a rival school.

Sample Projects

1. Look into a local bus system. What do the users think of the system? Suggest any minor and/or major changes and justify your suggestions.
2. Look at the current transit service(s) into your area. Who uses this service? How could the service be improved?
3. Suppose that a mass transportation system were to be implemented in your area. What type of system would you suggest? How would you implement and finance the new system?
4. Suppose that the present transit system is to be expanded into the surrounding communities. What routes, schedules, and stations would you recommend? How would you advertise the expanded system?
5. Devise a questionnaire to determine the travel requirements for a mass transit system in your area.

6. Investigate the problem of exhaust emissions from both gasoline- and diesel-powered buses. What simple methods could be used to reduce or alleviate this problem, and how would you apply these solutions to an existing bus system's equipment?
7. Consider a demand type of mass transit system for a local area with which you are familiar. What size vehicles would you recommend? How many vehicles would be needed to begin operation? What kind of fee structure would be appropriate? What legal problems might arise? What special equipment would be needed?
8. Look into the present maintenance procedures for a local bus system and consider possible improvements.
9. Consider and recommend ways of utilizing more effectively a school or university bus system during off-peak periods such as evenings, weekends, holidays, and vacation periods.

4

energy and pollution

L. Ambs

J. McGowan

4.1 INTRODUCTION

Energy is of major benefit to mankind, but it is also a major threat to modern society. At same time that energy makes more comfortable living conditions possible, it also creates harmful environmental effects. As a result of rapid increases in energy consumption during recent years, coupled with a current concern to preserve environmental quality, it is apparent that the world will have to make changes in the way energy is produced, transported, and consumed if we are to meet the needs of the future. This chapter will serve as an introduction to the important subject of energy utilization and the resulting environmental pollution that it causes.

In the specific areas of energy conversion and pollution control engineers are engaged in a variety of functions ranging from creative design and applied research and development to management. Engineers involved in these fields conceive, plan, design, and supervise the manufacture of a wide variety of devices, machines, and systems for energy conversion, transportation, environmental control, and other related areas. With such a broad range of opportunities it is not surprising that engineers with an energy conversion background work in occupations in the aerospace, automotive, chemical, electrical power generation, and many other industries. If we are to meet the future's changing energy requirements we shall have to develop new technologies to improve our efficiency in the production, conversion, and consumption of energy and to reduce its adverse effects on the environment. Thus, engineers of the future will have a vital role in relating the world's energy needs to technological reality.

4.2 ENERGY RESOURCES AND USES

The development of society can be characterized by a progressive substitution of machine power for muscle power. As will be discussed in detail later, this machine power became available as engineering systems have been developed to convert heat energy into useful power. With the technological development of energy conversion devices, the only other factor needed to supply energy is the availability of useful energy resources. In this section we shall discuss these resources and their uses and supplies.

The first attempt to use a source of power, other than muscles, occurred in the first century B.C. when water power was used for irrigation purposes. As the size and efficiency of waterwheels increased they were used for grinding grains and later became important power sources of the early Industrial Revolution. Even today, water is an important source of power, especially in mountainous terrain where electricity is generated in hydroelectric power stations. In such systems the kinetic energy of the flowing water drives complex and efficient hydraulic turbines instead of simple waterwheels.

With the development of engines driven by heat generated from combustion (heat engines) during the Industrial Revolution, the emphasis on energy sources was shifted from water power to fossil fuels. Other than the obvious fact that water power was restricted to a few geographical areas, one of the important factors that gave impetus to the development of fossil-fuel-fired steam engines was their potential as mobile power sources. Thus, although the steam engine was first used as an auxiliary waterwheel pump, by the middle of the nineteenth century the steam engine became the principal power source for the manufacturing industry of the world. In the present century, a steadily increasing number of energy conversion devices whose chief advantage is mobility have been introduced. The automobile powered by an internal combustion engine is an excellent example.

The interest in energy consumption, energy reserves, and the ability to deliver energy where it is desired can be tied to industrialization. Thus, the great demands to be placed on the energy reserves of the earth can be explained by the fact that almost every country in the world is trying to industrialize—and industrialization takes energy. For example, in the United States, the annual per capita energy consumption is approximately equal to the energy that can be obtained from 10 tons of coal. One can graphically picture the large amount of energy resources necessary to keep our industrial machines working. As Figure 4.1 shows, if we compare the gross national product (a measure of industrialization) and energy consumption per capita for various countries, it is apparent that a key difference between an underdeveloped society and an advanced society in today's world is the amount of energy consumed per person. In Figure 4.1 we have used a common unit of energy, the Btu. One Btu is the amount of energy needed to raise the temperature of 1 lb of water 1°F.

Because gigantic amounts of energy are used, one needs a large unit to talk about the world's total energy consumption; thus, we use the Q ($1Q = 10^{18}$ Btu). About $15Q$ have been used during the past 2,000 years, but one-half of this was used in the last 100 years. At the time of the Industrial Revolution the world was consuming only about $\frac{1}{100}$ Q/year; by 1960 the rate was $\frac{1}{10}$ Q/year. Thus, we had a 10-fold increase in total energy consumption while only doubling the world's population. Taking into account increasing industrialization and population growth, by the year 2050 the world will have spent about $75Q$ if the rate of energy use increases at a 3% annual rate or $275Q$ at a 5% rate of increase.

If one were to look at energy consumption on a regionalized basis, they would discover that the United States with 6% of the world's population

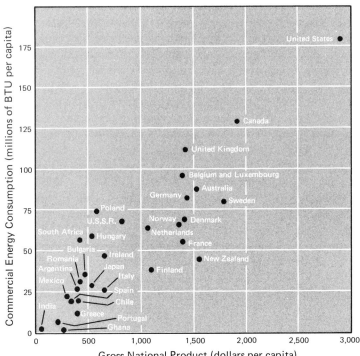

Figure 4.1
Relation between per capita energy consumption and gross national product.

accounts for about 35% of the world's energy use. By the year 2000 the United States' share will probably drop to 25% and the world average per capita energy consumption will have increased from the present one-fifth of the United States' average to about one-third of the United States' average (see Figure 4.2).

Not only are the demands for energy sources increasing, but the relative demands on various sources are also changing rapidly as needs increase. For instance, petroleum has been known for centuries, but until the nineteenth century the common energy sources were wood, water power, animals, and humans. With the development of the steam engine, coal became the source of energy for the Industrial Revolution. Even in modern times, the trends in the changing energy scene are striking. This is graphically illustrated in Figure 4.3, which shows the past and projected consumption of various energy resources in the United States. Note the projected use of nuclear energy in this figure: By the year 2000 it will amount to one-fifth of our total energy supply.

To estimate how long the various energy sources of the world will last, one faces the doubly difficult task of estimating the reserves of each source as well as the consumptive demands on each reserve. Many complex factors are involved in determining the energy consumption patterns, and, as was pointed out, an energy need of $200Q$ could be involved by 2050 just based on the predicted rate of energy consumption growth. Political, sociological, and technological factors are involved in growth patterns. To further compound

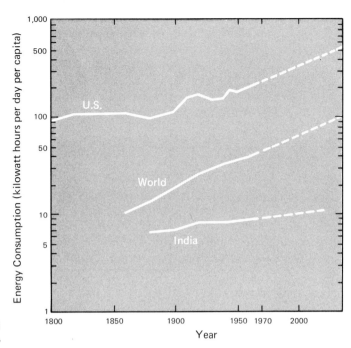

Figure 4.2
World and U.S. energy demand.

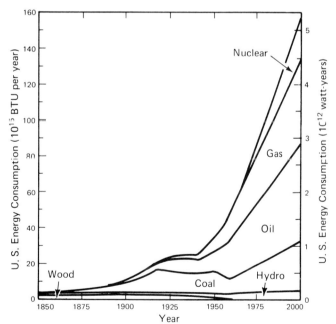

Figure 4.3
Breakdown of U.S. energy consumption. (*Courtesy Scientific American, 39, Sept. 1971.*)

the difficulty in prediction, different energy use sectors are growing at different rates. For example, it is convenient to divide the total demand into household and commercial, industrial, transportation, and electrical generation segments. Figure 4.4 illustrates the changing pattern of energy consumption for

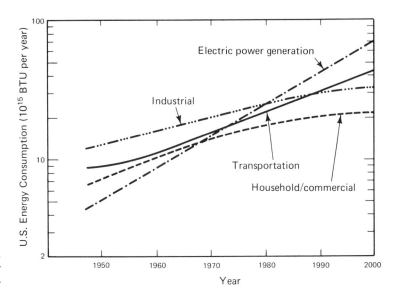

Figure 4.4
U.S. energy consumption by major consumptive sector.

the United States. It can be seen that the largest growths are expected in the electrical generation and transportation areas of the economy. This type of information is needed when predicting the demand on any energy reserve. For example, at the present time, the transportation industry is almost entirely dependent on petroleum as an energy source. Thus, any large-scale increases in transportation energy requirements will place immediate demands on petroleum resources.

When one attempts to estimate the total world energy resources he can classify them as capital (nonrenewable) or income (renewable) sources. In the past, energy capital has been limited to the fossil fuels including coal, oil, and natural gas, which were created several hundred million years ago. We now include nuclear fuels in this category, but the amount of energy available from nuclear fuels is variable. For example, the current type of nuclear reactors extract only about 1% of the available energy from nuclear fuels, while proposed breeder reactors may be able to extract close to 100% of the energy from the uranium fuels. When estimating energy capital resources it is also convenient to divide them into proved and potential resources. The proved resources are those that are known to exist and that can be used economically with present technology. Potential resources refer to those resources that may or may not be technically or economically feasible to utilize (oil shale deposits are an example of this type).

The energy income category refers to continuously available energy resources such as water power, geothermal power, farm wastes, wood, and solar energy. Presently, about 85% of the world's needs is supplied from capital reserves. Although this large dependence on energy capital is likely to remain in the near future, it is interesting to speculate on the possibility of exploiting the tremendous potential of our solar energy input. This can be seen in Table 4.1, which gives the energy budget of the United States for the year 1968. In the United States our consumption of energy from fossil fuels exceeds our consumption of energy from food by 2 orders of magnitude, but it is still much less than the available supply of solar energy.

Although it is beyond the scope of this section to go into a detailed estimate of our energy resources, we shall present the results of one study in Table 4.2. (A more detailed discussion is presented in Reference 1.) In this table estimates of conventional and nuclear fuel resources for the world and the

Table 4.1
Energy Budget of the United States in 1968

ENERGY SOURCE	ENERGY CONSUMPTION	
	Conventional Units (*per year*)	*Btu/yr*
Electric power	1.32×10^{19} kWh	4.5×10^{15}
Fossil fuels		
Crude oil	3.33×10^{9} barrels	1.6×10^{16}
Natural gas	1.93×10^{13} ft^3	1.94×10^{16}
Natural gas, liquids	5.5×10^{8} barrels	$.27 \times 10^{16}$
Coal	5.6×10^{8} tons	5.0×10^{16}
Food consumption	2000 kcal/person/day	5.8×10^{14}
Solar energy	Based on solar energy input of 2 cal/cm^2 over area of U.S. (3.55×10^{6} square miles)	1×10^{20}

Table 4.2
Energy Resources of the United States and the World (Q Units)

DEPLETABLE SUPPLY	WORLD	UNITED STATES
Fossil fuels		
Coal	20–30	5–7
Petroleum	3–6	.6–1.0
Gas	2–5	.5–1.0
Subtotal	25–41	6.1–9.0
Nuclear (ordinary reactor)	100	10
Nuclear (breeder reactor)	10,000	1,000

United States are presented. This table just includes the economically recoverable fuel supplies (those that are available at no more than double the current costs). If we were to compare these data with previous consumption figures, one might conclude that, at least for the United States, fossil fuel reserves are adequate to the year 2000 (with the possible exception of natural gas supplies). Beyond the year 2000, as the supply decreases, fossil fuel prices will begin to rise significantly. Hopefully, by then nuclear energy or solar power will be playing a major role by allowing us to conserve fossil fuels, providing technological options for energy resources, and yielding major savings on energy costs.

4.3 ENERGY CONVERSION SYSTEM ANALYSIS

In this section we shall introduce the physical laws that are required in order to analyze energy conversion systems. Some understanding of these laws is essential in order that the range of application and the limitations of the energy conversion process can be seen. We shall begin by considering conservation of mass and then examine the important laws of thermodynamics: the first law, which is commonly referred to as the law of conservation of energy, and the second law, which describes some of the limitations that must be placed on energy conversion devices. The impact of these laws will be seen not only in the engineering application of energy conversion systems but also in the environmental impact of energy conversion systems on the surroundings. For an interesting introduction to this subject the reader is encouraged to read Reference 2.

Before proceeding, it will be useful for us to introduce some of the terms that are necessary to describe energy conversion systems precisely. In engineering analysis it is very important to identify clearly whatever it is that is under consideration. We shall use the term *system* to identify in a very broad sense the subject under discussion. The system that we choose may consist of the expanding gases inside of a cylinder of an internal combustion engine. In this context we may be considering how much power we can obtain from the expansion of these gases. Alternatively, the system may be a nuclear power generation station, and we might be concerned with the amount of electrical power that can be generated for each unit mass of fissionable material. A second term that we shall find useful is *surroundings*. The surroundings consist of everything not in our system. In some contexts, it is perhaps more useful to term the surrounding the environment, particularly when we are considering the air pollution or thermal pollution aspects of energy conversion systems.

Another term, *energy*, is one we sometimes take for granted. The term energy is familiar to us from many places. We have seen it in physics and chemistry, and we use it in our everyday lives. But here, the technical definition or the technical implication of energy will be important to us. Energy is a system's capability to do work. Energy is present in many forms. *Potential energy* is the form of energy possessed by a system in a force field. Generally, we are most familiar with gravitational potential energy, such as that possessed by water before running over a dam. *Kinetic energy* is the energy that a system possesses as a result of its relative motion. *Electrical energy* is based on the fact that electrical potential or voltage can do work by causing a current to flow. *Chemical energy* refers to the energy liberated when a substance takes part in a chemical reaction.

If we were to examine all the various forms of energy, we would discover that all forms of energy have the units of force times distance. Thus, to put a numerical value on the amount of energy available from a certain process we must be careful to make sure we are on the same dimensional basis. For example, in the United States, a convenient unit of mechanical energy is the foot-pound, defined as the energy required to raise 1 lb vertically through 1 ft. However, at times it is more convenient to represent a certain amount of energy in British thermal units (Btu). For this case one would have to know that 1 Btu = 778 ft-lb. We could go on and on describing the multitude of units used in thermodynamics, however, Table 4.3 serves as a short guide to various conversion factors, depending on the system of units one desires.

Table 4.3
Conversion Factors for Energy and Power Units

Energy Units
1 Btu = 778 ft-lb = 107.6 kgm = 252 cal
1 J (W-sec) = .7376 hp = .2389 cal = 1 × 10^7 ergs (dyne-cm)
1 kWh = 3,413 Btu = 1.341 hp-hr = 8.6 × 10^5 cal

Power Units
1 hp = 2545 Btu/hr = 550 ft-lb/sec
1 kW = 1.341 hp = 1 × 10^{10} ergs/sec

The forms of energy we have not yet considered are work (W) and heat (Q), both of which are energy in transit. Work and heat are only considered relative to the system we choose. Work is the equivalent of a force acting through a distance. We do work if we lift a weight, pick up our arm, or push something across the table. All of these are equivalent to a force acting through a distance. Work is a transfer of energy from our system to the surroundings. In that sense it is transient energy. Energy is moving from one system to another or from a system to the surroundings. If we consider our body as a system, we require energy to lift a weight from the floor to the table. How efficiently we convert the energy of our body to the raising of a weight or doing work can be important to us. Heat or heat transfer is also energy in transit. The sense of warmth we feel when placing our hand next to a hot radiator is due to the heat transfer from the radiator. Technically, the definition of heat transfer is energy transfer from our system caused by a difference in temperature between the system and the surroundings. Both heat and work are energy transfers. With the first law of thermodynamics, heat and work are related to the energy of our system in a conservation law. The second law distinguishes between heat and work.

Once the concept of energy is recognized, it becomes important to consider power. Power is defined as the rate at which energy is expended in a process. The first unit for power, the horsepower, was chosen by James Watt, a great English inventor and steam engine pioneer. Watt found that an average horse could do work at a rate of 33,000 ft-lb/min, thus defining the standard mechanical power unit. It is interesting to note that in measuring electrical energy, power is usually expressed in terms of watts or kilowatts. (1 kW = 1.341 hp.) Again refer to Table 4.3 for various conversion factors for power and energy.

A final term of importance to energy conversion processes is the term *efficiency*. Efficiency is usually defined as the fraction of one form of energy that is converted to another form. For example, the efficiency of a fuel-burning electrical power generating plant is given as the ratio of the electrical energy produced to the chemical energy of the fuel consumed. Typically this efficiency is about 30%.

The law of conservation of mass is important to us because most of the energy sources we use come from mass in which the energy is being converted from one form to another. An example of this is the fossil fuel used in power plants. Conservation of mass says that the amount of mass remains constant, although its form may change. The mass of matter that 100 billion years ago was vegetation has slowly been converted to coal. This coal represents a useful (capital) energy source. In an electric power generating plant, the energy in this coal can be rapidly converted from one form to another, producing electricity. The carbon in the coal is oxidized with oxygen from the air to form carbon dioxide with a large release in chemical energy. The resulting carbon dioxide returns to the environment (surroundings) where it interacts with sunlight through photosynthesis to produce vegetation again. At this point, the cycle begins anew. This biological cycle has been very useful to man. The primary difficulty is that the slow portion of the cycle, the conversion of vegetation to coal, takes much longer than the conversion of the coal to the gases in the energy conversion process of the power plant.

Conservation of mass can be simply illustrated by the air pollution aspects of this system. During the natural coal-forming process, sulfur can be introduced to the coal. When the coal is burned, the sulfur appears in the surroundings in the form of sulfur dioxide (SO_2). Sulfur dioxide is harmful to life, and we as engineers must now modify our energy conversion processes or otherwise limit sulfur dioxide production. Some simple numbers will tell us what is involved.

If the United States uses 5 billion tons of coal to produce 12 billion kWh of electrical energy and this coal contains 3% sulfur, then 150 million tons of SO_2 will be produced. This corresponds to approximately $\frac{3}{4}$ ton of SO_2 for each person in the United States. If we were each asked to dispose of our share, we think you can see the type of environmental impact this could have in a short period of time. As the yearly consumption of energy increases and the use of high-sulfur fossil fuels continues, the problem could become more severe.

A similar problem exists with the solid exhausts from a power plant, fly ash. Coal consists of solid carbon plus other minerals. After burning the coal in a power plant or furnace, the coal is converted to carbon dioxide and water vapor and the remaining materials are left to be disposed of.

A final consequence is that all mass contains energy or can possess it. Mass has internal energy, and when it has motion it possesses kinetic energy. If a change in chemical identity is involved, chemical energy is involved. Other forms of energy that might be considered are potential energy, nuclear energy, electrical energy, and radiant energy. This brings us to the second of our physical laws, the first law of thermodynamics, which is usually referred to as the conservation of energy. A more precise statement of this law states that the energy of an isolated system remains constant. In an isolated system neither mass nor energy crosses the boundaries to the surrounding. Consequently, no change in total energy occurs.

At this point, we can invoke some of the other terms that we have introduced previously. Most of the systems that are of engineering significance are not isolated. By this we mean either matter and/or energy can cross the boundary into the system. Let us consider only those systems in which energy is allowed to cross the boundary. For our nonisolated or open system, energy can cross the boundary in two forms: first as a work transfer (W) and second as a heat transfer (Q). This allows us to write the following expression for the first law of thermodynamics, which states that the change in energy (ΔE) of the system is equal to the sum of the heat transfer and work transfer across the boundaries into the system. Algebraically this is given as

$$Q - W = \Delta E$$

By convention, work is positive when transferred out of the system, and heat transfer is positive when transferred into the system. Applications of the first law then become a matter of bookkeeping. We must keep track of the energy transfers into and out of our system. In addition, we shall also be concerned with the form of the energy within the system. Although the first law describes the total energy of the system, we know that the energy within this system, the total energy E, can be in a number of different forms. It can be in the form of potential energy, kinetic energy, and internal energy, as well as the others that we have described previously. For a system in which there is no heat transfer or work transfer, the first law then tells us that the total energy must remain constant. This does not mean, however, that the form of the energy must remain fixed. It is still possible for us to convert the form of the energy from one type to another.

The conversion of chemical energy to internal energy by burning a fuel is probably familiar to you. The head of a match contains chemical energy.

When we strike it and begin the chemical reaction, we are converting chemical energy to thermal internal energy. We then make use of this thermal energy to transfer heat to some paper, cigarette, or other substance in order to make further use of this energy. But if the match is defined as our system, initially what occurs is the conversion of energy from chemical form to internal form followed by a heat transfer from our system. At this point, we should note that the reverse of this process is not quite so easy. That is, it is not easy to utilize thermal energy as heat transfer into a stick of wood in order to create a matchhead. This type of natural process direction will be considered further when the second law of thermodynamics is discussed.

As an additional example of the first law, let us consider the automobile. The chemical energy available in the gasoline is converted in the engine of the automobile into the high-temperature internal energy of the gases. These high-energy gases then expand, forcing down the piston, which in turn turns the crankshaft and produces useful work. In this process, the chemical energy of the fuel has been converted into mechanical energy that can be used to move our cars. This process, however, is not 100% efficient. Some of the energy that was available in the fuel is not converted into useful work and is discarded as lost thermal energy. Also, after the gases have expanded and pushed the piston to the bottom of its stroke, the exhaust value opens and the extra energy remaining in the gases is discharged into the atmosphere. This lost energy is not available to do useful work. In current internal combustion engines, only about 25% of the energy available in the fuel results in useful work, with the remaining 75% lost. This loss of available energy occurs whenever energy conversion processes using heat energy occur.

This example can be explained by the third physical law, the second law of thermodynamics. The second law of thermodynamics can be stated as follows: *In a thermal energy converter only part of the available energy can be converted into useful work.* The thermal efficiency of an energy conversion process is defined as

$$\text{efficiency} = \frac{W}{Q_{\text{in}}}$$

where W is the net useful work and Q_{in} is the thermal energy used by the energy converter. In an electrical power plant, energy from the combustion of a fuel is transferred as heat (Q_{in}) into the water to produce the steam that drives the turbines. The expanding steam in the turbine does work (W) in powering the electrical generators. This expression for efficiency can be further developed by utilizing the first law of thermodynamics. That is, the heat transfer into the water in the boiler and the net work produced by the power plant are related to the heat rejected in the cooling system, Q_{out}:

$$Q_{\text{in}} - Q_{\text{out}} = W$$

This modifies the efficiency expression to

$$\text{efficiency} = \frac{Q_{\text{in}} - Q_{\text{out}}}{Q_{\text{in}}}$$

or

$$\text{efficiency} = 1 - \frac{Q_{\text{out}}}{Q_{\text{in}}}$$

Thus, thermal efficiency of a system is just an indicator of how well we make use of the energy available. As we reject less and less of the energy that is input to our system, the efficiency rises.

The important point to be made from this discussion is that the second law of thermodynamics limits the amount of energy that we can obtain from one source. From thermal sources, the efficiency that we can achieve is usually considerably less than 50%. This means that less than half of the energy that is available to us from our source is converted to useful work. This lost energy is dumped into the environment and manifests itself in a number of ways. It may be the energy that is rejected from the power plants that heat up our lakes and rivers or it may be energy that is disposed of in the atmosphere that could contribute to atmospheric problems. This thermal pollution, which we shall look at later in more detail, is just one of the environmental burdens that energy utilization can place on man.

4.4
ENERGY CONVERSION SYSTEMS

Before discussing various types of energy conversion systems, it is important to realize that many systems or sources of energy have been developed for convenience. Sometimes we require energy in a specific form at specific locations at a specific time—and are willing to pay for it. Thus, we sometimes support costly sources of energy if the convenience factors are high. Many examples exist such as battery-powered devices, electric home heating, and decorative lighting. As summarized from calculations by various authors, Table 4.4 presents relative costs of energy in different forms. One can note that the costs listed can vary over an exceedingly wide range, and sometimes the price paid for this energy may not be significant. The best example of this fact is the last item in Table 4.4. However, it should be pointed out that the yearly cost of operation of an electric watch is small compared to its total cost.

As shown in Figure 4.5, there are a multitude of ways to convert energy from one form to another. However, on a worldwide basis more than 99% of the energy presently consumed involves the combustion of fossil fuels, in which chemical energy is converted into thermal energy. Likewise, almost all the world's electricity is generated through the use of thermal energy and heat engines. By their nature, and governed by the law of thermodynamics, these indirect (requiring more than one step) energy conversion devices have efficiencies of less than 50%. As indicated in Figure 4.5, there are some direct energy conversion devices that can bypass this limitation. For example, solar energy converters and fuel cells are devices that operate at constant temperature and do not convert their primary source of energy to heat before generating electricity. At the present time, however, these devices have not

Table 4.4
Cost of Energy in Various Forms

Energy Source	Cost (cents/kWh)
Fuel oil	.5
Gasoline	1.0
Electricity (central station)	2.0
Gas Turbine	3.0
Food (sugar, bread, butter)	10.0
Lead-acid storage battery	20.0
Thermoelectric generator	30.0
Fuel cells	400
Candlelight	1,000
Flashlight battery	4,000
Mercury battery	50,000
Electric watch battery	7,000,000

reached their theoretical performance goals of about 100% efficiency. Solar cells are still only about 20% efficient, and fuel cells have actual efficiencies on the order of 50 to 60%.

Before looking at the various conventional (indirect) and unconventional (direct) energy conversion systems, we should note that one of the prob-

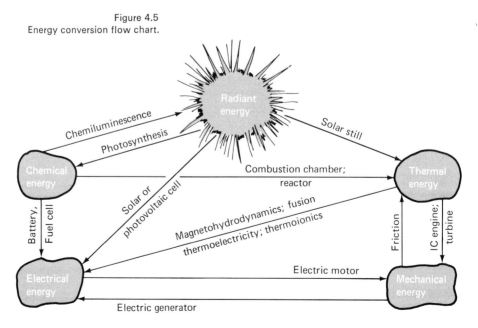

Figure 4.5
Energy conversion flow chart.

lems that engineers have to keep in mind is that there may be more than one way to measure the performance of an energy conversion system. Thus, efficiency alone cannot be the sole index of how well a device performs. For example, sometimes we measure the performance of a system by its power per unit volume, or per unit of fuel. Such performance criteria become important factors in many of our mobile energy conversion systems.

4.4.1 Conventional Energy Conversion Systems

The type of systems that deliver practically all the world's energy needs are based on heat engine systems. Although particular components in such systems have changed with time, in this type of energy conversion system, thermal energy released by combustion or nuclear reaction is converted into mechanical energy by an appropriate heat engine. A schematic of a typical heat engine system, a Rankine cycle, is shown in Figure 4.6. In this system fossil fuel is burned in the combustor and energy is transferred in the form of heat to produce vapor (steam if water is the working fluid). The vapor then expands through a reciprocating engine or a turbine to produce mechanical work. The work output can be used to drive a vehicle or, as shown, to produce electricity by driving an electrical generator. After passing through the expander, heat is rejected as the working fluid passes through the condenser (a necessary condition imposed by the second law of thermodynamics). The fluid then passes through a circulating pump and is returned to the vapor generator to repeat the same power-producing cycle. It is important to point out two characteristics of these systems that can contribute to environmental problems (a subject to be discussed in greater detail in a later section). First it should be recognized that the exhaust products from the combustor may present a potential air pollution problem. Second, the heat rejected (which is a

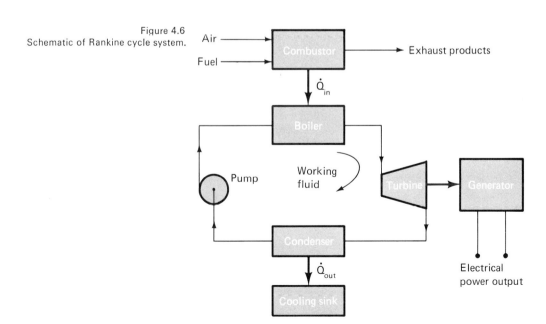

Figure 4.6
Schematic of Rankine cycle system.

form of low-grade energy) may pose another potential harm to the environment, so-called thermal pollution.

Historically, a system similar in design to that shown in Figure 4.6, using a reciprocating engine, provided the energy to drive the early power sources that became important during the Industrial Revolution. In practice reciprocating steam engines were used in large numbers as locomotive power sources until replaced by more compact and efficient diesel engines in the 1950s. For electrical generation applications, the reciprocating steam engine was replaced by large steam turbines in the early twentieth century. Replacement was due not only to economic and efficiency considerations but also to the ability to build larger power units. We again emphasize that engineers are continually aiming to increase the efficiency and power output of energy conversion machines while decreasing their costs. In Figure 4.7 one can see that the maximum power output of the steam engine and its successor, the steam turbine, has increased by more than 6 orders of magnitude from less than 1 kW to more than 1 million kW. For comparison, the maximum power output of other basic machines are also shown in this figure. It is interesting to note that all these machines are surpassed in power output by the largest liquid fuel rockets, which can deliver more than 16 million kW of power for brief periods.

In general, three types of thermodynamic power cycles account for the vast majority of power produced from heat engine systems. These include the previously mentioned Rankine cycle used in large steam-electric generating stations, the reciprocating internal combustion engine in vehicles, and the gas turbine in aircraft or peak power electric generating plants. The use of these

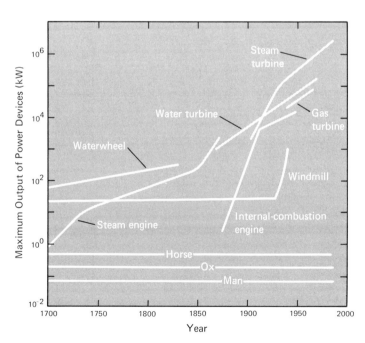

Figure 4.7
Power output of basic machines.

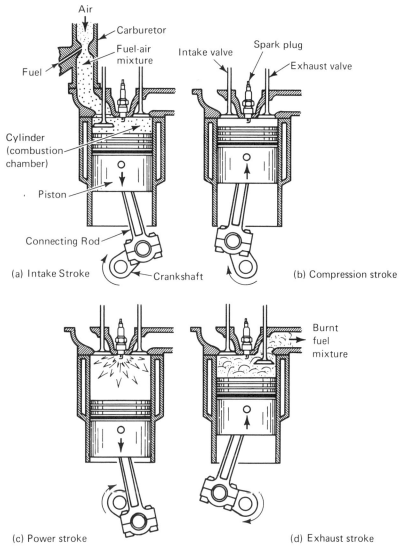

Figure 4.8
Combustion in an automobile engine.
(One cylinder of a typical automobile engine shown.)

cycles will continue to grow, even as more nuclear power plants are built in the future. (In nuclear power plants the fission of uranium releases energy, which is used to make steam, which then goes through the same cycle as in a fossil fuel power plant.)

A number of energy conversion systems are in use in transportation. We shall now briefly describe two major types: the internal combustion engine and the gas turbine. Figure 4.8 shows a schematic of a spark ignition internal

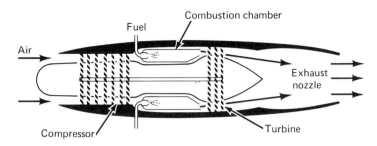

Figure 4.9
Schematic of a gas turbine engine.

combustion engine cycle. This cycle is usually referred to as the Otto cycle after Nikolaus Otto, who first built this four-stroke engine in 1876. The cycle begins when the intake valve opens and a mixture of fuel and air enters the cylinder during the intake stroke. The intake valve closes when the piston reaches the bottom of the stroke, and as the piston moves upward, the mixture is compressed. As the piston nears the top of the cylinder, a spark plug ignites the fuel-air mixture. The thermal energy from the combustion of the fuel-air mixture forces the piston down, and the rotating crankshaft does work. After the piston reaches the bottom of its travel, the exhaust valve opens, and the piston again moves upward, displacing the gases from the cylinder. At the top of the cylinder, the exhaust valve closes, the intake opens, and the cycle repeats. During the cycle, the chemical energy of the fuel is converted to thermal energy during combustion. This thermal energy is converted to useful work by the piston. Most of the thermal energy is lost from the cylinder when the exhaust valve opens. Typically, only 25% of the chemical energy available in the fuel-air mixture ever produces useful work.

A schematic of a gas turbine is shown in Figure 4.9. Gas turbines develop their power by expanding a high-pressure high-temperature gas as in an Otto cycle. Air enters the engine from the left and is compressed in a rotating compressor. Air leaving the compressor at high pressure and temperature has fuel injected into it in the combustion chamber. The fuel burns, and the high-energy gases expand through a rotating turbine, which is used to provide power for the compressor. The gases then further expand in the nozzle to a high exit velocity leaving the engine. The change in momentum of the gases passing through the engine produces net thrust, which is used to proper the aircraft. In a shaft power machine, instead of expanding in a nozzle, the gases are further expanded in another turbine, producing rotary or shaft output power.

4.4.2 Unconventional Energy Conversion Systems

One reason for the present interest in direct or unconventional energy conversion systems is the potential they offer for higher-energy conversion efficiencies—thus reducing waste energy. Also, many times particular configurations of these systems have been developed as the only possible energy-producing device available for a particular job. For example, the Apollo space mission's power requirements were uniquely suited for fuel cell applications. In the next few paragraphs we shall confine our brief discussion to three direct energy conversion systems that have received much notice: fuel cells, magnetohydrodynamic (MHD) generators, and thermionic energy converters.

A fuel cell is a direct energy conversion device that continuously converts chemical energy into electrical energy. A schematic of a typical fuel cell is shown in Figure 4.10. This device converts the energy in hydrogen or other fuels directly into electricity by a chemical reaction (of the same form as a combustion reaction) inside porous electrodes. The oxidizer is usually oxygen. However, current research is directed toward the development of fuel cells that use hydrocarbon fuels and air. Fuel cells are especially attractive since they are not theoretically limited to lower conversion efficiencies like heat engines and could conceivably have efficiencies as high as 90%, although today their actual efficiencies range between 50 and 60%. Recently, engineers have proposed the use of fuel cells for central power generation; however, the economic and technological problems to be overcome for this type of application appear formidable. As an interesting side note, in many ways the human body itself is a fuel cell system: Food in blood, which is an electrolyte, is oxidized catalytically by enzymes to produce energy, part of which is electrical.

In MHD energy conversion systems, thermal energy is directly converted to electrical energy by using the scientific principle that a liquid metal or ionized gas will generate electric power when flowing through a magnetic

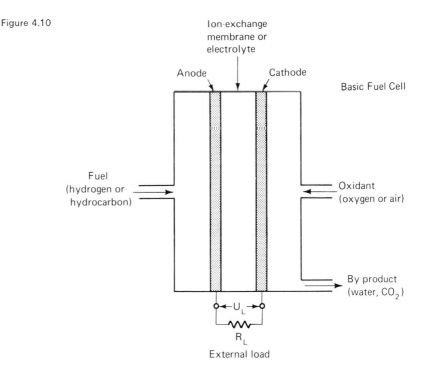

Figure 4.10

Basic Fuel Cell

field. Like the fuel cell, the MHD converter is an especially attractive system since it has no moving parts. In its simplest form, the MHD flow channel has electrical insulators on two opposite walls and power-removing conductors or electrodes on the other two walls (see Figure 4.11). The testing and development of MHD energy converters have been mainly confined to the laboratory with the exception of some fairly large power generation installations in Russia and Japan. Several U.S. power companies have funded design work for a peak-load MHD electrical generating station. The extreme temperatures required for efficient operation of MHD systems cause severe materials problems as well as some unique air pollution problems. If MHD technology can be developed, it should be possible to build fossil fuel power plants with efficiencies of 45 to 50% or higher if combined with conventional Rankine cycle systems.

Analogous to heat engine systems, the thermionic generator uses electrons as a working fluid instead of a vapor or gas. Electrons are driven by thermal energy across a voltage difference to produce electrical energy. In the thermionic generator, electrons are first evaporated from a heated cathode and are then condensed or collected on a cooler anode. These electrons flow through an external circuit back to the cathode. Thus, in a thermionic generator, thermal energy is converted directly into electricity through a process similar to that in a steam power plant where water is evaporated in a boiler and then condensed after doing useful work in an engine.

As with many of our so-called new energy conversion systems, the therminoic energy converter represents a new approach to an old discovery. It was Thomas Edison who first observed an electric current between an incandes-

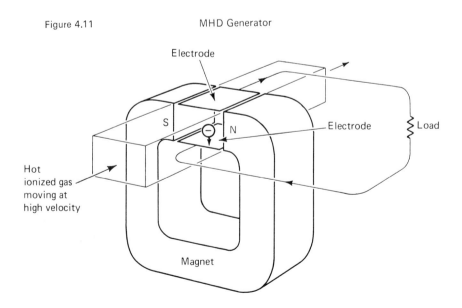

Figure 4.11   MHD Generator

cent filament and a cold electrode in an evacuated tube, but not until recently has there been active development of thermionic generators. With the discovery of materials that provide adequate electron emission rates without melting and by the addition of vapors to reduce the space charge effects, the performance of these systems have been greatly improved. Such systems have been designed for space power applications where high-temperature operation is advantageous. In addition, thermionic generators are well suited for use with nuclear reactors or radioisotope heat sources.

4.5 ENVIRONMENTAL ASPECTS OF ENERGY CONVERSION SYSTEMS

It is possible for energy conversion systems to impose health or safety hazards. The problems of environmental health should be faced today since they will require much more attention as space, air, water, and energy demands increase with the world's population and economic growth.

Whenever man uses energy, he pollutes his environment in some form. The severity of the problem is related to the type of pollution involved, the location of the pollution, and the quantity of pollutants emitted. Our increasing environmental burden is related to energy utilization in each of the following three ways, and we must address ourselves to each of them in order to alleviate the problem:

1. The extent of energy use is increasing.
2. The location of the energy use and its related pollution is critical.
3. The pollution per unit of energy used must be considered.

We have previously discussed the first point. Basically, this increase in energy consumption is due to two factors. First, the standard of living is increasing and this requires more energy consumption by increased industrialization. The second factor is that, in addition to increasing our standard of living, we are also increasing our population. A significant point when looking at both these factors is that, in most of the world, the total increase in energy use is much greater than just the population increase. For example, in the 10-year period from 1958 to 1968 in the United States, total energy consumption increased by 50.5%, while the population increased by only 15.5%.

Also, along with increasing energy consumption, the type of energy needs can have important effects on the environment. In the United States in 1970, household and commercial energy consumption amounted to 22% of the total; industrial consumption was 30%, transportation was 24%, and electrical power generation was 24%. Although all the energy modes will increase in the future, Figure 4.4 shows that electrical power is growing at the fastest rate. As we have already seen, less than half of the energy that is consumed for electrical power generation ever finds its way into doing useful work. This shift to electrical energy consumption will have a profound impact on the energy use and environmental pollution of the country.

This leads to an example of the second point to be considered—the location of the environmental pollution source. Under the provisions of the Clean Air Act of 1970 the Environmental Protection Agency has set national air quality standards for air pollution. Under these standards the atmospheric concentrations of various air pollutants will be limited to a level that is not hazardous to human health. With uniform standards throughout the United States it is safe to assume that those locations having the largest energy use will have the highest concentrations of pollution. It is these areas in which the most restrictive controls must be imposed. The location of pollution is as significant as the amount of pollution produced. Most energy utilization occurs near man, and consequently most pollution occurs near man. The greater the population density, the greater the energy utilization. Man has congregated himself in sprawling urban environments with the consequence that his own welfare is threatened. In these urban centers we find man's use of transportation increasing as he moves from his home to his work.

All energy utilization does not have to be near man. This is especially true of electrical power generation. In the past the economics of electricity distribution has dictated locating power plants near the areas where the power is to be used. This concept is changing. The Four Corners Power Plant in the southwestern United States was located far from the population centers it was built to serve. Efficient power transmission lines are used to transport the energy to the marketplace.

The impact of the National Air Quality Standards Act may also severely limit other energy utilization in urban areas. Transportation use in the urban areas may have to be restricted in order to meet the standards.

It is possible to foresee a time in urban areas when most energy used will be from electrical power that is generated far from the urban center. This may not be as efficient an energy use as it is today, but it would produce reduced local environmental consequences.

The third point is that the pollution that is produced per unit of energy use must be reduced also. At the present time, as can be seen in Figure 4.3, fossil fuel combustion accounts for over 95% of our energy resource. Combustion produces air pollution, and with most of the combustion involved in heat engine applications thermal pollution can be significant. Combustion processes must be cleaned up to reduce pollution. The air pollution potential of combustion will be discussed in Section 4.5.1. Following that thermal pollution will be discussed.

4.5.1 Air Pollution

Energy utilization produces most of the air pollution in this country. Table 4.5 gives the air pollution burden for the United States for 1969. Three subject areas—transportation, stationary sources, and industrial processes—are the primary energy utilization sectors of society. Seventy-five percent of the pollutants formed are a result of energy utilization.

Table 4.5
1969 Estimated Nationwide Emissions,
United States (millions of tons/year)

Source	Sulfur Oxides	Particulates	Carbon Monoxide	Hydrocarbons	Nitrogen Oxides
Transportation	1.1	.8	111.5	19.8	11.2
Fuel combustion in stationary sources	24.4	7.2	1.8	.9	10.0
Industrial processes	7.5	14.4	12.0	5.5	.2
Solid waste disposal	.2	1.4	7.9	2.0	.4
Miscellaneous	.2	11.4	18.2	9.2	2.0
Total	33.4	35.2	151.4	37.4	23.8

Transportation with its total reliance on fossil fuel consumption is the largest single contributor to atmospheric emissions. It contributes about 51% of the total tonnage emitted annually. Following transportation are stationary sources with 16% of the total emissions and then industrial processes with 14%. It is quite evident from looking at these figures that sources should be controlled based on the tonnage emission rates.

We can get some insight into the emissions problem if we consider the combustion process itself. In normal combustion processes, oxidation is the primary reaction. In many reactions, however, intermediate products are produced, some of which remain when the combustion process is complete. Theoretically, combustion reactions should go to completion perfectly. In practice this seldom occurs because of incomplete mixing, high-temperature chemical equilibrium, improper fuel-air ratio, or flame quenching. The net result is that intermediate products and/or incomplete combustion products are formed in most combustion reactions. This process is schematically illustrated in Figure 4.12.

Combustion reactions are usually categorized as being either premixed or unmixed, depending on whether the fuel and air are mixed before or during the combustion process. Most energy conversion devices utilizing combustion are unmixed, and therefore mixing plays a very important role in the extent of the completion of the reaction. For example, in the combustion of coal as in a power plant, the fuel is solid and reacts with oxygen in the air as mixing makes it available. Equilibrium considerations then enter the picture because as carbon is oxidized the product is initially carbon monoxide rather than carbon dioxide. Later, further oxygen, if available, completes the reaction of carbon monoxide to carbon dioxide. These reactions are strongly temperature-dependent and the rate of the reaction also depends on other chemical species present. It should be pointed out that even in complete combustion reactions some carbon monoxide remains in the products.

$$\text{Hydrocarbon fuel} + \text{Air} \longrightarrow \text{Combustion} \longrightarrow \text{Products of combustion}$$

$$C_xH_y + O_2 + N_2 \longrightarrow CO_2 + H_2O + N_2 + O_2 + N_2 + CO + NO + NO_2 + C_AH_A + C_BH_B + \cdots$$
$$\text{(polutants)}$$

Figure 4.12
Fuel combustion process.

Other chemicals present in fossil fuels can also present problems. For example, sulfur present in the input fuel converts to about 98% SO_2 and 2% SO_3 during normal combustion. In gaseous form, these are ready to combine with water vapor in the exhaust gas or outside atmosphere to form acids. In fuel oils most of these sulfur oxides leave the stack, whereas in coal combustion a large percentage of the sulfur remains in the ash. Most of the particulates obtained from coal combustion are from the minerals present in the parent coal. Presently, both solid particulates and sulfur oxides can be removed from the exhaust gas with varying degrees of success.

Nitrogen oxides (NO, NO_2) form under conditions that we usually think of as perfect combustion, that is, complete mixing, high gas temperatures, and adequate excess air. NO_2 does not exist at these conditions, but combustion gas equilibrium at high temperatures with excess air forces the

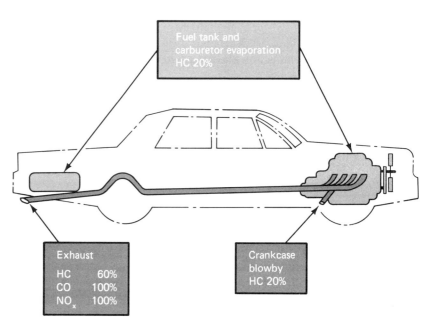

Figure 4.13
Automobile emission inventory (prior to 1961).

formation of NO. As the gas temperature increases, the concentration of NO increases. Power-plant furnaces are designed for high temperatures, and, under these conditions, the NO concentration may be as high as 1000 ppm. After the NO leaves the stack, atmospheric air and lower temperatures convert the NO to NO_2.

As you might have observed from this discussion, poor combustion produces excessive carbon monoxide, whereas good combustion produces nitrogen oxides. Finally, the sulfur in the fuel produces sulfur oxides. Current research in combustion processes is directed at reducing all three of these air pollutants simultaneously.

Although autos do not give the appearance of being as serious a polluter as power plants, their contribution to the air pollution problem is significant. The source of pollutants in cars is not completely from combustion of the fuels. In Figure 4.13 is shown the precontrol inventory of automotive emissions. In addition to the exhaust, fuel evaporation and crank case blowby have contributed to the hydrocarbon emissions. We should point out that the combustion process in internal combustion engines is different from the power-plant or space-heating furnace. Internal combustion engines are perhaps better characterized as intermittant combustion engines. A fuel-air mixture is produced and distributed usually nonuniformly to the cylinders of the engine. In the cylinder after compression the mixture is ignited and combustion proceeds. With changes in load produced by acceleration and deceleration the fuel-air ratio changes, and less than ideal conditions are found in the engine for combustion. The walls of the combustion chamber quench the combustion reaction, and unburned or partially burned hydrocarbons appear in the exhaust.

The internal combustion engine exhaust produces carbon monoxide, hydrocarbons, nitrogen oxides, and some particulates. One difficulty is that different modes of operation produce different emissions. Carbon monoxide is produced throughout the driving cycle, but oxides of nitrogen are generally produced during high power and acceleration conditions. Hydrocarbons are dominant during deceleration conditions or when the ignition system is malfunctioning. This type of characteristic makes the problem more difficult to correct. Since 1960 engine controls have been required by law to reduce emissions. Current legislation is directed at reducing the emissions level for cars to one-tenth the level produced by cars in 1970. The impact of this legislation can be seen in Figure 4.14. With present controls, the emissions would reach a minimum in the early 1980s and would then increase again because of a predicted increase in the number of vehicles. With 1976 controls this minimum would be lower and would occur about 1990, after which it could be expected to increase again.

Diesel exhausts present similar problems except that the emissions are not so high. A more esthetic problem exists with diesel smoke and odor. Controls are also in progress here and some degree of control is being obtained. In the case of aircraft, the emissions problem is related to the location of the source—the airport—as well as the strength of the source. In combustion modification on jet aircraft, the smoke problem has been sharply reduced.

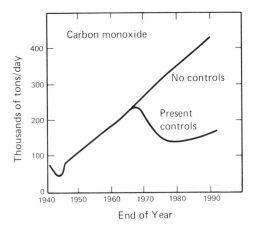
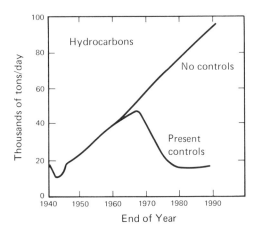

Figure 4.14
Impact of automobile emission control.

Emissions from other stationary sources and industrial processes are strongly dependent on the type of process that is involved. Reduction of emissions from these processes necessitates an inventory of emissions based on ambient sampling and a careful analysis of the process in order to affect a degree of control.

4.5.2
Thermal
Pollution

The second law of thermodynamics limits the amount of energy that can be obtained from conventional electrical generating power plants based on the Rankine cycle. Thus, the problem of thermal pollution arises when we burn chemical or nuclear fuel and convert, on the average, one-third of the fuel energy into electric power and return two-thirds to a cooling source (such as a stream or a river). As society has expanded its production of electric power and increased its discharge of heated water into the aquatic environment, severe ecological problems have sometimes shown up. We should remember that the term thermal pollution applies only to situations where heated water conditions adversely effect the quality of water and aquatic life. In some situations, the addition of heat to water can have beneficial effects. For example, rejected power-plant heat has been used to grow oysters in Long Island Sound.

The water temperature is a major consideration in determining the use of water. It plays a key role in the ability of the ecological system to maintain desirable characteristics throughout all biological stages. In addition to the direct lethal effects that temperature increases may have on biological systems, there may be effects on other potential water users, such as industries or municipalities. For example, higher temperatures lower the capacity of the water to hold oxygen and increase sedimentation (settling) rates. Chemical composition and acidity of the water may also change. In many ways, discharges of heated water are equivalent to placing organic wastes in a stream;

that is, the absorbing capacity of a stream is reduced. It is beyond the scope of our brief discussion here to go into all the ecological effects of thermal pollution; however, we must be aware that many scientists are presently engaged in research in this area.

The major users of water include farmers (irrigation), municipal governments (sewage treatment), industrial organizations, and electric power plants (see Figure 4.15). Electric power production accounts for more than four-fifths of the total cooling water use and for about one-third of the total water use. In general the problems associated with cooling water discharges from electrical generating power plants far exceed the problems from industrial sources. It is estimated that by 1980, primarily because of increased power demands, the power industry will use one-fifth of the total freshwater runoff of the United States for cooling. Also, the increase in size of individual plants and the lower efficiency of present nuclear power plants have made a large difference in the intensity of the problem. (For example, the maximum size of steam-electric power plants has risen from about 200 to over 1,000 MW in the last 15 years.) Even though the total amount of heat rejected to the environment may not increase when a single plant replaces two or more, the ecological impact is much greater locally.

At the present time, technological solutions to thermal pollution problems have not kept pace with the increased production of power. The solutions of which most people are aware fit into the following six categories:

1. Methods to minimize the effect of waste heat on the environment.
2. Means to reduce waste heat in conventional power plants.
3. Uses for waste heat.
4. Shifting the path of heat rejection directly to the atmosphere.
5. New, nonpolluting methods of power generation.
6. Less consumer use of electricity.

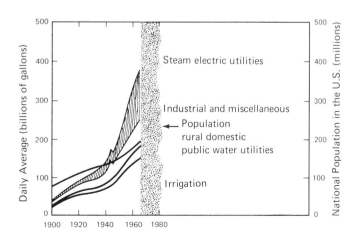

Figure 4.15
Water uses in U.S.

The possibilities opened up by this list are boundless, and it should be noted that many engineers are engaged in various phases of all categories. Another key point to be made in concluding this section is that we cannot continue to double the generation of electricity every 10 years or so without taking into account the possible worldwide climate changes caused by heat rejection. Presently, we are probably safe since the solar energy input to earth is about 200,000 times greater than the worldwide energy conversion for electric power. However, as we have pointed out, local concentrations of reject heat can become significant.

4.6 SUMMARY

In this chapter we have shown that demands will be placed on our energy supplies by increasing population and industrialization of all the countries of the world. The governing laws (conservation of mass and the first and second laws of thermodynamics) that engineers use to analyze energy conversion systems were introduced. Environmental consequences of energy conversion were discussed, and it was pointed out that these problems are influenced by the increasing use of energy, the location of the energy use, and the pollution per unit of energy for the specific energy conversion process.

REFERENCES

1. *Scientific American*, Sept. 1971 (Energy and Power issue).
2. Angrist, S. W., and L. G. Hepler, *Order and Chaos—Laws of Energy and Entropy*. New York: Basic Books, Inc., Publishers, 1967.

EXERCISES

1. In recent years, to stimulate the U.S. economy, economic measures have been enacted that will stimulate the sales of new U.S. automobiles. Your problem is to determine the amount of energy required to produce an average-sized U.S. automobile. You may express the energy required in consistent natural resource units such as tons of coal and barrels of oil. Be sure to state all your references and assumptions as you develop the problem solution.
2. The following list of projected technological breakthroughs during the next 25 years was published in the July 1970 of *Power*. Using available literature (be sure to cite your references), write a brief description (not more than 1,000 words) of any two subjects on the list.
   1970—Economic atomic power generation
   1971—Activated carbon absorption for water pollution control
   1972—Inexpensive soundproofing materials
   1973—Inexpensive sulfur removal from flue gases
   1974—Inexpensive reuse of water from municipal and industrial discharges

1975—External combustion engines for automobiles
   Nonpolluting automotive emissions
   Cheap catalyst preventing release of noxious gases into atmosphere
   Molten-salt absorption for air pollution control
   Inexpensive water desalting
   General use of cryogenics methods
   Inexpensive method for extracting oil for oil shale
1976—Inexpensive method for extracting sulfur from coal
   Manned permanent sea-bottom station
1977—Scientific method for removing solid waste from cities
   Fuel cells for utilities
1978—Selective chromatic recovery for industrial air pollution control
   New alloys—stronger, lighter, and more corrosion-resistant
   Petroleum-base fuel cell
1979—Widespread use of underground cables, except for HVC cross-country
1980—Permanent lunar base
   Electric car and gas-turbine-powered car
   Compact total-energy plant for the home
   Steam turbine-generators of 2,000-MW capacity
1983—Controlled thermonuclear power
1985—Commercial application of fast-breeder reactors
1988—Electrogasdynamics applications
1989—Cryogenic electric cables in and out of cities
1990—Magnetohydrodynamics applications
1995—Earth-weather control
   Commercial application of fusion reactor

3. During the past year there has been considerable interest in the problem of recycling waste materials, such as glass. Your problem is to determine the relative merits, on an energy basis, of glass recycling. Therefore, for a representative sample, say a quart bottle, compare the energy required to manufacture the bottle with the energy required to recycle the bottle or the energy required to reuse the bottle. Be sure to state all your assumptions and sources of data in your problem solution.

4. You, as a university student, have been asked to participate in the environmental cleanup program. As your first contribution, you are asked to calculate the amount (by weight) of sulfur dioxide (SO_2) given off as a result of your use of energy on the campus. For purposes of calculation you assume that your share of the energy is supplied by the burning of coal (from the university power plant and outside electrical companies) divided by the total number of students. Using sound scientific principles, calculate the number of pounds of SO_2 that you are responsible for during the first semester here.

5. An interesting alternative to the problem of thermal pollution is the use of systems that use low-grade energy. For example, one might propose the use of such energy to grow aquacultural or agricultural crops. Your problem is to define (using a schematic drawing) a system that you might propose to use the reject heat from a typical electrical power

plant, with a 1-MW power output. Be sure to include a basic mass and energy balance in your analysis.

6. In the text the Four Corners Power Plant was mentioned as an example of an efficient energy conversion system. However, its construction and operation have not been free from controversy. Discuss some of the environmental burdens that it imposes.

7. Rest rooms are found to contain both paper towels and heaters for drying hands. Find an electric hand dryer, and determine the amount of energy required to dry hands during one cycle of operation. Compare this figure with your estimated energy requirement to produce the paper towels. Discuss which hand dryer should be used.

8. In examining Figure 4.1 it is seen that a correlation seems to exist between G.N.P. and energy consumption per capita for nations of the world. However, if only the highly developed countries with a G.N.P. around $1,500/capita are considered, it is seen that a wide range of energy consumption levels are found. Discuss some of the reasons for the variation observed.

9. Verify the statements given in the text pertaining to energy growth prediction; i.e., at a 3% annual increase in energy consumption $75Q$ will be consumed by 2050 and at 5% $275Q$ will be consumed.

10. The predicted coal reserves for the United States are presently (1968) 2×10^{11} tons. If energy consumption increases at a rate of 3%/yr above the current level of 9×10^{16} Btu/yr, when will this coal be depleted? Coal has a heating value of 14,000 Btu/lb.

5

air pollution concepts

J. R. Kittrell

W. L. Short

5.1 INTRODUCTION

Throughout the history of civilization, advances in the business of living have been accompanied by increases in the complexity of problems of the environment. Centuries ago, the growth of cities led to the pollution of water by bacterial organisms from the intestinal tract of man. Fifty years ago the rivers and lakes of the United States began deteriorating from pollution by human and industrial wastes. Within the last 20 years, pollution of the atmosphere by man's wastes has become increasingly important. Some of these historical problems have been largely erradicated through combined social, economic, political, and technological solutions. Even though these historic problems periodically reappear, greater urgency must be placed upon the need for similar solutions to the more recent pollution problems.

Perhaps one of the most important challenges to our society is the growth of cities and industrial production presently leading to an increase in the discharge of both harmless and polluting substances into our atmosphere and our water. These discharges come from liquid effluent from a variety of industries, from the chimney stacks of industrial plants, from domestic and community heating, and from the exhaust gases of motor vehicles. Furthermore, the pollutants produced by man's activities combine with other substances in our atmosphere or in our waterways to produce compounds as undesirable as the original substances were, or more so. For example, the original pollutants may combine directly with oxygen in the air (thermal oxidation) or with oxygen in the presence of sunlight (photooxidation) to form a noxious substance. Sulfur dioxide from chimney stacks of industrial plants is oxidized to sulfuric acid aerosol, or sulfate. Nitric oxide from automobile exhaust is converted by thermal oxidation and particularly by photooxidation to nitrogen dioxide, and eventually to particulate nitrates. Hydrocarbons are oxidized to aldehydes, ketones, carbon dioxide, and organic aerosols. The joint oxidation of hydrocarbons and nitric oxides results in products which react to produce peroxyacetyl nitrates (PAN) and peroxybenzoyl nitrates, which are extreme eye irritants. Thus, gaseous oxidation reactions constitute a most important type of pollutant reaction in community atmospheres.

At the outset, we must define air pollution to provide a common basis for discussion. For example, do automobile exhausts constitute air pollution if the wind is blowing hard enough to disperse all the exhaust gases away from the city into the air over the Pacific Ocean? Do they represent air pollution if there is no wind? Do the substantial emissions of sulfur dioxide and solid particles from volcanoes constitute air pollution, regardless of whether man inhabits the region in the vicinity of the volcano? We shall consider air to be polluted if it cannot or will not be suitable for the specific purposes deemed desirable by man for the air. Hence, if pollutants in the air from either natural or man-made sources detrimentally affect a specific air resource (either in the physiological or esthetic reactions of man or the effects on plants and animals deemed undesirable by man), then the air is polluted. However, these same pollutants being emitted in the same quantities in some other region of the earth, such as the South Pacific, may not have a detrimental affect on man. The latter air resource, then, would not be polluted. (In some circles, this is viewed "as a license to pollute.") Air pollution results from a condition in which the rate of introduction of foreign substances into the atmosphere exceeds the self-purification capability of the atmosphere, i.e., its ability to cleanse itself of the various pollutants.

We are all aware that our attempts to combat air pollution are limited by economic and political constraints. For example, suppose that we could devise an attachment to automobiles that would completely eliminate exhaust emissions of pollutants and that each such device would cost $100. This is, at first glance, a relatively cheap price to pay for eliminating automobile pollution problems. However, the total cost of applying this air pollution abatement device to each of the 100 million automobiles in the United States would be $10 billion dollars, to be paid by the U.S. car buyers. Also this enormous sum of money excludes maintenance and training costs. For example, it represents 1% of the gross national product of the entire United States. Out of every $100 in goods and services produced by the entire population of the United States, $1 must go into exhaust converters. If we were to lay one dollar bills down once per second, it would require 300 years to accumulate $10 billion dollars. Any expenditure of this magnitude simply must be considered a real constraint. Yet another way to eliminate significant quantities of automobile-initiated air pollution in large urban areas such as the Los Angeles basin is to restrict the freedom of the people to drive for the benefit of the entire populace. Such actions restricting individual rights for the benefit of everyone have considerable precedent in U.S. history. Most northern cities, for example, restrict parking on the streets in winter nights to allow snow removal. This is an attractive alternative to the voters, for the inconvenience of restricted parking is acceptable in view of the importance of snow removal. Could we allow Los Angeles citizens to drive only every other day to cut the air pollution in half? Not likely. This represents a constraint that we wish to place upon the solution to pollution, and the result of this constraint will likely be a longer time period required to solve the problem. To be successful in attacking these problems, we must recognize the existence of political and financial constraints as well as the technical problems involved.

We can certainly consider air pollution to be one of the pressing national problems of today. However, that should not imply that the nature of air pollution is identical in all cities. For example, in Table 5.1 are reported some characteristics of smog, as the term is used in Los Angeles and London. We can readily see striking differences in the conditions under which smog forms in these two cities. Indeed, in spite of the common use of the term *smog*, the chemical nature of the air pollutants is quite different in these two cities, as are the specific atmospheric conditions. New York has many more characteristics of London smog than of Los Angeles smog. Hence, the specific regional air pollution problems in the United States are quite different, owing to varying climates and sources of pollutants. Our national problem, then, is actually a summation of regional problems. It is a national problem because (1) all areas have some degree of air pollution, (2) there are many common sources of air pollution such as automobiles and power plants that entail interstate transportation, (3) the pollutants themselves can be transported from state to state by winds, (4) the resources of the state and local governments are limited, and (5) continued growth of pollution would adversely affect the United States as a whole.

Table 5.1
Characteristics of London and Los Angeles Smog

	LONDON	LOS ANGELES
Fuel burned	Coal and oil	Oil and gas
Smog peaks at	Early morning	Midday
Temperatures	30–40°F	75–90°F
Humidity (air moisture)	High (fog)	Low (clear sky)

The air pollution problems of New York City and Los Angeles have been widely publicized for years. However, air pollution is not at all restricted to these two cities, or even to the United States. Air pollution has become a serious problem in Japan and Europe during the last decade. Consider, for example, Figure 5.1, which portrays the growth over time of German oil consumption, electric power generation, and motor vehicle registration. Also shown in this graph is the increase of population with time for Germany. It is apparent that there is a heavy increase in industrial production that has completely outstripped the linear growth in population. Consequently, it is not surprising that we find accumulation of trace substances in the air over German cities and frequently the formation of haze. If a haze builds up over densely populated areas, it can be dispersed only by sufficient winds. Without dispersion, haze and dust caused by pollutants have been reported at heights over over $\frac{1}{2}$ mile above the ground at Ludwigshafen, Germany.[1]

The story portrayed by Figure 5.1 is certainly one of the causes of increased pollution. There must be a contribution caused by population growth,

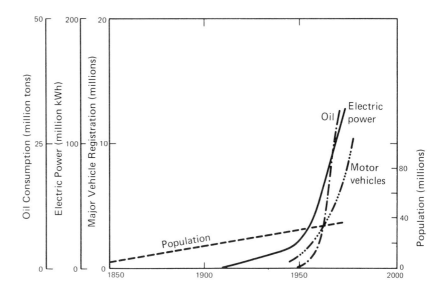

Figure 5.1
Growth of oil consumption, electric power consumption, and motor vehicle registration relative to population growth in Germany.[1]

which becomes more severe as we pack more and more people into each square mile of living space. However, another significant factor is the increased affluence of society, resulting in the growth of electrical power, oil, and automobile consumption far exceeding the population growth. Since electrical power generation and automobiles are already significant sources of air pollution and since their growth likely will accelerate with increasing affluence, they must represent primary targets if we are to attack pollution. Note, in passing, that we use statistics such as shown in Figure 5.1 as one of the primary measures of progress as a nation. Whereas economic indexes of progress are numerous—shares of stocks bought and sold, automobile and steel production, housing or heavy industry construction—and easily understood, we have no comparable environmental indexes that tell us if we are improving the livability of the nation or even holding onto the status quo. It would appear we need an index for gross national pollution as well as gross national product. While the city of San Francisco enjoys an increase in shipping tonnage in their port, raw sewage is still being discharged into the San Francisco Bay during severe rain storms.

Progress does not, however, demand that we accept pollution as an inevitable consequence. Pollution has spread because of factors that accompany progress: (1) population growth as a result of medical and sociological advances, which reduce the death rates both of the aged and infants; (2) expansion of industry because of technological advances, making more and more consumer goods available at prices the public can afford; (3) increased urbanization because of real or imagined advantages of living in metropolitan areas; and (4) increased prosperity of society, leading to more motor vehicles, more leisure time to use these vehicles, more "no deposit, no return bottles," etc. Pollution is promoted by many technological, social, *and* medical advances, not just technological advances. However, now that pollution is upon us, technological advances or applications of existing technology are required to solve this problem.

There are, of course, other factors that must also have contributed to the problems that accompany those items listed above, for these trends have been present for hundreds of years. Although we have never been able to accurately predict the consequences of our intervention in nature in the name of civilization, our capacity to intervene massively has existed only during the last few decades through technological advances. The earliest Spanish explorers of the Los Angeles basin reported accumulations of smoke from Indian campfires.[2] We were unable to generate a serious air pollution problem in Los Angeles, however, until we introduced millions of automobiles into the basin. Also, we have been inclined to view the benefits of technological, medical, and sociological advances as unmixed blessings, believing that damages to the environment were incidental and could be ignored. The side effects of DDT, now recognized to have an extremely long lifetime in nature, illustrate this. The accumulation of DDT in fatty tissues of animals and the side effects on the loss of strength of the shell of certain bird eggs were not fully recognized at the time of introduction of DDT. The effect of agricultural fertilizer on the eutrophication of our water resources was not anticipated. We must recognize that, although these advances were not unmixed blessings, they *were* nevertheless blessings. The impact of DDT on malaria and the necessity of fertilization to provide food to the growing U.S. population must be considered along with the environmental damage. It is equally wrong to ignore either end of this spectrum. Finally, we have often tended to be more concerned with whether something *can* be done and less concerned with whether it *should* be done. Recognizing the difficulty in evaluating such a question, it is nevertheless encouraging that such evaluations are now being attempted, as demonstrated in the supersonic transport (SST) debate in the U.S. Congress and the fate of some of the projects of the U.S. Corps of Engineers.

The difficulty in attacking air pollution problems is further compounded by the extremely low concentrations of substances causing pollution. A comparison of compositions of pollutants in pure and in polluted atmospheres is shown in Table 5.2. Even though some of these pollutants are in apparently low concentration (for example, sulfur dioxide in a polluted atmosphere ranges from .02 to 2 ppm parts of air), their presence is still quite obvious and quite objectionable to anyone breathing air containing these levels of sulfur dioxides.

These problems of the environment—water pollution, air pollution, transportation, and urban development—are difficult to solve. No amount of enthusiasm or sorrow, however sincere and profound it may be, can compel a problem to be solved. Ideas occur to us when they please, not when it pleases us. The necessary tools of engineering must be added to political, social, medical, and biological contributions to solve these problems. To

Table 5.2

Comparison of Pure and Polluted Atmospheres (ppm = parts per million, mg/m³ = milligrams per cubic meter)

	NORMAL	POLLUTED
Particulates	.01–.02 mg/m^3	.07–.7 mg/m^3
Sulfur dioxide	.001–.01 ppm	.02–2 ppm
Carbon dioxide	301–330 ppm	350–700 ppm
Carbon monoxide	1 ppm	5–200 ppm
Nitrogen oxides	.001–.01 ppm	.01–.1 ppm
Total hydrocarbons	1 ppm	1–20 ppm

Source: Reference 1.

effectively attack these problems, we must (1) isolate the specific pollutants of concern, (2) identify the important sources of these specific pollutants so that they can be reduced first (within our limited economic constraints), (3) define the standards specifying the harmful levels of these pollutants so we will know how far the level of pollutant must be reduced, (4) implement these standards by the most economic means available, and (5) police the standards to ensure that these harmful levels of pollutants are not being exceeded.

5.2 EFFECTS OF POLLUTION AND CLIMATE

As we shall see later, most of the pollutants listed in Table 5.2 have been emitted into the atmosphere in significant quantities long before man began adding large amounts of these pollutants to the atmosphere. However, natural scavenging or removal forces in nature have been sufficient in the past to prevent *accumulation* of these pollutants to levels detrimental to plants and animals. The additional pollutants generated by man's activities have been sufficient to raise the total level of *production* of pollutants (from both man-made and natural sources) above the level that can be *removed* by nature's purifying mechanism. Since this production level exceeds the removal capabilities, an *accumulation* of pollutants must occur. Different geographical localities in the United States have different levels of production of pollutants, which, in part, cause greater pollution problems in Los Angeles than in Oklahoma City. However, the atmospheric conditions in Los Angeles are such that the pollutants remain in the immediate vicinity at which they were originally emitted. Hence, the continued residence of these pollutants over a small area of the earth greatly overloads the natural scavenging action that takes place. Obviously, we cannot consider the effects of various pollutant levels without also considering the atmospheric conditions or the meteorology of that region.

The presence or absence of wind and its strength and direction is a key force in determining the immediate effects of adding pollutants to the atmosphere. If wind is present, it can disperse the smoke from leaves burning in the backyard or the products of combustion of 5 million automobiles in the Los Angeles Basin. In the absence of wind, accumulation of pollutants in these cases can be severe. Once pollutants are dispersed by winds over a wide geographic area, natural scavenging mechanisms are often adequate to remove these pollutants from the atmosphere. If so much pollutant is being produced that the scavenging mechanisms cannot naturally remove them, then accumulations of these pollutants in the atmosphere occur; the rate of accumulation in the rather dispersed environment is substantially less noticeable (but of no less ultimate importance) than that which occurs in localized areas such as the Los Angeles Basin.

Wind even has a large effect on the amount of pollution on a single block in the city. Figure 5.2 demonstrates the relationship between the concentration of carbon monoxide from automobile exhausts in streets to the wind speed. It is seen that at wind velocities below 6 ft/sec, ventilation was very poor and fresh air that entered the space between buildings from above did not even reach the street level. The carbon monoxide concentration increased considerably near the ground. At higher wind speeds the turbulence that formed between the buildings reached the street level, but fairly complete ventilation of the street and dispersion of the pollutants requires a wind velocity of 15 ft/sec or more. The accumulation of pollutants in streets also obviously depends on the height of the buildings and the width of the streets. Hemispheric pollution must be increasingly considered on a much wider scale than the simple concentration plot of Figure 5.2 for a given street. Stagnation periods of several days occur over large regions of the United States and lead to significant accumulations of atmospheric pollutants.

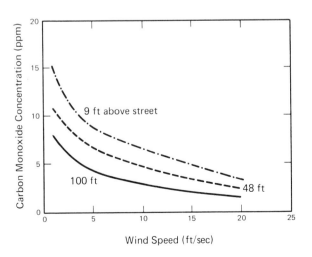

Figure 5.2
Effect of wind speed and distance above street on carbon monoxide concentration.

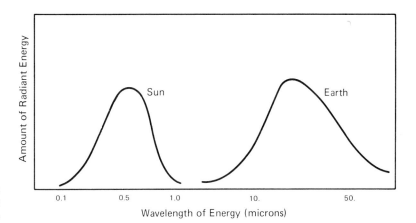

Figure 5.3
Energy distribution of sun and earth radiation.

Evidence also exists of recirculation of pollutants from a previous day over areas such as the Los Angeles Basin and the San Francisco Bay.

The source of energy for the earth is, of course, the sun. The energy from the sun is a primary influence on the earth's temperature and, ultimately, on its meteorology. One of the fundamental differences of the sun's heat from the heat emitted by a steam radiator is the high energy or intensity of the heat from the sun relative to that of the radiator. It is this property that allows us to get sunburned or tanned by long exposure to sunlight. The higher energy of the sun's radiation is due to the higher temperature of the sun and is measured by the wavelength of the sun's radiation, where smaller wavelengths are associated with higher energy radiation. In Figure 5.3 is plotted a graph of the energy distribution of the sun's radiation.

As this energy from the sun approaches the earth's atmosphere, it can begin heating up the molecules in the air. For the purposes of our discussion, we shall divide the atmosphere into two portions, the troposphere and the stratosphere. The lowest layer of the earth's atmosphere is the *troposphere* ("mixed layer"), continuing from the earth's surface up to a height of 6 to 12 miles and containing 95% of the earth's atmosphere. It is in this layer that the weather activity takes place, including winds, cloud formation, and precipitation. As is familiar to anyone having flown in an airplane, the temperature drops rather rapidly with increasing height above the ground in the troposphere, at an average rate of about 3.5°F/1,000 ft. Hence, an airplane flying at 30,000 ft would experience an outside temperature of about $-35°F$ if the ground temperature were 70°F. The layer of air above the troposphere is termed the *stratosphere*, wherein the temperature rises slowly with increasing height above the earth. In this region, there is very little air, very little moisture, and very little vertical movement of air. In Figure 5.4 is shown a graph that summarizes the temperature of the atmosphere at various heights above the ground.

The sun's energy is absorbed by the atmosphere and the surface of the earth, causing them to heat. However, the atmosphere and earth also radiate

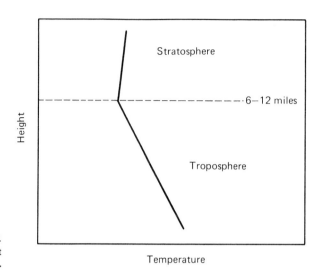

Figure 5.4
Temperature variations with height in earth's atmosphere.

energy back into space. To prevent an accumulation of energy and hence a continual heating of the earth, the amount of energy received from the sun must on the average equal the amount of energy emitted by the earth. At night, the solar energy is greatly decreased, so the earth cools until its lost energy decreases to match the incident energy.

Of the total sun's energy approaching the earth, about one-third (34%) is reflected back into space, by the clouds and dust in the earth's atmosphere. This reflectivity is called the earth's albedo. Hence, this portion does not contribute to heating the earth. About 3% of the sun's energy directed toward the earth is absorbed in the stratosphere by ozone. Ozone is a form of oxygen that is capable of absorbing energy in the ultraviolet wavelength, less than 1 micron in Figure 5.3. Most of the molecules in the atmosphere, such as water, are incapable of absorbing such highly energetic radiation. This absorbance of the sun's energy in the stratosphere causes the increasing temperature with height shown in Figure 5.4. About 15% of the sun's incident energy is, however, absorbed in the troposphere by water vapor, at wavelengths greater than about 2 microns in Figure 5.3.

Thus far we have accounted for about 52% of the sun's incident energy (34 + 3 + 15 = 52). The remaining 48% of the energy is absorbed very strongly by land, less strongly by water (more reflection), and even less strongly by snow. All this absorbed energy must eventually be reradiated into space to prevent the earth's surface temperature from rising. However, since the earth temperature is only about 70°F compared to the sun's temperature of about 10,000°F, the earth's radiation is much less energetic (longer wavelengths). As shown in Figure 5.3, the earth emits radiation in the 4- to 80-μ range, with a peak intensity near 10 μ. Carbon dioxide and water vapor, which are plentiful in the troposphere, are capable of absorbing much of the earth's radiation (in the range of 4 to 80 μ), except for a small region between about 8 and 11 μ. Thus, much of the emitted radiation from the earth is absorbed by the surrounding atmosphere and reradiated back to the earth. In fact, if it were not for the "greenhouse" effect of the carbon dioxide and water vapor in the atmosphere, the average temperature of the earth's surface would be closer to $-50°F$ rather than the comfortable $50–100°F$ temperature range to which we are accustomed. There must be, of course, a net loss from the upper atmosphere corresponding to the 48% of the incident sun's energy absorbed by the earth.

Another noticeable characteristic of atmosphere as we proceed upward from the surface of the earth is the decrease in pressure encountered. The pressure at the earth's surface is due to the weight of air (caused by gravitational attraction of the air molecules) acting downward on the earth's surface. The amount of air remaining above an observer at higher elevations is less, so the weight of air or the pressure of the air is less. It is this factor, of course, that requires that airplane cabins be pressurized for flights above about 10,000 ft.

To relate these concepts of temperature and pressure changes with altitude to air pollution, we need two additional concepts. First, hot air must rise; hot air is less dense than cold air, so a packet of hot air will be pushed upward by buoyant forces, much like a beach ball is pushed upward when submerged under water. Second, as a packet of air is forced upward into lower pressure regions (at higher altitudes), it expands and cools. This cooling effect is *not* due to lower temperatures at higher altitudes, but rather is due to the work done by the gas upon expansion. Air escaping from a balloon is slightly cooler than the air in the balloon because of the energy lost from the air accomplishing work on expansion (conversely, air that is compressed becomes heated). If no heat from a parcel of rising air is being interchanged with the surrounding air, then this process is termed *adiabatic*. The decrease in temperature of this dry parcel of rising air with increasing altitude is called the *dry adiabatic lapse rate*. It is approximately $5.5°F/1,000$ ft. The actual lapse rate is a measure of the stability of an air packet, and hence its ability to disperse pollutants.

Suppose that the actual temperature of the atmosphere above a source of pollutant is as shown by the solid line in Figure 5.5(a) and that the adiabatic loss of temperature of a rising parcel of air is shown by the dotted line. As this parcel of pollutant-containing air rises, it is always hotter than the surrounding air (read the temperature at any given altitude). Since it is always hotter than the surrounding air, buoyant forces tend to accelerate this parcel of air upward like a balloon, and the vertical motion of these parcels of air causes substantial turbulence. This disperses the pollutant over a large volume of the atmosphere. On the other hand, suppose the actual temperature of the troposphere does not decrease very rapidly with height, or even increases with height as shown by the solid line in Figure 5.5(b). Here a rising air parcel becomes relatively cooler than the surrounding environment and, hence, more dense. The parcel of air thus tends to return to its starting point, resulting in an accumulation of pollutants on the ground. This effect, called a *temperature inversion*, accumulates all emitted pollutants in a layer of air from 300 to 3,000 ft in depth. There can thus be a strong accumulation of

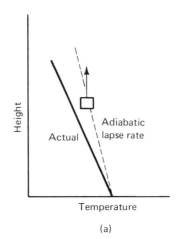 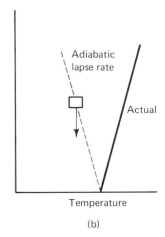

Figure 5.5 Effect of thermal inversion on air pollution.

smog around populated areas, because the parcels of smog-containing air cannot rise from the cool environment to the warm environment as required. In Figure 5.6c, the pollution-clearing effect of a strong ocean breeze is shown to relieve the smog inversion problem.

Thermal inversions can originate from a variety of sources. For example, an inversion can arise from a mass of warm air moving slowly across a relatively cooler land surface. A high-pressure region often is present off the coast of California, near Los Angeles. This air is heated as a result of being compressed in this high-pressure region. As this air flows or subsides away from the high-pressure region it can be hotter than the surrounding terrain, leading to a subsistence inversion. *Subsistence inversions* represent most of the temperature inversions at Los Angeles. *Radiation inversions* can be caused by fog reflecting the early morning sunlight away from the earth, slowing its rate of heating relative to the upper atmosphere. Such a cooler earth and warmer high-elevation air leads to the problems typified by Figure 5.6(b). The earth's topography (or its surface structure) can also contribute to temporary radiation inversion. For example, in Figure 5.7 is shown a housing and industrial complex at the base of a valley. Early in the morning, as the sun comes up over the mountain, the sun's rays shine and warm the air far above the valley floor. Since the valley floor still contains the cooler night air, a thermal inversion starts for the same reasons as described in Figure 5.6. This phenomenon is well known in Yosemite Valley, where the smoke of numerous campfires in the valley is held at the valley floor in the early morning hours, owing to a thermal inversion. However, in all these cases, as the sun comes higher into the sky, the valley floor is gradually heated and the thermal inversion disappears, resulting in a disappearance of the smoke as well. Of course, winds in excess of 5 to 7 mph can also disperse accumulations of pollutants in this case.

It is quite apparent that climatic conditions can have a gross effect on the tendency of emitted pollutants to accumulate in the atmosphere. How-

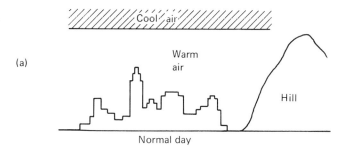

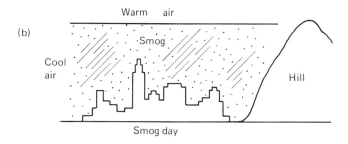

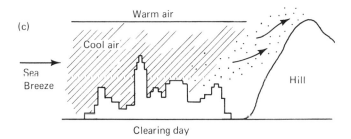

Figure 5.6
Thermal inversion and smog formation.

ever, it is interesting to note that the presence of atmospheric pollutants can indeed *alter* the climate that would be present in the atmosphere if the pollutants were not present.

Decreased visibility is one of the first items noted in the presence of large amounts of pollutants. Visibility is, of course, decreased when smog is being accumulated. In these cases, visibility is primarily altered by an increase of particulate materials suspended in the atmosphere. However, there are a variety of other factors that can also decrease the visibility but that are not pollutants. For example, some particulate materials are hygroscopic in nature; that is, they increase in size as they soak up moisture from the atmosphere. As these particles increase in size as a result of increasing humidity of the atmosphere, they become more effective in reducing visibility even though

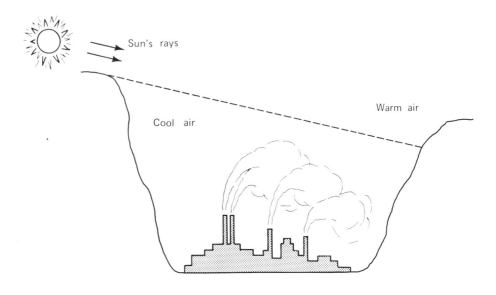

Figure 5.7
Effect of topography on smog formation.

there is no increase in total pollutants. Also, the transition of sea salt from a crystal to a droplet occurs at about 75% relative humidity. At Los Angeles International Airport a sharp decrease in visibility at 67% relative humidity was found. This airport is located within about 2 miles of the Pacific Ocean, a source of sea salt particles. Hence, it is not surprising to find this decrease in visibility with the increasing humidity. Although both humidity and particulates are important factors in determining the loss of visibility in polluted atmospheres, there is also considerable interest in visibility losses caused by sky color effects. This is caused particularly by the brown discoloration that might occur as a result of excessive nitrogen dioxide concentration in smog.

In addition to the effect of particles in the atmosphere on visibility, these particles can provide sites for water to condense in the atmosphere and thereby can lead to fog. This explanation has been given for the observation of central London being far more foggy than the outlying urban areas, even though air temperatures and relative humidity should lead to a lower fog level in central London. Under appropriate atmospheric conditions, particulates can result in increased precipitation. It has been found, for example, that when automobile exhaust gases containing lead aerosols are exposed to minute quantities of iodine vapor in the atmosphere, freezing nuclei in large numbers are produced. The production of large quantities of freezing nuclei, of course, is a basis of weather modification and the production of rain.

Atmospheric pollutants also considerably reduce the amount of solar radiation reaching the earth in areas of high industrial activity. Certain portions of solar radiation are more seriously affected than others. Since ultraviolet radiation is a major factor in the body's generation of natural vitamin D, one of the effects of air pollution of concern has been the excessive loss of ultraviolet radiation from sunlight penetrating polluted air. In some cases in Pasadena, California, smog has been shown to reduce the intensities of solar radiation in the ultraviolet region by 90% or more. In Table 5.3 the average duration of bright sunshine is shown to decrease considerably toward central London where there is more industry as well as more heating of homes. The loss of sunshine in central London (compared to surrounding country) can be seen to be greatest during the winter months when atmospheric pollution was heavy due to the higher level of domestic heating. Furthermore, there is generally a difference in temperature between central city and the rural surroundings. This temperature difference between the central part of the city and the suburbs can be from 1 to 7°C. It is thought that this is due to (1) the thermal capacity of buildings, which radiate heat during the night; (2) the decrease of outgoing radiation from the city during the night as

Table 5.3
Average Duration of Bright Sunshine in London (hours per day)

	JANUARY	JULY
Surrounding country	1.7	6.6
Outer suburbs	1.4	6.5
Inner suburbs	1.3	6.3
Central London	.8	6.2

Source: Reference 1.

a result of increased amounts of pollution in the atmosphere; and (3) the heat contribution from the burning of various fuels in the city. Owing in part to higher temperatures in the center of Cologne, the proportion of apple trees in blossom increased from the suburbs toward the center of the City. When 50% of the trees were in blossom in an area close to the densely built-up parts of the city, only 10% of the apple trees were in blossom in the surrounding villages. In more distant parts, the trees were not blossoming at all. Hence, there is a relation between the effects of pollution, urban temperature distribution, and development of vegetation.

5.3 EFFECTS OF POLLUTION ON PLANTS AND ANIMALS[3,4]

It is quite difficult to determine the precise effects of air pollution on animals or plants; humans are no exception to this generalization. This is due, in part, to the fact that a sample of air taken at one spot will generally be quite different from a sample of air taken several hundred feet from that place. The sample of air will also be different at a different time of the day or year. Furthermore, the evaluation of the effect of pollution is complicated because this air is inhaled by different people. Various people have different responses to pollution, because of differing time of contact with the air as well as differing responses of their own bodies to pollution. Through *synergism*, the body can have a far greater reaction to a combination of polluting chemicals than the total reaction expected from the effects of individual chemicals. For example, because of the synergistic effects of sulfur dioxide and benzpyrene aerosol, the combined effect of cigarette smoking and air pollution on the lungs may be more serious than either individual problem. However, there may also be a subtraction process where one pollutant chemical cancels out the effects of another. For example, small amounts of sulfur dioxide in the air appear to lessen the effects of ozone on plants. Furthermore, the body may build up either a resistance or a sensitivity to various polluting chemicals.

However, in spite of the difficulty in precisely measuring the effect of polluting chemicals on plants or animals, there are several obvious demonstrations of these effects through a variety of accidents that have happened throughout recorded history. In Poza Rica, Mexico, in November 1950, an accident at a factory producing sulfur from natural gas spilled hydrogen sulfide into the atmosphere during a foggy atmospheric inversion. This accident killed 22 people and hospitalized 320 others. In addition, unknown numbers of chickens, geese, ducks, cattle, pigs, and dogs were made ill. All the canaries and many of the birds and animals died.

In the Meuse River Valley of Belgium,[4] where a heavily industrialized area extends for some 15 miles, the terrain has a valley topography typically lending itself to inversions. When all of Belgium was blanketed by a thick cold fog in the first week of September 1930, the pollutants from the local industry were trapped by the inversion and gathered day after day. Within a few days, thousands fell ill and 60 persons died from the poisoned air, well above normal mortality rates. The dead were mostly the elderly and those already ill from diseases of the heart and lungs. Those who were merely sickened complained of coughing and shortness of breath. Autopsies showed only congestion and irritation of the tracheal mucosa and the large brochi. There was also black particulate matter in the lungs. The specific chemical pollutants causing this disaster were never isolated. It is suspected that sulfur dioxide, dissolved in water droplets and in the presence of several other pollutants, was oxidized to sulfuric acid mist with a particle size sufficiently small to penetrate deeply into the lungs.

Donora, Pennsylvania,[4] a town in the Monogahela River Valley, was markedly affected by a foggy inversion that covered a wide area of the northeastern United States in October 1948. The city contained a large steel mill, a sulfuric acid plant, a large zinc production, and other industries. Among the population of only 14,000 close to 6,000 fell ill. Instead of the normal 2 deaths for the period, the fog and pollution produced 20. Again the elderly were hit harder. Deaths were often accompanied by existing heart or lung disease. Most people coughed, although they also suffered from sore throats and chest constriction, headaches, breathlessness, burning eyes, running nose, vomiting, and nausea. One out of five birds and one out of six dogs were sickened. Specific chemical pollutants causing this disaster were not isolated, but sulfur dioxide and particulates were known to be present. Calls for medical assistance in Donora stopped abruptly on Saturday night, despite the dense fog that remained. It has been suggested that an increase in the size of the polluted fog droplets caused them to be deposited in the upper airway instead of penetrating deep into the respiratory tract.

The fogs of London are as notorious as is its air pollution. For centuries Londoners heated their homes with soft coal in open fireplaces. Resultant smoke in the presence of fog (the old original smog) combined with the modern contaminants of an industrial city to leave its mark on its inhabitants. During a 5-day inversion in December 1952, the deaths of 4,000 people were attributed to the noxious air. London has had other such incidents. Six cases have been recorded earlier, as far back as 1873, with deaths totaling over 2,500 people. It is possible, furthermore, that only the precautions brought about by these earlier calamities kept mortality numbers down in a similar inversion in 1962, when 750 people died.

According to air pollution probers, New York City has had severe episodes in 1953, 1962, 1963, and 1966. By comparing mortality figures over a number of years, it was concluded that in 1963 during a period of heavy air pollution, Asian flu, and extreme cold, 405 people died because of air pollution alone. In 1966 they concluded that a 3-day period of inversion with air pollution contributed to the deaths of 168 people. The fatalities in this 1966 episode were probably much lower than they might have been because the inversion occurred during the Thanksgiving weekend.

It has been concluded that (1) the effects of these incidents were anatomically localized and limited to the respiratory track, (2) the people most vulnerable were those who were elderly and those with preexisting disease of the cardial respiratory system, (3) meterological conditions were an important factor, and (4) not one but two or more interreacting pollutants were responsible: sulfur dioxide, sulfur trioxide, particulate solids, and droplets of fog were carriers.

Numerous instances in which specific contaminants have crippled livestock have been recorded. In the copper smelter at Anaconda, Montana in 1902, a loss of arsenic from the chimney stacks deposited on grass and poisoned large numbers of cattle, horses, and sheep. At a distance of 15 miles from the smelter, 3,500 sheep were grazing and 625 of them died. In Sweden in 1954, a steel plant spread molybdenum poisoning to many of the cattle grazing a half a mile away. In Germany in 1955 lead and zinc losses from foundaries killed many cattle and horses. Fluorine, which is produced from the fertilizer industry or from the aluminum industry, can also accumulate in cows and make them lose weight, give less milk, and eventually become so crippled that they must be killed.

For healthy growth, plants depend on sunlight, air, water, and soil, all of which can be affected by air pollution. The visible injury symptoms on plants are (1) leaf tissue collapse, (2) color changes in the leaf, and (3) alterations in the growth of the plant. In many cases, the injury pattern in highly characteristic of the toxic agent. However, disease, insects, nutrition, and other factors can produce leaf patterns very similar to those produced by air pollutants.

Sulfur dioxide fumes can destroy foliage for miles around the source of these fumes. Sulfur dioxide enters a growing plant through the stomata, the openings on the underside of the leaf, just as does carbon dioxide in the normal plant life cycle. The injury it causes may show up as markings along the edges between the veins of the leaf, with the damaged area appearing dried and white or ivory in color. Long exposure to sulfur dioxide may either stunt the development of the plant, resulting in chronically injured areas that never recover, or cause plant cells to die. In this case the tissues between the veins would collapse and the leaf would slowly take on the scars of sulfur dioxide. The effect of sulfur dioxide varies tremendously among species and varieties within the species. Plants with thin leaves, such as alfalfa, barley, cotton, and grapes, usually suffer most. Plants with fleshy leaves or needles, such as citrus and pine, can be resistant.

Fluorides enter the leaf through the stomata also. From there they move to the edges and tips of the leaf. Continued exposure to fluorine spreads the killing flouride compounds inward from the edge and tips of the leaves. The leaves exposed to fluoride generally have burnt, dried-out edges with a narrow reddish-brown line of dead tissue distinctly marking off the healthy part of the plant.

Photochemical smog, as encountered in Los Angeles, for example, is a relatively recent hazard to vegetation. Two different types occur from the smog. One is caused by ozone, which is a dominant constituent of photochemical smog. Ozone, which is a substance that oxidizes a selected compound not oxidizable by atmospheric oxygen, is the major (up to 90%) component of the photochemical oxidant. Although ozone is also disagreeable to man, it affects the plants by attacking the upper surface of the leaf, making the upper surface appear sploched or stippled. The tissues of some of the spots collapse if the concentrations of ozone are high enough. The other serious effect of smog is probably caused by PAN (an organic nitrogen compound, peroxyacetyl nitrate), which damages the lower side of the leaf. This makes the underside of the leaf of susceptible plants (citrus trees in particular) silver or bronze. Doses that are too low to cause irreversible damage will make the underside of the leaf seem temporarily water-soaked. Ozone appears to be most responsible for smog damage in the eastern part of the United States. In southern California, PAN appears to play the larger role.

5.4 SPECIFIC ATMOSPHERIC POLLUTION

As one of the obvious steps in attacking pollution problems, specific atmospheric pollutants and their primary sources must be identified. Table 5.4 represents a summary of the major contaminants, their man-made sources, their natural sources, and estimated annual pollution levels. It should be recalled that these pollutants are often being emitted in extremely low concentrations, shown in Table 5.1. However, the tonnage figures of pollutants are readily generated when you consider that in 1967 there were approximately 90 million cars consuming 60 billion gallons of gasoline annually. This means that *each automobile* in the country will discharge over 1,600 lb of carbon monoxide, 230 lb of hydrocarbons, and 77 lb of nitrogen oxides each year; multiply these numbers by 90 million to arrive at the total amount of these pollutants contributed by automobiles.

Table 5.4
Sources and Annual Emission Levels of Atmospheric Pollutants

Contaminant	Major Pollution Source (Man-made)	Natural Sources	Estimated Sources (Man-made) (tons)	Emission Naturally Occurring (tons)
Sulfur dioxide	Combustion of coal and oil	Volcanoes	146,000	None
Hydrogen sulfide	Chemical proc. Sewage treatment	Volcanoes Biological action in swamps	3 million	100 million
Carbon monoxide	Auto exhaust and other combustion	Forest fires Oceans	274 million	75 million
Nitrogen oxides	Combustion	Bacterial action in soil	53 million	1 billion
Ammonia	Waste treatment	Biological decay	4 million	1 billion
Hydrocarbons	Combustion exhaust Chemical processes	Biological processes	88 million	480 million
Carbon dioxide	Combustion	Biological decay Release from oceans	10 billion	1 trillion

Source: Reference 5.

Once the pollutants are discharged to the atmosphere they may enter into many chemical reactions. Often these chemical reactions will take place very slowly. However, some will be accelerated by sunlight and are known as photochemical reactions. Hence, some pollutants can be generated directly by man, whereas other pollutants can be formed from those by photochemical reactions. Those pollutants formed directly by man are termed primary pollutants and include carbon monoxide, carbon dioxide, nitric oxide, sulfur dioxide, and hydrocarbons. Those produced by reactions in the atmosphere are called secondary pollutants and include peroxyacetyl nitrate and nitrogen dioxide. The important pollutants and their chemical formula or notation to be used are shown below:

Notation	Description
NO	Nitric oxide, a primary pollutant produced by reaction of nitrogen and oxygen in the air during high-temperature combustion operations.
NO_2	Nitrogen dioxide, a secondary pollutant produced by oxidation of nitric oxide in the atmosphere. This compound imparts a brownish haze to the atmosphere.
O_3	Ozone, a secondary pollutant produced in photochemical smog during reactions involving nitrogen oxides and hydrocarbons. It also occurs naturally in the upper atmosphere. It is a highly reactive form of oxygen, has a very pungent odor, and is quite harmful to plants.
HC	Hydrocarbons, a primary pollutant found in automobile exhausts, and many other sources. They result from incomplete combustion of the fuel.
CO_2	Carbon dioxide, a primary pollutant produced by reaction between carbon in the fuel and the oxygen in the combustion gas. It is chemically inert and nontoxic and might be classed as a pollutant only because of the greenhouse effect (described later).
CO	Carbon monoxide, a primary pollutant produced by reaction between carbon in the fuel and oxygen in the combustion gas. It is formed when relatively small quantities of oxygen are present relative to the amount of carbon in the fuel. Carbon monoxide is an odorless and colorless gas but is highly toxic.
SO_2	Sulfur dioxide, a primary pollutant formed by reaction of the sulfur in the fuel and oxygen, is highly toxic and has a characteristic sharp odor.
H_2S	Hydrogen sulfide, a primary pollutant that is highly toxic and has a smell like rotten eggs, is released from some chemical processes and from sewage plants.
NH_3	Ammonia, a primary pollutant released from various chemical processes.

The major sources of these pollutants are shown in Table 5.4. Note that in some cases man is an insignificant supplier. However, this table does not give due weight to small quantities of emissions that, when confined to a small area, can reach dangerously high concentrations.

Since many of the man-made sources from Table 5.4 are combustion processes, let us consider a typical combustion application in more detail— the burning of fuel to heat a home.

We have several different choices of fuel for this purpose, the compositions of which are listed in Table 5.5. It is the role of several of the chemical elements, such as sulfur, that we must consider here.

Table 5.5
Chemical Makeup of Fuels Analysis
(% by weight)

Kind	Carbon	Hydrogen	Oxygen	Nitrogen	Sulfur	Ash	Methane	Ethane
Pennsylvania anthracite coal	84.2	2.3	2.1	.8	.6	10.0	—	—
Fuel oil	80	16	1	1	2	.1	—	—
Natural gas	—	—	—	1.30	Negligible	—	80.5	18.2

Source: Reference 6.

To burn a fuel, we recognize that sufficient air must be supplied; also, because of the large amount of heat released, the temperature at which the combustion takes place will be very high. The combustion of coal, for example, involves the reaction of carbon in the coal with oxygen from the air to yield heat and carbon dioxide:

$$C + O_2 \longrightarrow CO_2 + \text{heat}$$

In this equation C is used as a short-hand notation to symbolize carbon, O_2 symbolizes oxygen, and CO_2 denotes carbon dioxide. It is apparent that there is no way to prevent the production of carbon dioxide if the combustion of fossil fuels is to be used as a source of heat. If only a limited supply of air is available, carbon burns to form the poisonous carbon monoxide:

$$C + \tfrac{1}{2}O_2 \longrightarrow CO + \text{heat}$$

Most combustion operations produce predominantly carbon dioxide with small quantities of carbon monoxide.

Note from Table 5.5 that coal and fuel oil also contain sulfur. Any sulfur present in fuel is burned simultaneously and unavoidably with the carbon:

$$S + O_2 \longrightarrow SO_2 + \text{heat}$$

This reaction of sulfur in the fuel with oxygen from the air produces sulfur dioxide and heat. The amount of heat generation is insignificant because of the small quantity of sulfur in the fuel. Because of the relative abundance of natural gas in Los Angeles, the sulfur dioxide pollution problem is minimal there. In New York City, by contrast, coal and oil are the major fuels and the sulfur dioxide problem is quite significant. Sulfur dioxide emissions can be controlled by restricting the sulfur content of the fuels being burned or by removing the sulfur dioxide from the flue gases before they are emitted into the atmosphere. Most coals have sulfur levels well above the .6% value of Table 5.5 and fuel oils can have sulfur levels of 3 to 6%.

Nitrogen oxides are another primary air pollutant evolved during combustion operations. Although most fuels contain some nitrogen, the main source of nitrogen oxides is *not* the reaction of this nitrogen in the fuel but rather the nitrogen in the combustion air. The air necessary to provide oxygen for burning the fuel contains only about 20% oxygen, and about 80% nitrogen; nitrogen *from the air* combines with the oxygen *from the air* to form nitric oxide,

$$N_2 + O_2 \longrightarrow 2NO$$

After nitric oxide is emitted to the atmosphere in the flue gases, it reacts further with oxygen in the air to form nitrogen dioxide, denoted by NO_2:

$$NO + \tfrac{1}{2}O_2 \longrightarrow NO_2$$

Nitrogen dioxide is reddish-brown in color and contributes heavily to the atmospheric haze. Control of the nitrogen level of the fuel will only partially control nitrogen oxide emissions. These emissions may, however, be controlled by decreasing combustion temperatures or by removal of nitrogen oxides from the flue gases before they are emitted into the atmosphere.

Particulates come from a variety of sources. Smoke emitted when burning paper, wood, or coal consists largely of particles of carbon and tarry residue; the color of the smoke depends on the size of these particles, which in turn depends on the efficiency of the combustion process. Proper combustion control can eliminate smoke; if you hold a lighted match directly above a smoking cigarette, the smoke will be eliminated. Note from Table 5.5 that coal also contains large amounts of ash. This is an inorganic material such as a fine sand or metalic dust that is bound into the carbon of the coal. As the carbon is burned away, this ash is released and much of it is carried out into the atmosphere with the flue gases. Control of particulate emissions is achieved by proper combustion techniques for smoke abatement, by limiting the ash content of the fuel, or by removal of particulates from the flue gases before they are emitted into the atmosphere.

Thus far we have considered only the pollutants emitted by burning fossil fuels for home heating. Such a source can be a significant problem in the center of large cities, because of the large number of heaters per city block and the short stacks that emit the flue gases and pollutants. There are other important combustion operations, however. The automobile burns gasoline in the internal combustion engine, thus providing these same pollutants described above. However, SO_2 production is not a problem because of the negligible sulfur content of gasolines; the high temperatures to which air is exposed in the internal combustion engine produces large quantities of

NO. Electrical power generation plants convert heat energy into electrical energy. In nuclear power plants, the heat is produced from nuclear reactions, producing none of the pollutants of Table 5.4. In fossil-fueled power plants, this heat is produced by burning coal, oil, or gas. Hence, the pollutants produced are the same as those discussed above for combustion operations. Let us now consider the sources of these specific pollutants in more detail.

5.4
specific atmospheric pollution

Carbon dioxide is so common to all of our activities that air pollution regulations typically state that carbon dioxide emissions are not to be considered as pollutants. Indeed, as discussed above, these emissions cannot presently be controlled unless combustion is replaced by another power source, such as nuclear energy. However, carbon dioxide has long been theorized to be of global importance as a pollutant, with dramatic suggested effects of this pollutant on the earth's temperature.

5.4.1
Carbon Dioxide

Carbon dioxide is an essential ingredient in the life cycle of both animals and plants. Plants rely on the carbon dioxide and moisture in the atmosphere to produce plant cells. About 60×10^9 tons of CO_2 are used annually in this photosynthesis process. These carbon compounds are then used by animals for food and to provide both the energy and cell-building materials required by them. Carbon dioxide is returned to the atmosphere in processes of respiration by both plants and animals and by decay of organic matter. Carbon dioxide is also consumed by dissolving in the oceans and by formation of inorganic materials such as calcium carbonate.

Over the centuries, carbon dioxide levels in the atmosphere have probably been quite stable, until the activities of man became important. The probable result of increased emissions of CO_2 from combustion of fuels is shown in Figure 5.8 as an increasing rate of accumulation of CO_2 in the atmosphere. More than half the total contribution to carbon dioxide emissions results from coal and other solid fuels. Petroleum combustion produces about 30% of the total and natural gas less than 10%. On a relative basis carbon dioxide emissions in the year 2000 are predicted to be almost three times higher than those of 1965. It is expected that if a large quantity of this

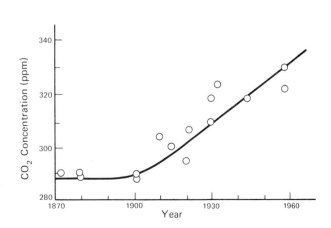

Figure 5.8
Average carbon dioxide concentration in North Atlantic region.[7]

carbon dioxide formed from combustion were to remain in the atmosphere, the concentration of carbon dioxide in the atmosphere would certainly increase.

Carbon dioxide in the atmosphere does not absorb solar radiation because of its short wavelength, but it both absorbs and reflects radiation of the wavelength emitted by the earth. Hence, in much the same manner in which a greenhouse functions, the resulting "greenhouse effect" increases atmospheric and surface temperatures.

This theory has been used to explain the continued climatic oscillations that have occurred in the past.[7] In these studies, a period of glaciation will sufficiently decrease the amount of water in the ocean so that the absorbed CO_2 will be evolved, thus increasing the atmospheric concentration of CO_2 and bringing on a warming trend and a recession of the glaciers. This trend will then continue until the ocean volume increases sufficiently to absorb the excess carbon dioxide and start a cooling trend.

With the CO_2 trend of Figure 5.8, brought about by man's activities, one would thus expect a continued increase in global temperatures. On the basis of estimates from Figure 5.8, the average atmospheric temperature increase because of CO_2 alone will be expected to be .5°C by the year 2000. If this increased temperature were used entirely to melt the polar ice caps, the ice would disappear in 400 to 1,000 years. Melting ice caps, if they occurred, would obviously result in inundation of coastal areas and major global climatic changes. However, when the details of the total atmospheric conditions were considered, it was not at all obvious what the results of this added carbon dioxide in the atmosphere might be. In fact, studies have been conducted that indicate that changes in atmospheric temperature have a very minor effect on the polar ice caps and that such other variables as moisture content and particulate content of the atmosphere have the dominant effect on global temperature. For example, particulates in the atmosphere, from volcanoes and man-made pollution, reflect solar radiation of all types back into space but are largely transparent to the longer-wavelength earth radiation. These particulates exist in the atmosphere in sufficient quantities to produce a "halo effect." This halo can be seen from great heights, such as from high-flying aircraft or spacecraft, and is caused by light being reflected from the particles.

Thus, the effects of other atmospheric contaminants might be opposite from that caused by increasing levels of carbon dioxide in the atmosphere. A large-scale cooling trend has been observed in the Northern Hemisphere since about 1955. This may be due to the disturbance of the radiation balance by fine particles and may have already reversed any warming trend attributable to carbon dioxide.

5.4.2 Aerosols or Particulates

Aerosols present in the atmosphere may be organic or inorganic in composition, and they may be in either the liquid or solid physical state. These particulates are perhaps the most widespread of all the substances that are considered pollutants. Although a variety of particle sizes originates from a variety of sources, the particles must generally be of very small size or they will not remain suspended in the air. Over recent decades, an increase in atmospheric fine particle concentration has been observed by measurements over the oceans. The Atlantic air masses have been observed to carry greater loads of fine particles than do Pacific air masses. This is not unreasonable considering the fact that the Atlantic is both smaller than the Pacific and is generally dominated by the heavily industrialized North American continent because of prevailing winds from the west, while the Pacific is primarily influenced by the much less industrialized Asian continent.

Particles larger than 10 μ in diameter come mainly from mechanical processes such as erosion, grinding, and spraying. Those between 1 and 10 μ are more numerous in the atmosphere and originate from mechanical processes and include industrial dust, ash, etc. Particles in the size range between .1 and 1 μ tend to contain more of the products of condensation. In this size range are found ammonium sulfate, products of combustion, and aerosols formed by photochemical reactions in the air. All size ranges of particles can grow by condensation, adsorb and absorb vapors and gases, coagulate or disperse, and absorb and scatter light. Table 5.6 summarizes the particulate emission levels from several man-made sources. Natural sources, such as volcanoes and winds, contribute as much particulate matter as all those sources combined.

Table 5.6
Emission Levels of Particulates, 1965

Source	Level (tons/yr)
Automobiles	1
Industry	6
Electric power plants	3
Space heating	1
Refuse disposal	1

Source: Reference 8.

Probably the most commonly emitted kind of particle anywhere is carbon. Carbon particles are nearly always present in the products of combustion of all types of fuel. However, these particles occur in greater quantities when burning fuel oil than when burning natural gas. The burning of coal provides even greater quantities, as well as large amounts of inorganic ash. Carbon particles are probably one-third to one-half of the total aerosol emissions.

Small droplets of oily or tarry materials are frequently found in combustion products from many types of sources. The most common sources are probably the emissions associated with the operation of motor vehicles, particularly crankcase emissions and exhaust emissions. The total quantity of these materials probably comprises 10 to 20% of particulate emissions.

Small droplets of acid, both organic and inorganic, are emitted from stacks of power plants, from industrial operations such as certain metal-working and cutting operations in storage battery reclamation, and even from motor vehicle exhaust. Like the other kinds of particulate matter, the total emitted quantities of these droplets are small, probably 5 to 10% of the total particulate emissions.

Emissions of inorganic dust consist primarily of silicates, carbonates, and oxides and probably are associated most commonly with sand and gravel plants and other phases of the mineral industry. The quantity emitted represents about 5 to 10% of the total particulate emissions.

In most situations, small particulate emissions represent a major proportion of the total particulate emissions. Many particulates are removed from the atmosphere by gravitational settling, in rain water, by impinging on buildings, etc. Nevertheless, those that remain have been suggested to cause global drops in temperature and to harm man, as discussed earlier. There are indications that the toxic effect of the sulfur dioxide and sulfur trioxide in the combustion of coal is enhanced by the accompanying particulate matter. This kind of effect has been noted in other cases involving aerosols and toxic gases or liquids; it has given rise to the theory that contaminants can adsorb on the surface of the particles and thus come into contact with deep inner surfaces of lungs and mucous membranes in greater concentrations than would otherwise be possible. Particulate emissions are also associated with reduction of visibility. There is evidence that the presence of minute particles promotes the photochemical reactions that produce smog. Furthermore, small aerosol particles are among the products of these reactions, adding to the visibility reduction produced by the emitted contaminants. Particulates thus represent a substantial form of air pollution.

5.4.3 Carbon Monoxide

Carbon monoxide has been considered an important atmospheric pollutant for many years because of its prevalence in automobile exhausts and in the gases from combustion with insufficient oxygen. In spite of widespread knowledge of the toxicity of carbon monoxide, there have been many fatalities resulting from exposure to excessive carbon monoxide. Because of the hazards it posed, initial air pollution studies of carbon monoxide in the urban atmospheres were carried out mostly to determine whether special segments of the population, such as traffic police and tunnel toll takers, were exposed to toxic levels of carbon monoxide. However, recently carbon monoxide measurements have been studied as an indication of urban air quality. It has been assumed that the only sources of carbon monoxide were combustion sources. However, some natural sources of carbon monoxide seem to be present (see Table 5.4).

The toxic effects of carbon monoxide arise from CO reacting with components of the blood in such a way that it can no longer assimilate fresh oxygen from the lungs. However, probably due in part to its lack of odor, long-term records have not been kept on the concentration of CO in the atmosphere. The sketchy data that exist do not indicate that carbon monoxide levels in the atmosphere are increasing, at least over the last 15 to 20 years. The average level of carbon monoxide varies greatly with location in the city. In a residential area in Detroit, 29 ppm has been recorded, but much higher levels of CO are present where vehicular traffic is heaviest. By measuring carbon monoxide level from an automobile moving in the traffic stream of Los Angeles, peaks as high as 120 ppm have been recorded.

The primary source of carbon monoxide as a pollutant is from the internal combustion engine. This source represents 70 to 80% of the total carbon monoxide emissions of 274 million tons/yr.[4,9] Combustion of coal, wood, and other incineration make up the remainder. Some marine life, such as the siphonophores, must also emit CO, for ocean water has been found to be supersaturated with CO relative to atmospheric levels of CO. About 95% of the CO formed by man's activities occurs in the Northern Hemisphere. This has resulted in about three times higher concentration of carbon monoxide in the Northern Hemisphere than in the Southern Hemisphere.

5.4.4 Sulfur Dioxide

Sulfur dioxide is one of several forms of sulfur that exists in the air; the others are hydrogen sulfide, sulfuric acid, and sulfate salts. The most common sulfide is hydrogen sulfide, which is emitted from natural sources and which is oxidized to sulfur dioxide in the atmosphere. As indicated in Table 5.4, however, pollutant sources of SO_2 are also quite important.

In the mid-1960s the estimated total worldwide sulfur dioxide pollutant emissions were 146 million tons. Of this, 70% resulted from coal combustion, 20% from petroleum combustion and refining, and the remainder from nonferrous smelting. Of the worldwide sulfur dioxide emitted, 93% was emitted by pollutant sources in the Northern Hemisphere.

Sulfur dioxide has been a major pollutant throughout the history of industrial processes—at least since the first smelting of copper sulfide or the beginning of the widespread burning of soft coal. Figure 5.9 shows the total annual sulfur dioxide emissions over the last 100 years. There is clearly a major increase of sulfur dioxide emissions between about 1950 and 1965. About 55% of this increase is related to increased coal combustion, and 31% is due to increased petroleum usage. Since SO_2 emissions have doubled between about 1905 and 1940 and then doubled again between 1940 and 1965, further large increases can be expected without some changes in combustion practices. Without making allowances for the introduction of additional emission controls, there will be a 226% increase in sulfur emissions by the year 2000 compared to the year 1965. This will increase the sulfur dioxide emissions up to 333 million tons annually.

The effects of SO_2 in the various air pollution disasters have already been suggested. It has been theorized that the SO_2 is oxidized to SO_3 in the air, which then combines with water droplets to form sulfuric acid. In combination with fine particulates, a sulfate aerosol (often ammonium sulfate)

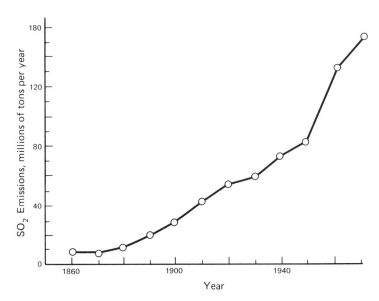

Figure 5.9
Estimated historical global sulfur dioxide emissions.[7]

can be carried deep into the lungs of man. In the early 1960s the increased concentration of sulfur dioxide in the air (due to combustion of high-sulfur fuel oil) was accompanied by increased acidity of rain and snow in Europe. The pH of Lake Vanern in Sweden fell from 7.3 in 1965 to 6.8 in 1967, and that of the Morrumsan River fell from 6.8 to 6.2 during this period (i.e., both became more acidic). Most aquatic organisms cannot survive at pH values below 4.0 and some fish, such as salmon, cannot survive at pH values below 5.5. It is possible that factors other than SO_2 in the atmosphere could have caused this pH decline (such as variations in temperature, biological activity, and mineral composition of the water), but the atmospheric sulfur pollution must be regarded as a possible contributing factor.

5.4.5
Hydrocarbons[10]

Hydrocarbon emissions have recently received substantial attention as an atmospheric contaminant. Although hydrocarbons can themselves be toxic or carcinogenic in sufficient concentrations, of major concern is their participation in photochemical reactions in the atmosphere to produce objectionable compounds causing eye irritation and plant damage. Man produces only about 15% of the total global emissions of hydrocarbons, but he produces far more than half the hydrocarbons in the atmosphere of urban areas.

Gasoline is one of the most important sources of hydrocarbon emissions, coming from evaporation at storage facilities and evaporation from the automobile gas tank. Hydrocarbons are also emitted from automobile crankcases and unburned hydrocarbons escape in the automobile exhaust. Other hydrocarbon sources are evaporation of industrial solvents and combustion of coal and wood. Natural sources of hydrocarbons include forests and vegetation, which emit terpenes, and bacterial decomposition of organic matter, which produces large amounts of methane (swamp gas). More than

80 different hydrocarbon compounds can be found in the Los Angeles atmosphere, at levels up to 10 ppm. Fifty to 80% of these hydrocarbons are methane, with many paraffins, olefins, and aromatics making up the balance.

Hydrocarbons, in combination with sunlight and nitrogen oxides (automobile emission), are important in the formation of photochemical smog. Such ingredients are available in profuse supply in Los Angeles. In addition, the dispersion of pollutants there is limited horizontally by low winds and surrounding mountains and vertically by the inversion that lies over the area some 320 days a year.

5.4.6 Nitrogen Oxides

The cycle of nitrogen compounds in the environment is complex, consisting of generation of nitrogen compounds from pollution and natural sources and consumption of nitrogen compounds through natural mechanisms. There are at least five gaseous compounds and two solid nitrogenous compounds. These materials—nitrous oxide, nitric oxide, nitrogen dioxide, ammonia, ammonium compounds, and nitrates—are for the most part interrelated in a complex environmental cycle. The major pollutant in the nitrogen group is nitric oxide, approximately 50 million tons/yr. Nitrous oxide and ammonia appear to be generated primarily from natural sources.

Oxides of nitrogen from automobile exhausts or from other combustion equipment constitute one of the main problems in air pollution in the United States. For instance, exhaust gases from automobile engines contain between 300 and 3000 ppm of nitric oxide. This nitric oxide undergoes photochemical transformation to form the brownish nitrogen dioxide and other compounds that are eye and plant irritants. Nitrogen oxide emissions from several combustion sources are summarized in Table 5.7. This table des-

Table 5.7
Nitrogen Oxide Emissions from Combustion Assuming No Control
(thousands of tons per year expressed as nitric oxide)

YEAR	HOUSEHOLD AND COMMERCIAL	INDUSTRIAL	ELECTRIC GENERATION	TOTAL STATIONARY SOURCES	TRANS- PORTATION	GRAND TOTAL
1950	724	1,877	1,274	3,875	3,557	7,432
1955	657	2,086	1,842	4,585	4,643	9,228
1960	664	2,856	2,303	5,823	4,871	10,694
1965	707	3,267	3,162	7,137	5,509	12,646
1970	765	3,380	3,966	8,111	6,529	14,640
1975	849	3,526	5,045	9,420	7,768	17,188
1980	966	3,768	6,049	11,143	9,193	20,336
2000	1,450	3,579	8,879	13,908	18,488	32,396

Source: Reference 11.

cribes nitrogen oxide emissions from combustion sources as an overall U.S. average. Note that about half the nitrogen oxides as pollutants are formed from stationary sources such as electric generation plants and that about half are formed from transportation sources such as motor vehicle exhausts. By the year 2000 the total nitrogen oxide emissions (without controls) will approximate 32 million tons. It is important to recognize, however, that in nonindustrialized urban areas, the contribution from motor vehicles may be as high as 90% of the total nitrogen oxides present as pollutants in the atmosphere. The oxides of nitrogen emissions are related to fuel consumption and the number of motor vehicles. These vehicular emissions of nitric oxide in the Los Angeles Basin were approximately 450 tons/day in 1963 and are expected to almost double by 1980.

Let us consider the role of nitrogen oxides in the formation of photochemical smog. The amounts of primary and secondary pollutants during a typical smog episode in Los Angeles are summarized in Figure 5.10. The primary pollutants, NO and HC, are building up in the predawn hours from the increase in man's activities, such as freeway traffic. Small amounts of NO_2 are emitted directly, and some NO_2 is formed slowly by a reaction of the NO with oxygen in the air:

$$NO + \tfrac{1}{2}O_2 \longrightarrow NO_2$$

After dawn, the presence of sunlight becomes an important factor in the atmospheric reactions taking place. Any NO_2 that is present in the atmosphere interacts with the sunlight to produce NO and an oxygen radical:

$$NO_2 + \text{sunlight} \longrightarrow NO + O\cdot$$

This radical, designated by the dot in $O\cdot$, is an extremely reactive compound. It combines very rapidly with oxygen in the air to form ozone:

$$O\cdot + O_2 \longrightarrow O_3$$

Hence, the ozone concentration will increase after sunlight is available to decompose the NO_2. Ozone itself can also react with NO to form more NO_2:

$$NO + O_3 \longrightarrow NO_2 + O_2$$

Hence, small amounts of NO_3 present in the atmosphere at dawn form the oxygen radical. Ozone is produced immediately from this oxygen radical but is consumed in producing the NO_2. Hence, after dawn, we see the NO_2 accumulating, which produces the brownish atmosphere characteristic of the photochemical smog. Ozone only begins accumulating in the atmosphere after all the NO is depleted (so no more ozone can react to form NO_2), which occurs around 9 A.M. in Figure 5.10.

Hydrocarbons in the atmosphere can also react with ozone to produce a hydrocarbon radical, which is also highly reactive. This radical,

$$HC + O_3 \longrightarrow HCO\cdot + O_2$$

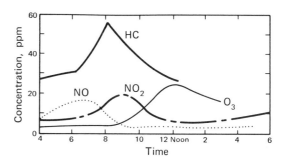

Figure 5.10
Change in pollutant levels during day.

can react with NO to form NO_2 *and another hydrocarbon radical*:

$$HCO\cdot + NO \longrightarrow HC\cdot + NO_2$$

In contrast to the previous reactions of the oxygen radical, the hydrocarbon reaction produces additional reactive radicals. Hence, hydrocarbons are particularly detrimental in the production of photochemical smog for they produce radicals that continually accelerate other photochemical reactions. Finally, some of these hydrocarbon radicals can react with NO_2 in the atmosphere to form peroxyacetyl nitrates (PAN):

$$HCO\cdot + NO_2 \longrightarrow PAN$$

We thus see a decrease in hydrocarbon content of the atmosphere in Figure 5.10 as ozone builds up. A late afternoon rush hour will repeat the cycle as long as sunlight is available.

We thus can see the importance of the role of sunlight, NO, and hydrocarbons in the formation of photochemical smog. An air pollution problem centering upon smoke and SO_2, as is prevalent in London and New York City, need not have components such as sunlight present (see Table 5.1). Because of the critical role of hydrocarbon radicals, attempts have been made to curtail photochemical smog by eliminating hydrocarbon omissions. The many sources of hydrocarbon emissions make such efforts very difficult, and present efforts are directed toward control of both hydrocarbons and nitrogen oxides.

The actual reactions taking place in a polluted atmosphere are much more complex than the set of seven reactions outlined above. The reaction sequence probably includes from 15 to 30 individual reactions.

5.5 ATMOSPHERIC SCAVENGING PROCESSES[5]

Atmospheric pollutants can be removed from the atmosphere by two routes, deposition and chemical conversion to normal atmospheric constituents. Deposition, of course, is a natural settling process that will take place when particles are removed from the atmosphere by gravitational forces. Gases or vapors and very small particles are not removed directly by deposition. They can, however, be adsorbed on the surface of dust or other particulates and thus removed by deposition. Chemical reactions of gases or vapors can convert pollutants to other normal atmospheric components; for example, CO can be converted to CO_2, and a small amount of NO may be converted into N_2 (nitrogen) and O_2 (oxygen). Rainfall greatly accelerates the removal of atmospheric pollutants. The rain can pick up small particulate matter and dissolve gaseous pollutants, which are then carried to the earth. Some pollutants, however, are not soluble in water and hence cannot be removed by rainfall. Figure 5.11 illustrates the effect of rainfall as a scavenging agent. The figure shows curves for concentration as a function of time for three different effluents: (1) a gas unaffected by rain fall, (2) a somewhat soluble gas (SO_2), and (3) a particulate material of uniform particle size. This graph shows that the effect of "washout" is considerable.

As shown in Figure 5.9, the total production of sulfur dioxide exceeds 100 million tons/yr; sulfur dioxide is the primary form of sulfur polluting the atmosphere. The level of sulfur dioxide in the atmosphere would increase continuously without scavenging mechanisms. However, most of the sulfur dioxide in the atmosphere is oxidized by reacting with oxygen in the atmosphere to form sulfur trioxide:

$$SO_2 + \tfrac{1}{2}O_2 \longrightarrow SO_3$$

Sulfur trioxide is highly water-soluble (forming sulfuric acid) and is almost completely removed from the atmosphere by precipitation. Sulfur trioxide can also react with water vapor in the air (which is present even without rainfall) to form a sulfuric acid mist or aerosol. Either this aerosol or sulfur

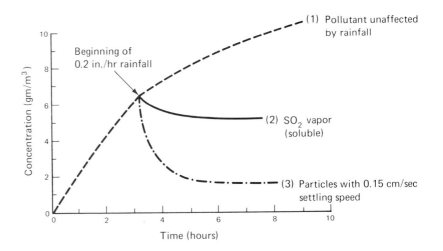

Figure 5.11
Effect of rainfall on specific pollutants.

trioxide can then react with suspended alkalies, such as ammonia or calcium compounds in dust, to form salts. These salts are then removed either by deposition or precipitation.

Carbon monoxide levels in the atmosphere are probably not increasing, although this is difficult to measure because of limitations of analytical instruments. Carbon monoxide is probably being converted by oxidation to carbon dioxide in the upper atmosphere:

$$CO + \tfrac{1}{2}O_2 \longrightarrow CO_2$$

Carbon dioxide, in turn, is converted into oxygen in the photosynthesis of plants; however, most of the plant material is subsequently reoxidized into CO_2 years later through the decay of dead plants. Carbon dioxide is also absorbed in the weathering of rocks and other solid materials. Large amounts of carbon dioxide are also dissolved in the earth's oceans, which are estimated to contain 60 times as much carbon dioxide as the earth's atmosphere. Because of the large capacity of the ocean to dissolve carbon dioxide, much of the CO_2 emitted to the atmosphere may well end up in the oceans.

Nitrogen oxides released into the atmosphere are rapidly converted by sunlight into nitrogen-containing hydrocarbon compounds and higher nitrogen oxides (such as nitrogen dioxide, nitrogen trioxide, and nitrogen pentoxide). These nitrogen oxide compounds are removed in the same manner as described for the sulfur oxides—by precipitation and deposition. The organic molecules are decomposed by oxidation into carbon dioxide and water.

Particulates are removed from the atmosphere primarily by gravitational settling. Small particles, which settle more slowly, are also removed by collisions with stationary objects, such as hills, buildings, and trees. Particulates normally remain suspended above the ground only for a few hours or a few days. However, particulates can be suspended for much longer periods of time, as evidenced by the particulate load over the Atlantic Ocean, thought to originate from the industrialized North American and European continents. In particular, large volcano erruptions have caused worldwide particulate suspensions for a year, and the early atomic bomb explosions suspended particulates for 7 years. Hence, extreme erruptions can generate very long-lived particulate suspensions, particularly if particles reach the stratosphere where rainfall is absent.

5.6 AIR POLLUTION LEGISLATION[4]

Many of the early attempts to deal with pollution problems depended on existing laws prohibiting a public nuisance. This approach has gradually been discarded as cumbersome and uncertain in its results, as well as inadequate to deal with the modern array of pollutants. Also, it makes no provision for the employment of preventative measures and techniques that can control emissions. Nevertheless, such regulatory agencies as the Los Angeles Air Pollution Control District have effectively used the *threat* of application of nuisance laws to accelerate compliance with air pollution regulations by individual companies. The nuisance accusation is difficult to contest by companies and, if upheld by courts, will require immediate and complete compliance with regulations or immediate cessation of plant operation. Because of the potential consequences, individual companies will usually try very hard to prevent the formal filing of a nuisance complaint by the Los Angeles Pollution Control District. Historically, of course, any person who suffers injury from pollution originating from activities on another person's property may sue to recover damages or to require cessation of the activity if it constitutes a continuing nuisance. Private litigation for relief from the effects of pollution has been largely unsuccessful, for it is both expensive and would close down an otherwise lawful business (which is so drastic an occurrence that courts have been reluctant to grant it).

The federal government did not become involved in air pollution control prior to 1955, leaving the problem to be attacked by individuals, cities, counties, or states. As a result, a vigorous policy of control was initiated by California. A number of cities, such as Pittsburgh, achieved a substantial reduction in smoke levels. However, local government programs were generally understaffed and without sufficient financial and trained manpower resources to meet their needs. By 1963, for example, only 34 local programs had annual budgets exceeding $25,000 and 7 of these were in California. Of the other 51 local air pollution control agencies in existence at that time, 21 tried to function on less than $10,000/yr.

In July 1955, Congress authorized a federal program of research and technical assistance relating to air pollution control, administered by the Secretary of the Department of Health, Education and Welfare. The assistance was directed toward state and local governments primarily; monetary appropriations under this law increased from $186,000 in 1955 to $11 million in 1963. This legislation continued the policy that state and local governments have the primary responsibility for air pollution problems, whereas the federal government provides overall coordination and support.

In 1960, the Division of Air Pollution of the Public Health Service was formed and Public Law 86-493 directed the Surgeon-General to study the problem of motor vehicle exhausts and their effects on human health. In this capacity, the Public Health Service rendered research support and technical assistance and generally tried to stimulate interest in pollution control, but the primary responsibility for regulatory control still rested with state and local governments. By 1963, governmental staffs had concluded that current pollution control efforts were not even satisfactory to control pollution at its 1963 level, much less control the anticipated growth of pollution.

5.6.1
1963 Clean Air Act

In December 1963, the Clean Air Act was passed by Congress, expanding the federal role in air pollution control. This act established a program of federal grants to state, regional, and municipal agencies to stimulate their air pollution control programs. It also greatly enlarged the federal research and development program carried out by the Department of Health, Education and Welfare (HEW), and specifically isolated motor vehicle emissions and sulfur oxide emissions for intensive investigation. The act, for the first time, gave the Secretary of HEW the powers to establish and publish air quality criteria defining the effects of air pollution on health and property, thus providing guidance for state, regional, and municipal agencies in establishing enforceable standards of air quality. Finally, it required the Secretary of HEW to provide procedures for control of air pollution from federal facilities.

The Clean Air Act was amended in October 1965 to direct the Secretary of HEW to regulate emissions from motor vehicles. The secretary issued regulations to be applied to the 1968-model-year automobiles. Air quality criteria for sulfur oxides were issued in March 1967, as directed by the 1963 Clean Air Act. These criteria were involved in a minor controversy for, as representatives of HEW conceded in later Senate hearings, the specified sulfur oxide controls were beyond the reach of existing control technology.

5.6.2 1967 Air Quality Act

With the continued increase in air pollution in the United States, the Air Quality Act of 1967 (amending the 1963 Clean Air Act) was adopted. This act required the Department of Health, Education and Welfare to define atmospheric areas or air basins in the United States, which are essentially homogeneous in topographical and meteorological factors dispersing pollutants. These areas are California-Oregon Coastal, Washington Coastal, Rocky Mountain, Great Plains, Great Lakes—Northeast, Mid-Atlantic Coastal, Appalachia, and South-Florida. Then, HEW was to designate numerous *air quality control regions*, which are smaller geographical areas containing communities that are affected by common air pollution control problems and that require a uniform air pollution control action. Approximately 100 of these regions have been designated, such as the greater New York City area, including adjacent counties in Connecticut and New Jersey. HEW was also directed to develop and publish *air quality criteria*, which define the relationship of individual air pollutant concentrations in the air to adverse effects to health and property. These criteria have been published for sulfur oxides, photochemical oxidants, carbon monoxide, hydrocarbons, and particulate matter. Air quality criteria for nitrogen oxides, polycyclic aromatic hydrocarbons, fluorides, lead, and odors (including hydrogen sulfide) were published in 1971. In the language of the act, "Such criteria shall... reflect the latest scientific knowledge useful in indicating the kind and extent of all identifiable effects on health and welfare which may be expected from the presence of an air pollution agent...." The act also directs HEW to develop and publish information on the prevention and control of air pollution. These publications describe specific technological methods for achieving the air quality levels described by the air quality criteria, including specific techniques, the effectiveness of these techniques, their costs, and the economic feasibility of achieving various levels of air quality with alternative control techniques.

Since the state and local governments still have the primary responsibility for preventing and controlling air pollution, the states must develop *ambient air quality standards*. These standards are the specific concentrations of pollutants that are to be achieved in the surrounding (ambient) atmosphere within the air quality control region. They are not the specific pollutant levels in a chimney stack of a furnace (which would be an *emission standard*). They reflect the feasibility of achievement within the immediate future (state-of-the-art of pollution control technology) as well as the federal air quality criteria. The state submits these standards as well as plans for implementing them to HEW within certain prescribed time intervals. The department reviews them and, when approved, the states are to control pollution in accordance with these plans. If any individual state fails to act or if its action is inadequate, HEW is empowered to initiate abatement action.

The act also prohibits state and local governments from regulating emissions from *new* motor vehicles; this prohibition can be waived in any state that had adopted vehicle emission standards prior to March 30, 1966, if that state requires more stringent standards (California is the only state meeting this qualification, and the prohibition has been waived for 1969 and 1970 vehicles). Hence, this act provides for federal preemption in vehicular emission standards, on the grounds that most motor vehicles are destined for interstate travel. For example, even though a farmer in Oklahoma may have little need for exhaust emission controls, his automobile may be owned later by a factory worker in Los Angeles. The task of seeing that control devices are kept in place and properly maintained remains the function of state and local governments, but the act specifies a program of grants to assist the states in developing suitable motor vehicle inspection programs (up to two-thirds of the development costs). As might be expected, specific inspection programs vary widely. Apparently, the regulation of air pollution control devices for *old* motor vehicles (prior to the 1968 model year) remains the realm of state and local governments.

Finally, the 1967 act further expanded federal research and development programs relating to air pollution; broadened the grant program to the state, regional, and local air pollution agencies; and established advisory groups to assist HEW in controlling and preventing air pollution. By 1970, the operating responsibilities of HEW in this field were shouldered by the National Air Pollution Control Administration (NAPCA), which was formerly the National Center for Air Pollution Control, which previously was known as the Division of Air Pollution.

As has been discussed above, air quality criteria have been issued for sulfur oxides, photochemical oxidants, carbon monoxide, hydrocarbons, and particulate matter. In gathering scientific evidence for such criteria, the many air pollution disasters mentioned earlier were evaluated. In addition, other evidence linking the level of air pollution to higher death rates from heart disease and stroke was evaluated. For example, a recent study correlated death statistics of Buffalo, New York, over a selected time span with data on that city's level of suspended particulate air pollution during the same period. The findings pointed to, but did not conclusively demonstrate, a statistical relationship between air pollution levels and death from chronic respiratory diseases. Another study investigated possible links between the number of deaths in New York City and the level of sulfur dioxide in the air. The data of this study covered the period between 1960 and 1964. There were 10 to 20 more deaths per day when the SO₂ readings were .4 ppm or more, compared to days when these readings were .2 ppm or less. This, once again, points to a possible relationship between sulfur dioxide in the air and death. Airborne particles have also been linked to lung cancer, mesophelioma, and cancer of the plura. Apart from being a potential health hazard by themselves, minute particles adsorb molecules of gases such as sulfur dioxide on their surface. When inhaled, the particles with the adsorbed gas can lodge deep in the air passages of the lungs, where they may cause severe irritation. It has been suggested that the length of time between the ingestion of the particles and the onset of cancer may be as long as 20 years. In the development of criteria, laboratory and field studies concerning the effects of pollutants on materials, vegetation, and animals are also considered. In addition to a complete discussion of all the studies evaluated, the published air quality criteria include a summary of the effects of the individual pollutant, as exemplified in Table 5.8 for sulfur dioxide. The air quality criteria for sulfur oxides also states "... adverse health effects were noted when 24-hour average levels of sulfur dioxide exceeded 200 micrograms/m³ (0.11 ppm) for 3 to 4 days. Adverse health effects were also noted when the annual mean level of sulfur dioxide exceeded 115 micrograms/m³ (0.04 ppm)... adverse effects on vegetation were observed at an annual mean of 85 micrograms/m³ (0.03 ppm). It is reasonable and prudent to conclude that, when promulgating ambient air quality standards, consideration should be given to requirements for margins of safety which take into account long-term effects on health, vegetation, and materials occurring below the above levels." For comparison to the levels of Table 5.8, Figure 5.12 represents measurements of sulfur dioxide concentrations for eight U.S. cities.

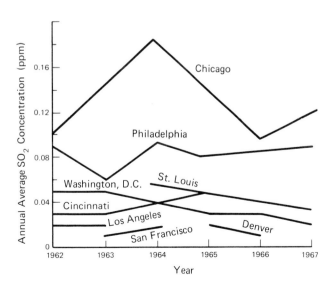

Figure 5.12
Trends in sulfur dioxide concentrations.

Table 5.8
Air Quality Criteria for Sulfur Dioxide

Concentration	Effect
1. .52 ppm and higher (24-hr average)	With particulates, increased mortality in man may occur
2. .25 ppm and higher (24-hr average)	With smoke, increased daily death rate in man may occur
3. .19 ppm (24-hr average)	With low particulate levels, increased death rates in man may occur
4. .11–.19 ppm (24-hr average)	With low particulate levels, increased hospital admissions of older persons for respiratory disease may occur; absenteeism from work, particularly with older persons, may also occur
5. .25 ppm (24-hr average)	With particulates, a sharp rise in illness rates for patients over age 54 with severe bronchitis may occur
6. .21 ppm (24-hr average)	With smoke, patients with chronic lung disease may experience accentuation of symptoms
7. .037–.092 ppm (annual average)	With smoke, increased frequency of respiratory symptoms and lung disease may occur in man
8. .046 ppm (annual average)	With smoke, increased frequency and severity of respiratory diseases in school children may occur
9. .040 ppm	With smoke, increase in mortality in man from bronchitis and lung cancer may occur
10. .12 ppm	With high particulate levels, corrosion rate for steel panels may be increased by 50%
11. .03 ppm (annual average)	Chronic plant injury and excessive leaf drop may occur
12. .3 ppm (8 hr)	Some species of trees and shrubs show injury
13. .05–.25 ppm (4 hr)	With ozone or nitrogen dioxide, moderate injury to sensitive plants

Although the legislation represented by the Air Quality Act of 1967 represents a clearly definable philosophy of attacking our air pollution problems, several alternative philosophies have been proposed that are believed by some people to provide a stronger attack. First, the usefulness of our air quality criteria, which are directed at *individual* air pollutants, has been questioned. These criteria attempt to relate the levels of one pollutant, such as sulfur dioxide, to detrimental effects on health and property. In fact, there are probably many important interrelationships between specific pollutants and their combined effects on health. Furthermore, the specific effects of a pollutant on individuals is also influenced by such additional factors as accompanying illness, cigarette smoking, age, and heredity. These observations are supported by the entries in Table 5.8. Hence, if these arguments can be extended to many other pollutants, it is hopeless to look for a single disease and a single cause as is implied in the development of air quality criteria. Also, the doses of sulfur dioxide, for example, applied to laboratory animals are many times higher than those associated with air pollution; doses of a magnitude comparable to air pollution episodes do not produce mortality in animals in spite of the statistical, epidemiological studies linking mortality in man to specific sulfur dioxide levels. The latter studies can be greatly influenced by other substances that increase in concentration by the same atmospheric conditions causing an increase in sulfur dioxide level and by synergistic interactions of these with sulfur dioxide. Also, measuring instruments that record the sulfur dioxide level can be made inaccurate by other compounds in the atmosphere, so the readings typified in Table 5.8 become unclear. In addition, a control population, which is like that exposed to sulfur dioxide in every respect except its exposure, is not available. Finally, critics of the air quality criteria and standards approach assert that this concept implies there is a certain air pollutant level below which we are "safe." In fact, we may be harmed by any level of pollutants above the natural background concentration and the degree of harm increases as the pollutants increase. The conclusion of the opponents of the air quality criteria approach, as elegantly argued by Cassell (12), is that health requirements are best met "by the control of emissions to the greatest extent feasible, employing the maximum technological capability." Interestingly enough, this also is the approach required to achieve the "reasonable and prudent" air quality criteria, given the present technological and economic constraints of feasibility. Hence, although different in philosophy, the conclusions of the two groups are identical.

The philosophy of our national air pollution control program has also been attacked in that it specifies *ambient* air quality standards, to be proposed and implemented by the individual states. The alternative proposed is, of course, national *emission* standards that would regulate the emission levels of pollutants at each source, such as the chimney stack of a furnace (the early versions of the 1967 Air Quality Act included such provisions, but these were deleted in the act as finally passed by Congress). The primary deficiency of ambient air quality standards, of course, is that they simply define the permissible levels of pollutants in the total air quality control region; they do not specify the source of the pollution and do not provide guidelines for how much each polluter must reduce his emission level to achieve the ambient air quality standards. Furthermore, state and local air pollution agencies will be discouraged from setting stringent emission standards by local industrial firms and communities that must meet these standards while competing with other companies in the nation not having such stringent emission standards. This competitive disadvantage will require that each state improve the health and welfare of its populace at the expense of the economic growth and employment of the individual communities, thus restricting decisive action by each state in setting emission standards. Finally, arguments against ambient air quality standards suggest that the placement of air quality monitoring devices throughout an air quality control region (to measure ambient levels of sulfur dioxide, for example) would be so expensive that the states cannot effectively police their own ambient air quality standards anyway. There are also, of course, arguments against the use of national emission standards. Suppose that we specify the emission levels of 5 industrial plants, all of which emit pollutants into an air basin. Then 30 years from now, when there are 20 plants in operation, the quality of air in the basin must be degraded (there are four times as many plants operating at these emission levels). If we set emission levels in anticipation of the pollutant load in 30 years, not only will money be wasted for overly stringent emission controls (assuming air quality criteria are valid) for the next 30 years but also the atmosphere will be degraded in 60 years. If we anticipate the pollutant load for the next 100 years in setting today's emission levels, air pollution control will be so expensive that the 5 plants now operating must be shut down. Giving the local inhabitants of the region the choice between this alternative and much less pollution control, they will almost certainly choose the latter. If we make emission standards more stringent each time a new plant starts up, we are not only requiring the new plant to pay more to operate than his competitors, but we are also requiring the existing plants to periodically replace all their control equipment with equipment meeting the new standards. In addition, the use of national emission standards does not take into account the local differences in air quality control regions, so some industrial firms are controlling their emissions to provide very low ambient levels and others to provide high ambient levels. The use of national emission standards also requires the small manufacturer to invest a higher proportion of his available capital in air pollution control equipment than a large manufacturer, even though he contributes less total pollutant to the atmosphere. Finally, the cost of sampling the chimney stack of every plant in the country on a periodic basis would be even greater than the cost of policing an air quality control region for ambient levels.

In conclusion, neither approach is without faults, and it is likely that future legislation will combine the two concepts. Probably, ambient air quality standards will be maintained, and national emission standards will be superimposed for some of the large sources of pollutants, such as electrical power generating plants.

In the above paragraphs, we noted a deficiency in air quality criteria in that they implied there is a certain level of pollutant below which we are safe. It has been suggested that a cost-benefit analysis leading to an *effluent fee* would be a more rational approach to controlling air pollution than air quality criteria and standards. (See Figure 5.13.) This concept is based on the premise that pollution is caused by companies and individuals because they do not have to pay the cost of the pollution that they create. If we consider the true cost of air pollution to be measured by the effects of air pollution on life, on materials of construction, and on cleaning and maintenance bills, then it is argued that the polluter should bear these costs along with his other costs of production. Governmental agencies, then, should charge an effluent fee that reflects these costs; the fee would increase as the amount of pollutant discharged into the atmosphere increases. The polluter, then, would decrease his pollutant emission level until the benefits he derived (lower effluent fees) just equals the cost of his added air pollution abatement equip-

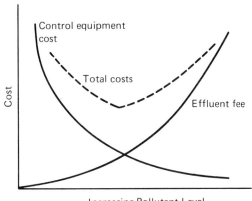

Figure 5.13
Cost-benefit analysis in air pollution.

ment. Being intimately familiar with his own operations, he could reduce pollution at a minimum cost passed on to the consumer. Although the concept of voluntary reduction of emissions by polluters is very attractive, the effluent fee approach is very difficult to apply in practice. Assessing the cost of pollution is very difficult; it requires that we place a fixed value on human life, that we isolate effects of individual pollutants and their costs from other forms of damage, that we measure long-term effects of pollution, and that we apportion these total costs between individual polluters. In addition, there is no guarantee that the individual polluters would invest the capital in abatement facilities to recover effluent fees as compared to investing the capital in new corporate ventures that return more money to the corporation than available through the effluent fees. Costly monitoring of pollutant streams would also be required to set the effluent fees. In short, the method is difficult and costly to implement and uncertain in its success in reducing air pollution.

5.6.3 1970 Clean Air Act

In 1970 President Nixon formed the Environmental Protection Agency (EPA) collecting the personnel responsible for our attack on all environmental problems. The operating responsibility for prevention and control of air pollution is carried out by the Air Pollution Office of the EPA, formerly the National Air Pollution Control Administration. Also in 1970, Congress passed the Clean Air Act, amending the Air Quality Act of 1967. This act authorized the EPA to continue designating air quality control regions for 90 days after enactment, but thereafter any undesignated portion of a state would be a single region (such regions could later be subdivided by the state with EPA approval). Under this act, the EPA is to set *national* ambient air quality standards. Both primary and secondary standards are to be set, where primary standards are designed to protect the public health with an adequate safety margin. Secondary standards are designed to protect the public welfare from any known or anticipated adverse effects. These ambient air standards have been set and are summarized in Table 5.9. The states are to provide plans for implementation, maintenance, and enforcement of these standards for each air quality control region (or portion thereof) within the state. These plans are to attain the primary air quality standards within 3 years and are to include *emission* standards needed to attain the national ambient standards as well as assurance of necessary funds to carry out these plans.

Table 5.9
National Primary and Secondary Ambient Air Quality Standards (all levels not to be exceeded more than once per year)

Pollutant	Primary Standard	Secondary Standard
Sulfur oxides	.03-ppm annual mean level	.02-ppm annual mean level
	.14-ppm maximum 24-hr level	.1-ppm maximum 24-hr level
Particulate matter	.75-μ/m^3 annual mean level	60-μ/m^3 annual mean level
	260-μ/m^3 maximum 24-hr level	150-μ/m^3 maximum 24-hr level
Carbon monoxide	9-ppm maximum 8-hr level	Same as primary
	35-ppm maximum 1-hr level	Same as primary
Photochemical oxidants	.08-ppm maximum 1-hr level	Same as primary
Hydrocarbons	.24-ppm maximum 3-hr concentration (6–9 A.M.)	Same as primary
Nitrogen dioxide	.05-ppm annual mean level	Same as primary

In addition, the EPA is authorized to set *emission* standards for all *new* stationary sources of pollutants. A stationary source is one that is fixed in a single location (such as electric power generation plants and cement plants), in contrast to a mobile source (such as an automobile). A new source is one for which the construction or modification is begun after the publication of these standards. The states are to provide plans for implementing and enforcing these standards, and the EPA may *delegate* enforcement authority to the states (although EPA maintains concurrent enforcement authority). The EPA is also to establish stationary source *emission* standards for hazardous air pollutants that are not covered by national air quality standards but that contribute to mortality or serious illness.

Title II of the 1970 Clean Air Act, which part is also known as the National Emission Standards Act, addresses itself to emission standards for moble sources. It requires all light-duty motor vehicles and engines (such as an automobile) to achieve at least a 90% reduction from 1970 emission levels of carbon monoxide and hydrocarbons by the 1975 model year and a 90% minimum reduction from 1971 vehicle emissions of nitrogen oxides by the 1976 model year. A 1-year extension of these restrictions is possible under certain conditions. To ensure that new vehicles conform to these regulations, the EPA tests prototypes of new vehicles as well as vehicles from the assembly line and issues certificates of conformity. In addition, manufacturers must warrant that vehicles are designed, built, and equipped to conform with emission standards and are free from defects that would lead to nonconformity during the vehicle's useful life (5 years or 50,000 miles).

REFERENCES

1. GEORGII, HANS-WALTER, "The Effects of Air Pollution on Urban Climates," *Bulletin W.H.D.*, No. 40, 1969, 624.
2. *Encyclopaedia Britannica*. Chicago: William Benton, publisher, 1971.
3. *Air Pollution*, Vols. I, II, and III, 2nd ed. (A.C. Stern, ed.). New York: Academic Press, Inc., 1968.
4. *Air Conservation*. New York: Report of the Air Conservation Commission of the American Association for the Advancement of Science, 1965.
5. ROBINSON, E., and R.C. ROBBINS, "Sources, Abundance and Fate of Gaseous Atmospheric Pollutants," *SRI Project PR-6755*, Feb. 1968.
6. *Chemical Engineering Handbook*, 4th ed. (J.H. Perry, ed.). New York: McGraw-Hill, 1963.
7. LOVELOCK, J. E., *Air Pollution and Climatic Change, Atmospheric Environment*, Vol. 5, p. 403–411, Pergamon Press, 1971.
8. "Air Quality Criteria for Particulates," *HEW Report AP-49*, U.S. Department of HEW, Washington, D.C., 1969.
9. "Air Quality Criteria for Carbon Monoxide," *HEW Report AP-62*, U.S. Department of HEW, Washington, D.C., 1970.
10. "Air Quality Criteria for Hydrocarbons," *HEW Report AP-64*, U.S. Department of HEW, Washington, D.C., 1970.
11. BARTOK, W., et al., "Systems Study of Nitrogen Oxide Control Methods for Stationary Sources," *Report PH-22-68-55*. National Air Pollution Control Administration, NTIS No. GR-1-NOS-69. Washington, DC, 1969.
12. CASSELL, E. J., "Law and Contemporary Problems", *33*, No. 2, 197, Durham, No. Carolina: School of Law, Duke University, 1968.

EXERCISES

1. Explain how a temperature inversion greatly increases the accumulation of pollutants at the earth's surface.
2. Explain the difference between a radiation inversion and a subsistence inversion.
3. Explain the difference between ambient air standards and emission standards.
4. How could increased carbon dioxide concentrations in the ambient air melt the polar ice caps?
5. Explain how pollution may have contributed to decreasing world temperatures over the last 20 years.
6. What are the two largest sources of nitric oxide emissions?

7. Why would hydrocarbon emission controls be used to help control photochemical smog?
8. We know that nitrogen dioxide causes the brownish haze associated with photochemical smog. Is nitrogen dioxide therefore a primary pollutant?
9. Since we use sulfur limits on combustion fuels to decrease sulfur dioxide emissions, why don't we use nitrogen limits on fuels to decrease nitric oxide emissions?
10. Differentiate between epidemiologic and toxicologic studies of the effects of air pollution on plants and animals. Why are each necessary, and what limitations are present in the use of each in specifying air quality criteria?
11. What is meant by a synergistic effect of two or more pollutants?
12. Describe the relative importance of population growth compared to other factors causing the rapid increase in air pollution in the United States. Comment on the relative magnitude of air pollution problems you would expect in India (having over twice the U.S. population) and in the United States.
13. Summarize the advantages and disadvantages of using effluent fees to limit the emissions of air pollutants.

6

engineering materials

J. E. Ritter, Jr.

6.1
MATERIALS IN
ENGINEERING DESIGN

Today, more than ever before, the engineer is faced with an unprecedented number of problems. He must design devices and structures that function over a vast spectrum of environmental conditions. These vary from the low pressures found in outer space to the very high pressures existing in the ocean depths and include temperatures ranging from below that of liquid helium ($-270°C$) to those encountered in nuclear reactors and rocket engines (up to $1,650°C$). It is part of the engineer's responsibility to select materials from which these structures and devices will be fabricated and to specify changes when materials have failed in their intended function. Also, there are many current technological problems that do not have the exotic image of the environmental extremes but nevertheless require new materials and new solutions. The technical solution to such problems as low-cost housing, mass transportation, and human implants will undoubtedly require new concepts and new materials. In addition, there are few industries today where materials are not the key to meeting increasingly severe service conditions, improving quality, and lowering costs.

The ultimate goal of engineering design is the fabrication and operation of devices or systems that will perform desired functions. Since performance, cost, and life depend on the characteristics of the materials from which the device or system is fabricated, selecting the requisite material becomes a significant aspect in the design process. If, for any given application, some material could be found that possessed all the right properties for that application, the consideration of materials could be postponed until the final stage in the design process and would simply involve identifying the material that possessed all the properties needed to meet the design specifications. However, this ideal situation does not yet exist, and designers do not have an unlimited choice of property combinations. Consequently, the choice of materials cannot be left until the end but must occur in at least a tentative way as the design proceeds in order for later steps based on intermediate calculations and decisions to be realistic. Before a final choice of materials can be made, trade-offs and modifications in materials requirements and/or in the design are generally required.

The design of an engineering component or structure must fulfill two main requirements: functionality and fabricability. In designing for functionality the selected material must be able to support the loads that will be applied in service and to operate under the service conditions of temperature, corrosive environments, etc. In general this aspect of design involves the "conventional" engineering properties of strength, stiffness, ductility, etc.

The second aspect of a design is just as important and follows logically from the first: One should be able to fabricate the design easily and economically from the material that has been selected on the basis of its functional properties. This involves the designer in a consideration of a very much wider range of qualities, such as the ability of the material to be machined, shaped, and joined. Very few of these qualities can be expressed in absolute terms, and the designer is forced into making some sort of a comparative assessment, often with experience as his only guide. We shall return to the selection of materials in more detail after discussing the testing and examination of materials.

6.2 MECHANICAL TESTING

The fabrication and use of materials depends in large measure on such mechanical properties as strength, hardness, and ductility. Numerical data describing these properties are obtained from standard types of tensile, hardness, bend, impact, creep, and fatigue tests discussed in this section. Although the results of these specialized tests are empirical in nature, they can still be most useful to the engineer if interpreted in a meaningful way and are, therefore, widely used for design purposes.

6.2.1 Tension Test

The tension test is the most common and useful means of evaluating mechanical properties. In the tension test (Figure 6.1) a specimen is subjected to increasing elongation until it fractures. The tensile load and elongation are measured at frequent intervals and the results are expressed in terms of stress and strain, which are independent of specimen size.

The *normal stress*, σ, is defined as the ratio of the load on the sample, P, to the original cross-sectional area, A_0:

$$\sigma = \frac{P}{A_0} \text{ lb/in.}^2 \text{ (psi)} \tag{6.1}$$

The *average linear strain*, ϵ, is defined as the ratio of the change in length of the sample, ΔL, to its original length, L_0:

$$\epsilon = \frac{\Delta L}{L_0} \text{ in./in.} \tag{6.2}$$

Typical tensile stress-strain curves are shown in Figure 6.2.

In the early stages of the tensile test the sample extends elastically; that is, the sample will return to its original length if the load is released. In the elastic region stress is proportional to strain and is described by Hooke's law:

$$\sigma = E\epsilon \text{ psi} \tag{6.3}$$

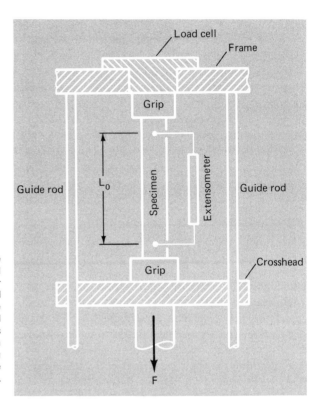

Figure 6.1
Schematic of a universal testing machine. The specimen is held by grips and a load is applied to the movable crosshead by a mechanical or hydraulic drive system. This load is transmitted to the specimen and causes elongation. The magnitude of the load is measured by the load cell, and the elongation of the specimen is measured with an extensometer. The specimen is in tension for this arrangement. Reversing the direction of the crosshead subjects the specimen to compressive stress.

where E is the *modulus of elasticity* and is a measure of the stiffness of a material. In engineering practice, the elastic modulus of a material is primarily of interest as a measure of the amount of deflection that will occur in a structure under a given load; the lower the modulus, the greater the elastic deflection will be. On comparing steel alloys with $E = 30 \times 10^6$ psi with aluminum alloys with $E = 10 \times 10^{10}$ psi, the steel alloys will deform about one-third as much as aluminum alloys for the same stress. Elastic modulus (or stiffness) is very important in certain designs where the deformation must be kept to an absolute minimum.

When the tensile specimen is stretched beyond the elastic region, it cannot return to its original shape upon unloading and the sample has now deformed *plastically*. For engineering purposes the limit of usable elastic behavior is described by the *yield strength*. The yield strength is defined as the stress that will produce a small amount of permanent deformation, generally a strain equal to .2% of the original length of the specimen. For low-carbon steels the *yield point* is used as the practical definition of the elastic limit.

As plastic deformation increases, the specimen becomes stronger (work hardening) so that the load required to extend the sample increases with further straining. The shape of the curve in the plastic region depends on the rate of work hardening. A low rate of work hardening signifies that only a small increase in stress beyond the elastic region is required to continue deformation. Eventually the stress reaches a maximum value, termed the *tensile*

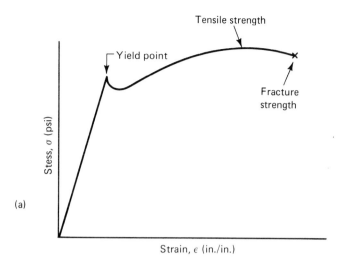

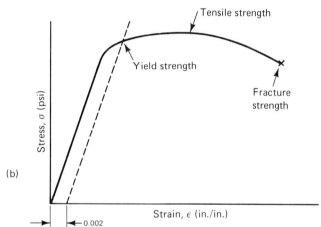

Figure 6.2
Typical tensile stress-strain curves.
(a) Low-carbon steel. (b) Nonferrous metal.

strength. At this point, the specimen develops a neck, i.e., a local decrease in cross-sectional area at which further deformation is concentrated (see Figure 6.3). After necking has begun, the stress decreases with further strain until the sample fractures. In materials that fracture without necking, the tensile strength and *fracture strength* are the same, but when necking occurs, the stress at fracture is always lower than the maximum stress.

After fracture, the overall elongation and reduction in area of the tensile sample can be measured. Both percentage *elongation* and percentage *reduction area* are used as a measure of ductility, i.e., the property of a material that represents its ability to deform. If L_f represents the gage length after fracture, then the percentage *elongation* is

$$\% \text{ El} = \frac{L_f - L_0}{L_0} 100\% \tag{6.4}$$

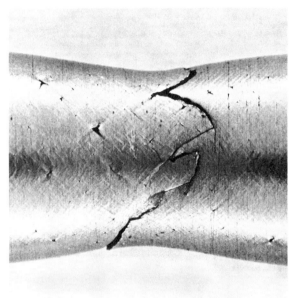

Figure 6.3
Tensile test specimen of a medium carbon steel illustrating necking and fracture.
(*Courtesy D. J. Wulpi, International Harvester Co.*)

The percentage *reduction* in area is given by

$$\% \text{ R.A.} = \frac{A_0 - A_f}{A_0} 100\% \tag{6.5}$$

where A_0 is the original cross-sectional area of the specimen and A_f is the cross-sectional area of the specimen at the fractured cross section. Figure 6.4 shows two bolts deliberately pulled to fracture in tension to illustrate brittle and ductile behavior. The brittle bolt failed with no apparent plastic flow, while the ductile bolt showed extensive plastic deformation.

The area under the stress-strain curve has the units of in.-lb/in.^{3}, which is the energy required to deform 1 in.^{3} of material. Thus, the total area under the curve is a measure of how much energy the material has absorbed before fracture and is referred to as *toughness*.

The results of a tension test are extremely useful to an engineer. In many structures only elastic deformation can be tolerated; therefore, a knowledge of the yield strength or yield point determines the maximum load that can be safely employed. Ductility is another important property of materials since it represents an insurance factor against excessive loads that may not have been considered in a design. In other words, a material with a high ductility will allow considerable deformation to occur before fracture takes place. This allowance is important in situations where unforeseen loads, exceeding the yield loads, are encountered. Finally, many techniques of fabrication, such as rolling, wire drawing, and forging, depend on the ability of a metal to withstand appreciable plastic deformation prior to fracture. Here it is important to know the ductility and rate of work hardening of the material being fabricated, as well as the toughness.

Figure 6.4
These two steel bolts, intentionally pulled to failure in tension, demonstrate brittle and ductile behavior. The brittle bolt, left, was hard (Brinell hardness of 600); the ductile bolt was soft (Brinell hardness of 205).
(*Courtesy D. J. Wulpi, International Harvester Co.*)

Laboratory Project: Measure and compare the tensile properties of a variety of materials such as mild steel, copper, aluminum, brass, nylon, and polyethylene.

6.2.2 Compression Test

Many brittle materials such as cast iron, concrete, stone, building brick, and wood are used in constructions subjected primarily to compression. Also, soft metals and plastics are used in such compressive stress applications as bearing linings, washers, and gaskets. The compression test, therefore, is directly applicable to design problems involving these materials.

The compression test is similar to the tensile test except that the load is applied in compression rather than tension. The stress-strain properties for compression can be defined for brittle materials as described above for tension. For ductile materials, however, it is not possible to determine the ultimate and fracture stresses since in compression the cross-sectional area increases with increasing load. This increase in cross-sectional area makes it difficult to fracture the specimen, and continued application of load increases the lateral deformation until a flat disc is produced (see Figure 6.5).

The elastic modulus and yield strength for ductile metals are approximately equal in tension and compression. For brittle materials, where the ultimate strength in compression can be determined, the strength in compression is much greater than in tension—hence, their wide use in compressive structural applications. For example, in concrete the ultimate strength in compression is more than 10 times the ultimate strength in tension (see Table 6.1).

The determination of accurate stress-strain curves in compression is considerably more difficult than in tension for two primary reasons. First,

Table 6.1
Comparative Tensile and Compressive Strengths of Brittle Materials

Material	Tensile Strength (psi)	Compressive Strength (psi)
Acrylic plastic (Plexiglass)	10,600	17,000*
Gray cast iron	24,000	120,000
Alumina ceramics	30,000	300,000
Concrete (28-day)	400	5,000
Window glass	10,000	500,000

*Stress at which excessive deformation or rupture occurs.

for longer specimens, the danger of failure by lateral buckling increases. Second, friction between the specimen ends and the compression plates hinders the expansion of the ends. This condition is known as end restraint. Its effect is to hold the ends of a compressive specimen near their original diameter while the center portion expands, resulting in a barrel-shaped specimen as shown in Figure 6.5. Because of these difficulties, compression specimens with specific ratios of lateral dimensions to length are selected so that on the one hand they are not too long, thereby avoiding buckling, or on the other hand too short so that the end restraints do not influence the stress-strain relations.

Large lateral deformations, hence barreling, are not produced in a brittle material; instead, failure results by shear and sliding along an inclined plane, as shown in Figure 6.5. The shear fracture plane is usually inclined at

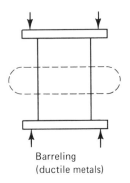 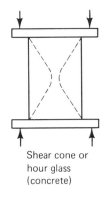 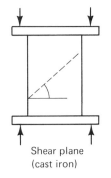 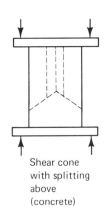

Figure 6.5
Failure of ductile and brittle materials in compression.

Barreling (ductile metals)

Shear cone or hour glass (concrete)

Shear plane (cast iron)

Shear cone with splitting above (concrete)

an angle of about 45° and for cast iron will tend to run from corner to corner of the specimen. The hour-glass fracture common in concrete is evidence of the end restraint effect, which strengthens the material in a cone-shaped region at each end and leaves a weakness around the edge of the specimen, from which the shear cracks spread, meeting in the center. Sometimes instead of the hour-glass fracture, splitting above the bottom shear cone will occur.

Laboratory Project: Test samples of brass, mild steel, cast iron, and cement in compression. The test specimens should be cylinders with length to diameter ratios of about 2 or 3 to 1. Calculate the yield strength for the ductile materials and the fracture strength for the brittle materials and sketch the mode of failure where appropriate. If possible, compare these compressive strength properties to those determined in tension for the same materials.

6.2.3 Hardness Test

Hardness measurements are easy to make and are used extensively for a variety of purposes. However, the hardness of a material is still an ambiguous term that has many meanings depending on the background of the person involved. To a person concerned with the strength of materials, hardness is most likely to mean the resistance to deformation, and to the design engineer it often means an easily measured and specified quantity that indicates something about the strength and quality of the material. In general, hardness usually implies a resistance to deformation, and for many materials the property is a measure of their resistance to abrasive wear. Figure 6.6 depicts the abrasive wear of a backhoe tooth. At the left is a new backhoe tooth, while center and right views show opposite sides of a worn tooth used in rocky frozen soil. In this type of service, the surfaces of the tooth can be deeply cut or gorged by rocks during the digging operation. Being of soft, low-carbon steel (AISI 1010, .1% carbon), the retaining shoe of the tooth (center) has worn considerably more than the opposite side of the same tooth at right that is made of a hardened, medium carbon alloy steel (AISI 8640, .4% carbon).

Various methods have been designed to measure the hardness of materials ranging from simple scratch tests to more elaborate and reproducible

Figure 6.6
At the left is a new backhoe tooth; the other views show both sides of a tooth worn by long operation on rocky, frozen soil. Shown in the center, the top of the tooth (AISI 1010 steel) wore considerably more than the opposite side (AISI 8640 a medium carbon alloy steel, hardened).
(*Courtesy D. J. Wulpi, International Harvester Co.*)

laboratory methods. The most common hardness tests used are the Brinell, Rockwell, Vickers, and Knopp. All these tests are similar in that a standard load is applied to an indentor or penetrator resting on the surface of the material being tested. Hardness is calculated by relating the load applied to the geometry of the indentation. For instance, in the *Brinell test* (Figure 6.7), a steel ball 10 mm in diameter is used with a load of 500 or 3,000 kg, the latter being used for harder materials. The Brinell hardness number (BHN) is defined as

$$\mathrm{BHN} = \frac{P}{\text{surface area of indentation}}$$

$$= \frac{P}{(\pi D/2)(D - D^2 - d^2)} \text{ kg/mm}^2 \qquad (6.6)$$

where D is the diameter of the ball and d is the diameter of the impression expressed in millimeters. A point of practical interest is the relationship between the BHN and tensile strength of a material. Empirically, it is found that for many metals the tensile strength in psi is approximately 510 times the BHN.

It should be kept in mind that all hardness measurements are made on or close to the surface as contrasted with other measurements of mechanical strength that are made on the bulk of the material. For this reason, the surface condition of the specimen is most important and should be free from any dirt or oxide scale. Also, to increase the accuracy of the hardness measurement, indentations should not be made near the edge of a sample or any closer than 3 times the diameter of a impression to an existing impression. Further, the thickness of the test specimen should be at least 10 times the depth of the impression.

Laboratory Project: Measure the BHN of various materials such as low- and high-carbon steel, copper, aluminum, and brass. Determine the relationship between the BHN and tensile strength for these materials.

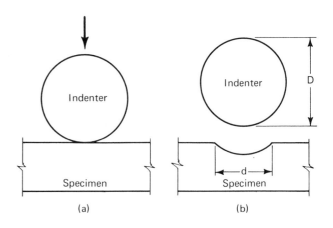

Figure 6.7
The Brinell hardness test. (a) A loaded indenter with diameter D is forced into a specimen. (b) The BHN depends on the size of the indentation, d, that remains when the indenter is removed.

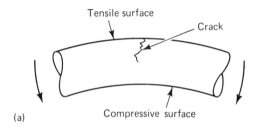

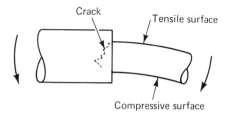

Figure 6.8
Bending fractures usually develop on the tensile surface (a). Sharp fillets or other geometric discontinuities concentrate the stress, causing cracks to develop more rapidly (b). Arrows indicate bending direction.

6.2.4 Bend Test

Bending is the term applied to deformation in which a change of curvature is produced in relatively long thin members. Most structural members in engineering are subject to some bending. Bending results in tensile stresses being set up on one side of the member and compressive stresses on the other.* Figure 6.8 shows how cracks are oriented in cylindrical and filleted shafts under bending stresses. Cracks always initiate on the tension side of the beam where they can then propagate to cause failure. Also, because the stress is highest at stress raisers such as sharp fillets or tool marks, the crack usually initiates at these points. Figure 6.9 shows a cylindrical shaft that failed in bending.

The bend test is widely used because it is probably the easiest stress system to apply. For example, brittle materials, such as rigid plastics, ceramics, and glasses, are generally tested in bending since the bend test eliminates the gripping problems inherent in the tension test and the end restraint difficulties inherent in the compression test. The actual test involves applying a load to a specimen in either three- or four-point bending (see Figure 6.10). The bending moments shown in Figure 6.10 can be thought of as a measure of the tendency of the applied loads to produce bending in the beam. Bend tests

---

*It should be noted that a *neutral plane* exists in a beam subjected to pure bending, the bending stresses along this plane being zero. The neutral plane is usually located along the centroid of the beam. Compressive stresses exist in those parts of the beam above the neutral plane, and tensile stresses in those parts below it.

Figure 6.9
This cylindrical shaft (1.6 inches in diameter) of 1046 steel was alternately loaded as a cantilever beam and failed in bending. The fracture's symmetry indicates that the stresses and the number of load applications were about the same on each side.
(*Courtesy D. J. Wulpi, International Harvester Co.*)

Figure 6.10
Schematic of bend test loading. In three-point loading a load (P) is applied in the center of a beam supported at both ends. In four-point bending, the load is applied at two points equal distances from the end supports. The variation in the bending moment along the length of the beam is shown for each case.

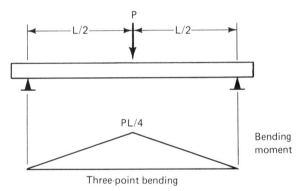

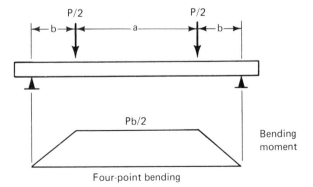

using four-point bending are generally preferred because of the constant bending generated between the inner loading points.

The results of the bend test for brittle materials are usually expressed in terms of the *fracture strength* or *modulus of rupture*. The fracture strength can be calculated from the formula[4]

$$\sigma_f = \frac{Mc}{I} \text{ psi} \tag{6.7}$$

where M is the maximum bending moment, c is the distance from the neutral axis of the specimen to the outer surface, and I is the moment of inertia. For round specimens of diameter D, $c = D/2$ and $I = \pi D^4/64$, the fracture strength in four-point bending is given by

$$\sigma_f = \frac{16Pb}{\pi D^3} \text{ psi} \tag{6.8}$$

where P is the load and b is the distance shown in Figure 6.10.

Laboratory Project: Measure the dependence of fracture strength of glass rods on their surface condition (as-received, abraded with emory paper, and acid-polished in a solution of 15% hydrofluoric and 15% sulfuric acid). Also, the dependence of strength on humidity can be demonstrated by testing the glass samples dry versus wet. Since there is generally large variation in these test results, a number of samples will have to be tested at each condition and an average of the results computed.

6.2.5 Impact Test

During World War II a great deal of attention was directed to the brittle failure of welded Liberty ships in which some ships broke completely in two with no prior warning. These tragedies focused attention on the fact that normally ductile mild steel can become brittle under certain conditions. As an example, the cluster gear shown in Figure 6.11 has two broken teeth. Tooth A

Figure 6.11
In this gear, tooth A broke in service and tooth B was fractured by a single blow in the laboratory. The similarity in fracture surfaces indicates that tooth A was also broken by a sharply applied load.
(*Courtesy D. J. Wulpi, International Harvester Co.*)

failed in service, while tooth B was fractured in the laboratory in which a 100-lb weight was dropped on the tooth from several feet. Both fracture surfaces indicate brittle failure and this similarity in appearances reveals that an impact load probably caused tooth A to break.

Three basic factors contribute to brittle or catastrophic failure in ductile materials (notably carbon steels and plastics):

1. Stress raisers in the material such as cracks, notches, or weld defects
2. Low temperature
3. Suddenly applied (impact) load

Various types of tests have been developed to measure the tendency of a material to behave in a brittle fashion. The *notched-bar impact* test is the most commonly used one. This test measures the energy necessary to fracture a standard notched bar with an impact load and as such gives a measure of the *notch toughness* of a material under shock loading. The test is performed by placing a specimen across parallel jaws of the impact machine and then releasing the pendulum from a known height (Figure 6.12). The pendulum strikes the specimen, fractures it, and moves upward. From knowledge of the mass of the pendulum and the difference between the initial and final heights, the energy absorbed in fracture can be calculated. The "tougher" the material, the greater this energy absorption will be.

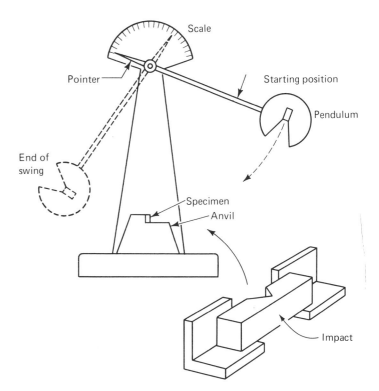

Figure 6.12
Schematic of a standard notched-bar impact testing apparatus.

Figure 6.13
Combinations of fracture modes are shown by fracture surfaces of three notched-bar impact test specimens that were broken at different temperatures. On the left, fracture is mostly ductile (150°F); in the center, combined ductile and brittle (50°F); and on the right, brittle (−20°F).
(*Courtesy D. J. Wulpi, International Harvester Co.*)

The notched-bar impact test is often used to assess the transition temperature from the ductile to brittle states that occurs in certain materials as the temperature is lowered. Figure 6.13 shows three samples of the same material (a carbon steel) as it reacted in notched-bar impact tests at different temperatures. From left to right, fracture modes are mostly ductile, mixed ductile and brittle, and brittle alone. These fractures correspond to high, intermediate, and low temperatures, respectively. In ductile fractures extensive plastic deformation and hence energy absorption occurs before failure. Figure 6.14 gives some typical notched-bar impact test results for a 4340, alloy steel containing .4% carbon, as a function of temperature. Unfor-

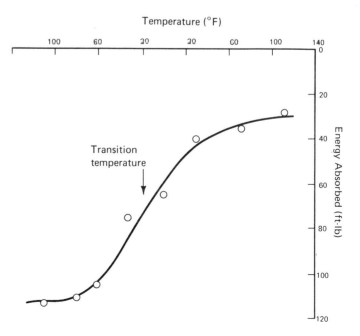

Figure 6.14
Typical notched-bar impact test results for a 4340 steel. The transition temperature is defined as the temperature at which the impact energy is midway between its values at high and low temperature.

tunately, the results of such tests are exceedingly difficult to interpret in terms of design requirements, and the correlation of service performance with test results is at best only qualitative. The main use of the notched-bar impact test is in the determination of the influence of metallurgical treatments on the transition temperature. This usefulness lies in the fact that notched-bar impact values vary considerably with small changes in metallurgical processes such as alloying and heat treatment.

Laboratory Project: Measure the impact values of mild steel as a function of heat treatment. Also, measure the impact values of mild steel and copper over a range of temperatures by placing the specimens in either some cool liquid (alcohol and dry ice) or a furnace just prior to testing. The test should be completed within 5 sec after removing the specimens from the coolant or furnace.

6.2.6 Fatigue Test

Fatigue fracture is a cumulative result of a large number (often many millions) of stress applications, none of which reaches the ultimate tensile strength and, usually, not even the yield strength. The number of fatigue failures in engineering practice is very large, and they probably account for more than 90% of all the mechanical fractures known. Breakage of suspension and valve springs in automobiles, propellers in ships and aircraft, steel cables in cranes and elevators, axles of railway rolling stock, and even huge rotors in steam turbines are just a few examples of fatigue failure. Figure 6.15 shows the fatigue fracture surface of a broken shaft. Fracture slowly progressed from the key way (upper right) through nearly 50% of the cross section before final rapid rupture (bottom left).

Figure 6.15
Fracture surface of a fatigue failure that started at a sharp corner of a keyway in a steel shaft.
(*Courtesy D. J. Wulpi, International Harvester Co.*)

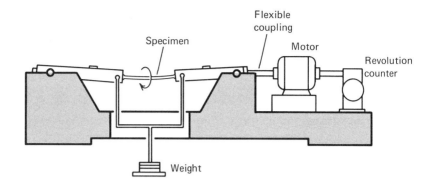

Figure 6.16
Schematic of a R. R. Moore rotating-beam fatigue-testing apparatus. The specimen is supported and loaded through ball bearings in the shafts so that it can rotate freely about its axis. The deflection of the specimen is exaggerated for purpose of illustration.

The fatigue test determines the stresses that a standard specimen can safely endure for a given number of cycles. One of the most popular types of testing machines is shown in Figure 6.16, where the sample is loaded in pure bending. As the sample is rotated, all points on the circumference pass from a state of compression to one of tension. Each revolution thus constitutes a complete cycle of stress reversal, which is repeated several thousand times per minute. Specimens are tested to failure using different stresses, and the number of cycles to failure is recorded for each stress. The results are plotted as stress versus the logarithm of the number of cycles to failure. Figure 6.17 shows some typical results for ferrous and nonferrous metals. Commonly, ferrous metals show distinct *endurance limits*, while nonferrous metals do not. The endurance limit is the maximum cyclic stress that can be sustained by a

Figure 6.17
Typical fatigue curves for ferrous and nonferrous metals.

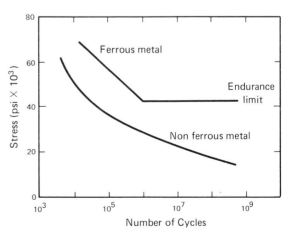

material indefinitely without failure. The *fatigue strength* of a material is defined as the stress that just causes failure after a certain number of stress applications has occurred.

Fatigue failures in metals begin at points of stress concentration such as abrupt changes of cross section, voids, nonmetallic inclusions, or small cracks or notches caused by poor machining or careless handling. The initial fatigue crack grows under the action of the cyclic stress until the remaining cross-sectional area can no longer support the applied stress and the sample fails. In Figure 6.15 the progress of the fatigue crack is indicated by a series of rings or bench marks away from the keyway and the rough-appearing region is where final rupture took place. Since surface imperfections account for many instances of stress concentration, surface finish is of great importance to parts subjected to cyclic stresses. Connecting rods for high-speed engines are always highly polished to prolong the service life, while the design engineer avoids geometric discontinuities such as sharp corners or radii since these are common points of stress concentration in structural components.

Laboratory Project: Measure the dependence of fatigue strength on surface finish by determining the number of cycles to failure at a given applied stress on mild steel specimens of various surface finishes. Because a large variation is generally found in test results on fatigue, a number of tests should be made for a given condition and an average tabulated.

6.2.7 Creep

In many applications materials are required to sustain steady loads for long periods of time, e.g., the components of the internal combustion and jet engine, gas turbine parts, steam pipes, and components of nuclear reactors. Under such conditions, the material may slowly deform until its usefulness is impaired. Although such time-dependent deformation, commonly termed *creep*, may be negligible during the first few minutes in service, over the lifetime of a structure it can become excessive and even result in final fracture without any increase in load. Figure 6.18 shows a high-temperature creep failure of a steel steam pipe. A local hot spot developed that caused the tube to deform or "thin out" until it eventually ruptured in that region.

Figure 6.18
High-temperature creep failure of a steel steam pipe. Note the thinning of the tube in the rupture region, which was caused by the development of a localized hot spot. (*Courtesy R. Huebner, Owens-Illinois, Inc.*)

A *creep curve* is a plot of the elongation or strain of a tensile specimen versus time for a given applied stress and at a given temperature. The typical curve in Figure 6.19 is usually considered in four parts: (1) initial elastic and plastic elongation following load application, (2) primary creep, (3) steady-state or secondary creep, and (4) tertiary creep. In primary creep the rate of work hardening is greater than the rate of softening, and thus the creep rate continually decreases. During steady-state creep the rates of work hardening and softening are about equal, and the creep rate is nearly constant and is a minimum. In tertiary creep the sample may begin to neck, and the creep rate increases with time until failure. The steady-state or minimum creep rate is the most critical since it continues over long periods of time to high strain values.

The essential information obtained from creep testing is *creep strength*. This is generally reported in one of two ways:

1. Stress that produces a given strain after some predetermined time at a specified temperature, e.g., the stress that produces .2% creep in 10,000 hr at 1500°F
2. Stress that produces a given minimum creep rate at a specified temperature

These data may be reported in tabular or graphical form.

In designing against creep failure, we normally have two requirements:

1. If the service life can be assigned, the total deformation must not exceed a certain limiting value within that time.
2. Within the assigned life and limiting deformation, fracture must not occur.

Because of the long times involved in getting meaningful creep data, we are often forced to extrapolate existing data in designing for a creep-resistant

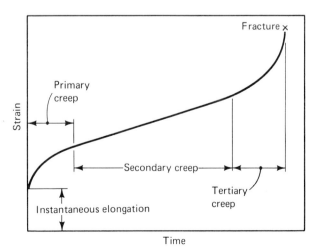

Figure 6.19
Typical creep curve showing the four stages of elongation for a long-time, high-temperature creep test.

material. When this is done, we must be extremely cautious in extrapolating existing data to longer times or higher temperatures since it is very easy to come to an erroneous conclusion from this practice. A further design difficulty is that nearly all creep data have been determined for simple tension. Many applications have stresses in shear, bending, compression, and combinations of these. Thus, the use of available data often requires extensive ingenuity on the part of the engineer.

Laboratory Project: Take short lengths of lead and lead-tin solder, suspend them from a rigid support, and hang various loads on them. Record the deformation versus time and plot creep strain against time for a given applied stress. Is there a relationship between the minimum creep rate and the applied stress?

6.3 EXAMINATION OF MATERIALS

A person who is curious about how a thing is made invariably tries to take it apart or to look inside it. Taking a material apart in a way that would permit you to see how it is put together is not a simple problem; however, a number of experimental techniques do exist that, in effect, will allow you to look inside. These examination techniques are most important for they enable you to relate the structural characteristics of a material to its physical or mechanical properties. In this section we shall discuss the three methods most widely used in examining materials.

6.3.1 Optical Microscope

Of the devices that have been used in studying the structure of materials, the optical microscope undoubtedly has yielded the most information. Since most of the materials that we are interested in are not transparent to light, we shall confine our discussion to the reflected-light microscope, shown in Figure 6.20. This microscope can operate over a range of magnifications from about 20 to $2,000\times$.

Most engineering materials are comprised of many small crystals or grains.* To observe the grain structure (or microstructure), the specimen surface must first receive some special preparation. The first step in preparing a specimen for observation in a microscope consists of grinding and polishing the surface with finer and finer abrasives until it has a mirror finish. In this condition all regions of the surface reflect incident light back into the lens system, and no microstructure is visible. To provide the necessary detail, the specimen is subjected to chemical attack by an etching reagent. The reagent attacks the surface beginning at the grain boundaries, i.e., the region where two grains meet. The chemical attack causes the grain boundaries to take the

---

*For an excellent, easy-to-read description of the crystalline structure of materials, the reader is referred to *Materials*, A Scientific American Book, W. H. Freeman and Company Publishers, San Francisco, 1967.

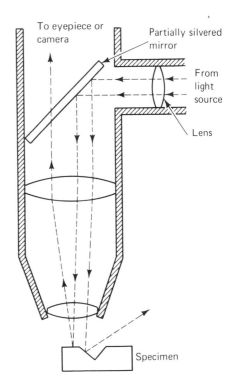

Figure 6.20
Schematic of a reflected-light microscope showing the effect of a grain boundary in the specimen surface. Light reflected by the grain boundary does not reenter the microscope, and the boundary appears as a dark line.

shape of grooves. Since the light reflected by the grooves does not go back to the eyepiece, grain boundaries appear as dark lines. Further etching results in the grain surfaces being attacked. Since some grain surfaces etch more rapidly than others, the result will be that some grains will become oriented so that they reflect less light than others and will appear shaded. The microstructure of copper is shown in Figure 6.21. The realignment of the grains by deformation is clearly apparent.

Laboratory Project: Examine and compare the microstructures of the various materials whose tensile properties were measured. Note that different materials require different etchants. An excellent treatment of metallographic technique is given in G. L. Kehl, *The Principles of Metallographic Laboratory Practice*, McGraw-Hill Book Company, New York, 1949.

6.3.2 Electron Microscope

The transmission electron microscope provides magnifications many times greater than that obtained with the best optical microscopes (up to $75,000 \times$), but it has some disadvantages. It is expensive and usually requires considerable skill and patience on the part of the operator. The principle of the transmission electron microscope is that a beam of electrons (instead of light) is focused on the surface of the specimen by electromagnetic lenses. As the electron beam passes through the specimen, certain regions of the specimen will either absorb or scatter electrons and will appear dark on the screen, thus providing the contrast necessary for producing an image.

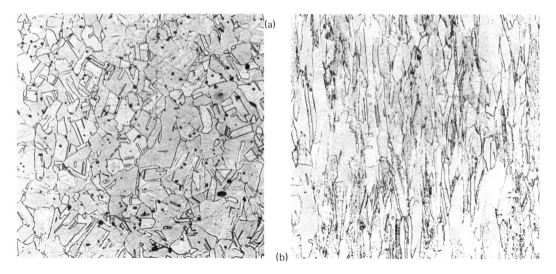

Figure 6.21
Photomicrograph of a high-purity copper sample showing its grain structure. (a) Prior to deformation, grains are equiaxed. (b) After deformation, grains are elongated in the direction of deformation. Magnification: 500X.
(*Courtesy J. Haggerty, Arthur D. Little, Inc.*)

Because the beam of electrons cannot pass through a piece of metal thicker than 10^{-5} cm, nearly all early work was done by the *replica* technique. Replicas of specimen surfaces can be made of a thin layer of a suitable plastic, and the thin replica can then be used as the specimen. This technique gives information about the surface detail only. In recent years transmission pictures of metal and ceramic samples made very thin by a combination of machining and chemical attack have been obtained. From photographs of these thin samples one truly gets a glimpse inside a material. Figure 6.22 shows a transmission electron photograph of a glass-ceramic material such as that used in the familar pyroceram cookware.

Figure 6.22
Transmission electron micrograph of a glass-ceramic material. This picture was taken after the partial transformation of the glassy phase to the ceramic phase (angular particles). Magnification: 19,500X.
(*Courtesy H. McCollister, Owens-Illinois, Inc.*)

The scanning electron microscope is particularly useful for characterizing large, irregular surfaces with high resolution since it permits a wide range of magnification (20 to 40,000×), extremely good depth of field, and a perspective that gives a good feeling for a position in the third dimension. With this microscope the surface of the specimen is scanned by a beam of electrons, and a picture of its surface is built up in much the same way as a television picture is constructed. Additional advantages of this type of microscope are that neither thin specimens nor surface replicas are required. Figure 6.23(a) shows a scanning electron photograph of the grain and void structure in a ceramic material and (b) how it is influenced with heat treatment.

6.3.3 X-Ray Diffraction

All the above techniques for examining materials do not reveal the details of the arrangement of atoms in solids. For this information, you must use X-rays. X-rays are a form of electromagnetic radiation that behave somewhat similar to light rays in that they can be reflected, refracted, diffracted, and transmitted under suitable conditions. Several X-ray diffraction techniques are available such as the rotating crystal, and the powder techniques. None of them yield photographs like the microscope but instead give a pattern of spots or lines from which calculations can be made regarding the position of the atoms and how far apart the atoms are. Of these techniques the powder method is preferred in many instances since it reveals the crystal structure accurately without requiring a single crystal of the material. In the powder

Figure 6.23
Scanning electron micrographs of the fracture surface of a ceramic material. (a) Before heat treatment. (b) After heat treatment for several hours at 2,200°F. Note the significant increase in coarseness of the structure and in void size. Magnification: 5,000X.

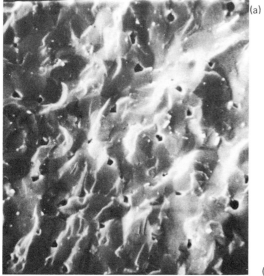

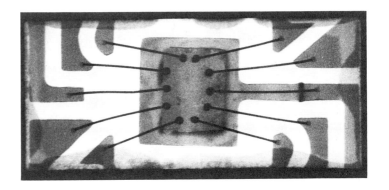

Figure 6.24
X-ray radiograph of an integrated circuit inside a hermetically-sealed alumina package. The lead wires (similar to those shown in Figure 7.29) are connected to metallized conductors on the alumina substrate. Defective areas in the braze joint between the silicon single crystal chip and the substrate are apparent. Magnification: 40X.
(*Courtesy J. Haggerty, Arthur D. Little, Inc.*)

method an X-ray diffraction pattern is obtained by the passage of a fine beam of X-rays through a powder of fine filings of the material.

Another important use of X-rays is in the rapid nondestructive evaluation of finished parts. X-ray examination is carried out by passing X-ray radiation through the part being examined and allowing the radiation to impinge upon sensitive film. Internal defects in the part will appear as dark areas on the film since they absorb less radiation than sound material. Figure 6.24 shows an X-ray radiograph of a defective integrated circuit.

6.4 SELECTION OF MATERIALS

Selecting the most suitable material for a given application is by no means a simple matter, for there are many factors to be considered. There is frequently no single answer since several materials, each with its particular advantages and disadvantages, may be about equally suitable. The engineer must use his best judgment based on his own experience and study and that of other engineers, scientists, and technicians. In general, the material most suitable for a given use will be that material that most nearly supplies the *necessary properties* and *durability* with a *satisfactory appearance* at the *lowest cost*. The factors that enter into the selection of a material are discussed below.

6.4.1 Materials Requirements

The requirements placed on materials are derived from a study of the desired performance of the product or system being developed, as well as the environment in which it will operate. However, to be useful, the materials requirements must be translated into properties that can be measured by an appropriate test. In some cases this is relatively easy, as, for example, in parts where unidirectional static loads are involved. Here, mechanical strength properties of materials can be measured in tension and compression and directly compared against the stresses encountered in application. Unfortunately, for many service conditions there is no direct or single correspondence between the requirement and a measurable materials property. In these cases the evaluation process may be quite complex and depend on predictions based on simulated service tests or on property tests most closely related to the service requirements.

6.4.2 Durability

If a material is to be satisfactory for a given use, it must be durable; that is, it must continue to function properly during the design life of the structure of which it is a part. It should be kept in mind that wear and corrosion probably account for more withdrawals from service than does fracture. Environmental factors, such as humidity or chemicals, can cause deterioration and subsequent failure of materials. Other environmental conditions, such as high or low temperatures, adversely alter the service performance of most materials. To be durable, a material must resist all forms of destruction to such a degree that the system or device will not be rendered unsafe or inefficient at any time during its prescribed life.

6.4.3 Appearance

The well-designed product, whether it be a domestic applicance, structure, machine part, tool, or any other device, should be in harmony with its environment. Both its lines and the material of which it is composed will influence its final appearance. For a machine part that is hidden from view appearance may not be vital, but for any exposed part the material employed should be chosen with due consideration to appearance as well as durability. For example, a domestic appliance must be designed for functionality and ease of mass production, while at the same time retaining a pleasing appearance for sales appeal. Materials that do not present a pleasing appearance may often be painted or metal-plated to improve their appearance as well as to provide a durable covering that will resist corrosion or decay.

6.4.4 Fabricability

The two main considerations in the fabrication of any engineering component are forming and joining. Forming includes casting, drawing and extruding, forging, machining, and sintering. Joining includes both welding, brazing, and soldering, on the one hand, and the use of fasteners such as screws and rivets, on the other. In principle, the fabricability of a material is a measure of its ability to be worked or shaped into a finished and useful form or part. Therefore, the physical embodiment of a design—its configuration, size, weight, finish, and dimensional tolerances—is directly related to the fabricability characteristics of materials. Fabricability is generally characterized in two different ways. One way is in terms of the various materials processing methods, that is, the ease with which a material can be cast or machined or welded (hence, the terms castability, machinability, and weldability). Another way is in terms of certain mechanical or physical properties such as ductility, fluidity, and hardenability. It is important to remember that processing operations will almost always have some effect on a material's functional or service performance. This effect may either improve or reduce performance; therefore, fabricability considerations are closely interrelated with service requirements.

6.4.5 Cost

Common sense advocates, and competition usually forces, the use of the most economical material that will satisfy the considerations of functionality, durability, and appearance. The items included in the initial cost of the product will be dependent on the quality, availability, and workability of the material, but the total cost should include such additional items as the cost of installation, maintenance and repair, the interest on the investment, and, in the event that the product must be used longer than the anticipated life of the material, the cost of replacement. Unless these additional items are considered, a material with a low initial cost may ultimately prove to be very uneconomical when compared with a more durable material of higher initial cost.

6.4.6 Ecology

An economic factor that deserves special attention is the use of materials in an ecological sense. Today, we can no longer afford not to take into account in our design such factors as the conservation of materials resources and the "disposability" of materials. All production operations produce some scrap such as turnings in metal-cutting processes and runners and risers in castings. Any change in process, component design, or stock form that can reduce this scrap not only will help conserve our limited materials resources but also will directly reduce costs. For example, a component produced by machining from an aluminum alloy bar may well be designed to be produced by impact extrusion with a far higher percentage of materials utilization. Also, after the usefulness of a product has ended we must consider how to dispose of the materials economically without adding to the pollution of our environment. For example, since many devices made of glass and thermoplastic materials can be economically recycled or broken up for reuse as raw material, many cities and industrial firms now have recycling centers for these materials.

6.4.7 Engineering Data

Before concluding this section, a word of caution about the use of engineering data. All engineering materials show a certain variability in mechanical properties, and often these properties are determined under conditions different from the proposed service; e.g., strength data are often obtained in a uni-axial test, whereas stresses encountered in service are generally complex. In these cases the engineer must project the properties of a material in service from the known experimental data. Further, uncertainities usually exist regarding the magnitude of the applied service loads, and approximations are necessary in calculating the stresses for all but the most simple member. Thus, to provide a margin of safety and to protect against failure from unpredictable causes, it is necessary that the allowable stresses be smaller than the stresses that produce failure. This can be achieved by dividing the strength of the material by a number called the *factor of safety*.

The value assigned to the factor of safety depends on an estimate of all the factors mentioned above. In addition, careful consideration should be given to the consequences that would result from failure. For example, the factor of safety is increased if loss of life would result from failure. Also, the type of equipment will influence the factor of safety, being lower for equipment where light weight is of prime consideration and where vibrating or fluctuating stresses are encountered. The factor of safety generally ranges from 1.5 to 4.0.

Library Reference Problem: Below are listed four applications and several materials. Rank the materials in order from best to worst for the application, and *explain why*. If you are unable to find the data you require, list the sources you have consulted and explain what additional data are necessary. Make judicious guesses, interpolations, and extrapolations where necessary.

1. Flashlight housing: polyethylene, 1020 steel, Bakelite, cartridge brass, aluminum.
2. Safety dash for automobile: sheet aluminum, polyethylene, rubber, 1020 steel, polystyrene foam.
3. Helicopter rotor blades: epoxy fiberglass composite, cast stainless 304 steel, forged stainless 304 steel, muntz metal, admirality brass.
4. Frying pan: pyrex glass, alumina, cast aluminum, sheet aluminum, cast iron, sheet copper, sheet titanium.

6.5 MATERIALS OF THE FUTURE

Modern technology is placing unprecedented demands on engineers in the development and utilization of materials. It is only necessary to examine the frontiers of such fields as aviation, atomic energy, and electronics to recognize the key role played by materials. Lightweight materials that retain structural strength and toughness well above 1,000°C could more than double the range of jet aircraft and make vertical takeoffs and landings practical for large planes. Applications of nuclear energy call for materials that do not deteriorate under heavy and prolonged radioactive bombardment. Designers of computers and communications equipment are looking for materials that will store and process information in still smaller spaces and will switch signals in smaller fractions of a millionth of a second. With so many needs now obvious, the hunt for new materials and the better utilization of existing materials is now underway on an extensive scale.

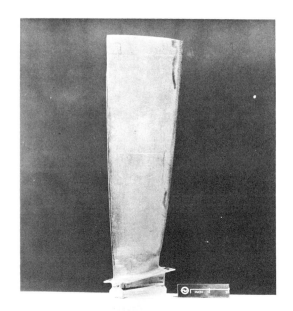

Figure 6.25
Close-up of Borsic^R-aluminum composite fan blade for use in aircraft gas turbine engine. This lightweight composite material is comprised of an aluminum alloy reinforced with boron fibers and has emerged as a higher strength/density and modulus/density ratio replacement for titanium.
(*Courtesy Pratt & Whitney Aircraft*)

6.5.1 Composites

Glass-reinforced plastics have already demonstrated the exceedingly high strength to weight ratio possible with composites and have found wide use in pressure vessels and in such consumer products as furniture and sporting equipment. Although composites reinforced with glass fibers are strong, they are not particularly rigid and many reserarch and development programs are now concerned with reinforcing with high-modulus fibers such as boron, silicon carbide, graphite, and sapphire. On a weight basis, a graphite filament is the strongest and stiffest material yet developed, being 3 times stronger than steel and 10 times stiffer than glass. The most ambitious composite development program is being conducted by the U.S. Air Force, which is evaluating boron-reinforced structures for aircraft wing and tail components, engine parts, helicopter rotor blades, and reentry vehicles (see Fig. 6.25). There is also keen interest in combining refractory metals such as tungsten, titanium and molybdenum and ceramic fibers in the same manner as plastics are reinforced with fiberglass, and many believe that this will be the new superstrength material of the future. Devising an assortment of superlative composite materials that can be manufactured at reasonable costs is one of the immediate challenges for the materials scientist.

6.5.2 Superalloys

Improving the strength of metals, particularly steels, is still a fruitful area of research. Right now we have reliable 200,00-psi yield strength steels (see Fig. 6.26) and before the end of the decade steels with a strength of 500,000 psi will be available. Titanium is now finding wide applicability in the aerospace industry and offers the properties of steel for little more than half the weight of steel, with a bonus by way of good resistance to corrosion. Unfortunately, titanium remains an expensive material because of the difficulties in extracting and forming it. Nevertheless, titanium and its alloys are used in almost every turbojet engine built today, and within a few years they may be the main structural material for the supersonic transports. Dispersion-strengthened superalloys now under development should provide the next major improvement in metals for high-temperature service. Small oxide particles are dispersed through these alloys to inhibit creep in the upper temperature ranges. The most promising alloys under development are based on thoria dispersed in alloys of nickel and cobalt. In jet engines these alloys may allow a 200°F increase in the operating temperature, which would result in about a 12% power gain with a corresponding fuel savings of about 10%.

6.5.3 *Plastics*

After many years of research and development plastics are now available that can replace many conventional materials. It is not surprising that plastics dominate the toy industry and are being widely used in the packaging industry. Polyethylene and polyvinyl chloride in the form of tubes and pipes for domestic and industrial plumbing are increasingly being preferred over copper, steel, and cast iron. Glass-filled nylon is used instead of zinc, magnesium, or aluminum in precision die castings. Glass-reinforced plastics have made the manufacture of small boats a big business and are fast making inroads into the housing market. Almost half of the furniture manufacturers today are

Figure 6.26
This trispherical pressure hull for the deep submergence rescue vessel was fabricated from hemispheres, hot-spun from 3-inch-thick plates of a 5 Ni-Cr-Mo-V alloy steel referred to as HY-140(T) steel. The hemispheres were heat-treated (quenched and tempered) to a minimum yield strength of 140,000 psi.
(*Courtesy U.S. Steel Corp.*)

Figure 6.27
Clear, impact-resistant Lexan^R polycarbonate with a thin layer of gold was chosen as the face shield to protect the Apollo astronauts. Lexan^R is tougher at −75°C than any other clear material at room temperature and is capable of withstanding the extreme high and low temperature of space exploration. In the foreground is a motorcycle helmet made of Lexan^R.
(*Courtesy General Electric Co.*)

using plastic parts and components in place of wood. The success of plastics in these uses in an indication of further developments. Research is now being conducted on the synthesis of new plastics with a unique combination of properties (see Fig. 6.27). It is thought that in the next decade a plastic will be developed with suitable mechanical properties for use up to temperatures of 150°C. This development could completely alter the competitive situation in such huge metal markets as automobiles and construction.

6.5.4 Ceramics

Traditional ceramics such as clay, flint, and quartz are familar in the form of pottery, sanitary ware, bricks, tiles, and windows. Through advances in theoretical understanding of the solid state and in engineering manipulation, an ever-increasing diversity of new ceramics is being developed. The chief advantages of ceramics in these new applications are abundant raw materials, low cost, and excellent high-temperature properties. The developments in missiles and space flight provided the financial support for much of the advanced ceramics research in the 1960s, the principal aerospace applications being reentry heat shields, rocket nozzles, and windows. However, ceramics are finding increasing use as components in such high-temperature applications as the gas turbine engine because of the need for higher operating temperatures to increase efficiency (see Fig. 6.28). In less glamorous fields, the properties of hardness and good wear resistance have led to an increasing number of applications for advanced ceramics such as thread guides in the textile industry, wire drawing dies, metal-cutting tools, and pump seals in the chemical industry. The trend in ceramics research is toward increasing simplicity and more rigorous control of the materials and toward purification of the component elements, simplifications of the internal structure, and

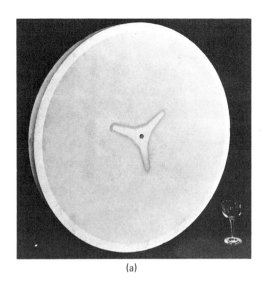
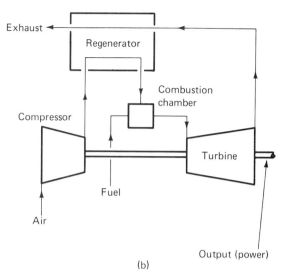

(a) (b)

Figure 6.28
(a) Glass-ceramic heat regenerator, 28-inch OD by 3 inches thick, with the rim and hub fused bonded to the matrix (a honeycomb-type structure). This heat regenerator is used in the gas turbine engine to preheat the incoming air with the heat gained from the exhaust gasses. This has resulted in over a 250°F increase in the operation temperature. (b) Air flow path in a gas turbine engine, illustrating how the regenerators are utilized.
(*Courtesy Owens-Illinois, Inc.*)

refinement of the production techniques. Through new and daring innovations in the future, we should see such unthinkable ceramic developments as glass submarines and ceramic armor.

6.5.5
Electronics

Some of the most sophisticated new concepts in materials are those involving electrical, magnetic, and optical materials. Transistors such as shown in Fig. 6.29 have provided the most dramatic materials breakthrough and have widely replaced the more cumbersome and complicated vacuum tube. The development of the ferrites, ceramics that are magnetic insulators as contrasted to conductive metals, has revolutionized the computer industry, with their greatest application being in magnetic core memory systems. Ferrites are also important in TV and microwave devices and in magnetic tapes for recording equipment. In equipment for lasers and energy converters we have complicated optical and thermomechanical systems being replaced by individual pieces of materials. For example, thermoelements made from high-purity intermetallic alloy can either pump heat or generate power. They produce electricity if exposed to a temperature differential or develop hot and cold junctions if supplied with electric current. The guiding objectives in future research and development in electronic materials will be smaller size, greater efficiency, and lower costs. Fruits of this research are already evident in making TV sets, radios, and electrical equipment smaller and lighter, more reliable, and less costly.

Figure 6.29
Scanning electron photomicrograph of an integrated circuit deposited on a single-crystal silicon chip. Conductors, capacitors, and other active components are developed by multiple vapor depositions of thin films through masks. An integrated circuit of this type is typically bonded to a high thermal conductivity substrate to dissipate heat and connected to a plug-in lead frame by gold or aluminum wires that are welded to pads on the integrated circuit. Magnification: 300X.
(*Courtesy J. Haggerty, Arthur D. Little, Inc.*)

6.5.6 Human Implants

No machine is more complex, no environment more hostile, but no challenge is more satisfying than the quest to prolong human life and usefulness with prostheses that range in complexity from contact lenses and artificial teeth to artificial organs. Simply stated, the problem is to find materials compatible with bone, blood, tissue, and body chemicals that do not upset the body's delicate balance and purity. Ceramics have successfully been used to replace teeth and to anchor high-load orthopedic prostheses to the skeletal system (see Fig. 6.30). Prosthetic noses, cheeks, etc., may be molded from various

Figure 6.30
(a) An alumina ceramic femoral condyle prosthesis positioned next to the natural condyle of a Rhesus monkey. (b) An X-ray of the condyle prosthesis implanted in a Rhesus monkey.
(*Courtesy W. Campbell, Ohio State University.*)

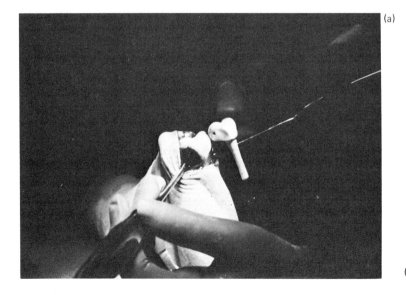
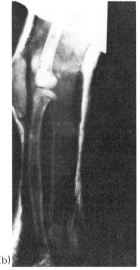

materials, including vinyl, latex, and silicones. Polytetrafluoroethylene has been used to build up jaw bones. Although much research is and will be devoted to developing and improving materials so that they are compatible with the body, dramatic advances in materials will also be made in another area. Almost all synthetic "parts" now being used or in development are usually mirror-image replacements for the real organism. In the future there will be little resemblance, and materials will be developed primarily for their body function. If the function of a kidney is to filter and purify, then the material that will perform this function may not be bigger than a golf ball, with a space-filling envelope that may not resemble a real kidney at all. The next decade holds much future and promise for bioengineering materials.

Library Reference Problems: (1) Write a brief essay concerning an advanced technological frontier in materials. For reference consult such journals as *Scientific American, Machine Design, Metals Progress, Plastics Age, Materials Engineering, Glass Industry, Industrial Research,* and *Production Engineering.* Possible essay titles would include "Application of High-Power Lasers to the Generation of Fusion Power," "Glass-Material for Deep Sea Submersible Vehicles," "Potential of Whisker-Reinforced Metal Composites," "Advantages of the New Glass-Ceramic Materials Over Conventional Glasses and Ceramics," "Future of Fiberglass-Reinforced Plastics," "Feasibility of Recycling Glass Containers," "···Plastic Containers," "···Automobile Tires," and "···Scrap Metal." (2) Write a book review of an introductory materials text. Some highly recommended introductory materials books include *The New Materials* by D. Fishlock; *The New Science of Strong Materials* by J. E. Gordon; *Metals, Atoms, and Alloys* by C. L. McCabe and C. L. Bauer; *Ceramics: Stone Age to Space Age* by L. Mitchell; *The Atomic Structure and Strength of Metals* by N. F. Mott; *Materials,* A Scientific American Book based on the September 1967 issue of *Scientific American; Strong Solids* by A. Kelly; *Metals in Service of Man* by W. D. Alexander and A. Street; *Materials and the Development of Civilization and Service* by C. S. Smith; *Ceramics in the Modern World* by M. Chandler; *Glass in the Modern World* by F. J. T. Maloney; *Crystals and Crystal Growing* by A Holden and P. Singer; *The Nature of Solids* by B. A. Rogers; *Man, Metals, and Modern Magic,* American Society for Metals publication; and *The Story of Metals,* American Society for Metals publication.

6.6 SUMMARY

The selection, fabrication, and use of materials depend in large measure on their mechanical properties. Numerical data describing these properties are obtained from the standard tests that we discussed in this chapter. Since the properties of materials are directly related to their structure, the various means of determining their structural characteristics were also described.

The problem of selecting the requisite materials for a given application is one of very great complexity and inevitably results in some form of compromise between those properties that are required and those that are merely tolerated. In this chapter we attempted to outline the major considerations involved in the selection of materials. We concluded this chapter with a discussion of some frontiers in materials research and development that currently challenge the materials specialist.

REFERENCES

1. FISHLOCK, D., *The New Materials*. London: John Murray, 1967.
2. MURPHY, G., *Properties of Engineering Materials*. Scranton, Pa.: International Textbook Comprny—College Division, 1957.
3. RUSKIN, A. M., *Materials Considerations in Design*. Englewood Cliffs, N.J.: Prentice-Hall, Inc., 1967.
4. SCHLENKER, B. R., *Introduction to Materials Science*. New York and Sydney: John Wiley & Sons, Inc., and Australasia Pty. Ltd., 1969.
5. SMITH, C. O., *The Science of Engineering Materials*. Englewood Cliffs, N.J.: Prentice-Hall, Inc., 1969.

7
fundamentals of electrical networks

R. V. Monopoli

The electrical engineer uses the terms electric current, electric circuit, electromotive force, active element, power, and many others in his daily work. He is not alone in requiring knowledge of these terms but shares this need with engineers and scientists of all types. They, too, must have a basic understanding of the electrical instruments that are being used more frequently than ever before in their professions. The design, manufacture, and application of electrical equipment will provide increasing numbers of engineers the opportunity to exercise their creative ability in helping to solve society's problems. Some of the most important terms and concepts used by electrical engineers will be defined and used in specific examples in order to give the student an understanding of their meaning and the situations in which they are applied. The problems we shall consider are those of electric charges in motion (electric current) within man-made electric networks. Laws governing the behavior of electric charges in this environment will be developed.

7.1
DEFINITIONS, CONCEPTS, AND CIRCUIT LAWS

There are positive and negative electric charges. The unit for electric charge is the coulomb (C). An electron has the smallest known electric charge. It is 1.6×10^{-19} C in magnitude and negative. Alternatively, 6.25×10^{18} electrons together have a charge of -1 C. A proton is a positively charged body with charge equal to $+1.6 \times 10^{-19}$ C. The smallest known electric charge is assigned the symbol e, and $e = 1.6 \times 10^{-19}$ C. Any electric charge may be written in terms of this minimum charge as $N \times e$, where N is an integer.

Electric current is basically the flow of electric charge in a conductor. We deal quantitatively with electric current by defining the *amount* of current as the time rate of change of charge passing through a specified cross-sectional area of a conductor. Let's examine this definition in detail. In Figure 7.1, protons are moving to the right and electrons to the left in the conductor. (Many conductors allow only electrons to move, but we shall consider a general case here.) If the number of protons moving to the right through the cross-sectional area A in time T (say $\frac{1}{2}$ hr) is N_p and the number of electrons moving to the left through the cross-sectional area A is N_e, then the net charge q moving to the right through A in time T is

$$q = (N_p + N_e)e \text{ C} \tag{7.1}$$

If the electrons had moved instead to the right during this period, then (7.1) would become

$$q = (N_p - N_e)e \tag{7.1a}$$

The *current* is the net amount of positive charge that has flowed by a given point, divided by the time T it took to do so:

$$i = \frac{q}{T} = \frac{q\text{C}}{30 \text{ min}} \tag{7.2}$$

By convention the direction of positive current flow is the direction of net *positive* charge. The unit for current is the ampere (A), which is 1C/sec. For the situation above, if $q = 1800$ C, then, since 30 min $= 1800$ sec, we have

$$i = \frac{1,800 \text{ C}}{1,800 \text{ sec}} = 1 \text{ A} \tag{7.3}$$

In (7.2), the current calculated is really an average value over the 30-min period. To develop the concept of an instantaneous current, the standard calculus approach is taken, and T is allowed to approach zero giving the "derivative of q with respect to time" (denoted dq/dt in calculus),

$$i(t) = \lim_{T \to 0} \frac{q(t + T) - q(t)}{T} \tag{7.4}$$

This equation is read as "the limit of q/T as T approaches 0 is dq/dt". If, for example, the charge passing through cross section A is linear with time, say $q(t) = Q_0 t$, where Q_0 is some number of coulombs per second, then

$$i(t) = \frac{dq}{dt} = \frac{dQ_0 t}{dt} = Q_0 \text{ A} \tag{7.5}$$

Figure 7.1
Moving charges in a conductor.

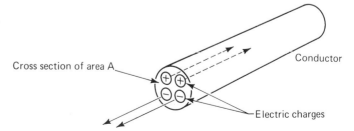

Some terms commonly used in discussing electric circuits are defined next. An *electric circuit* (or *network*) is a closed path composed of active and passive elements to which current flow is confined. An *active circuit element* is one that provides electrical energy to a circuit (a battery, for example). A *passive circuit element* is one that absorbs electrical energy from a circuit and either stores it or dissipates it in the form of heat (a light bulb, for example). The active elements in a circuit produce an *electromotive force*, which is the "pressure" in the circuit that causes a current to flow. The unit of electromotive force (abbreviated emf) is the volt. Alternative terms used for emf are *voltage*, or *potential difference*, or simply *potential*.

The *resistivity* of a material is a measure of its opposition to current flow. *Conductivity* of materials, on the other hand, is a measure of how well they conduct current. Resistivity and conductivity are reciprocal quantities. The symbol for resistivity is ρ, and for conductivity the symbol is σ. They are related by

$$\rho = \frac{1}{\sigma} \tag{7.6}$$

These quantities are properties of the material and do not depend on the size of the piece of material. The concept of *resistance*, denoted by the letter R, does depend on the size of the piece of conductor. It is a measure of the total opposition to current flow offered by a piece of material of a specific resistivity and size. In the case of a conductor of uniform cross-sectional area A and length l, the resistance R is computed from the equation

$$R = \frac{\rho l}{A} \Omega \quad (\Omega \text{ is the symbol for ohm}) \tag{7.7}$$

The unit for ρ in the metric system is the ohm-meter. Therefore, in (7.7) the units for l must be meters and for A, (meters)². For example, a conductor 1 m long and .001 m² in cross section has a resistance $R = 1\,\Omega$ if $\rho = .001\,\Omega$-m. From (7.7) it is seen that the total resistance R is directly proportional to the resistivity and the length and inversely proportional to cross-sectional area. The symbol used for a resistor in circuit diagrams is

The total *conductance*, denoted by G, of a piece of material of a specific size is computed from

$$G = \frac{1}{R} = \frac{\sigma A}{l} \mho \quad (\mho \text{ is the symbol for mho}) \tag{7.8}$$

The unit for σ is mho per meter, where the unit mho is ohm spelled backwards.

The resistance of several different types of conductors of a specified size is listed in Table 7.1, which indicates that silver is the best conductor. However it is too expensive for general use. Therefore, copper, while not quite as good as silver, is most commonly employed as an electrical conductor. It is important to have low resistivity, especially in long conductors (power lines

Table 7.1
Resistance Values for Conductors of Specific Material and Size

Conductor	Resistance (in ohms)
1,000 ft of copper wire of 1/10-in. diameter	1
1,000 ft of iron wire of 1/10-in. diameter	6
1,000 ft of silver wire of 1/10-in. diameter	.925

run for hundreds of miles), the keep the undesirable power losses as small as possible and to deliver as much power as possible to the load.

The concepts defined above may perhaps be best understood by using them all together, as is done in connection with the electric circuit shown in Figure 7.2. The small circles indicated on the connections to the active and passive elements represent the terminals of these devices. It is seen that these particular elements are *two-terminal devices*. It is assumed that the active element is connected to the passive one by perfect conductors, i.e., ones that do not offer any opposition to the flow of electric current ($R = 0$ or $G = \infty$). This assumption means that A and B are electrically the same point. This is also true of C and D. In circuits points such as these are called *nodes*. The active element has a terminal marked plus and one marked minus which means that the positive terminal has a net positive charge relative to the negative terminal and vice versa. Note that this is a relative relationship and that it is possible for both terminals to be positively charged but for one to be more positive than the other. The potential difference or emf is a measure of the difference in charge between two points in a circuit. In this case, the two points are A and C (or, equivalently, B and D).

In metallic conductors, it is electron flow that constitutes the current. Thus, since the positive terminal of the active element is more deficient in

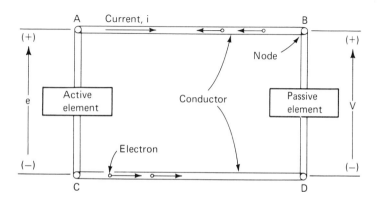

Figure 7.2
Electric circuit (ideal).

electrons than the negative terminal, there is an emf of e volts across the active element that draws electrons from its negative terminal, through the circuit, and into its positive terminal giving rise to a current flow in the direction opposite to the electron flow.

In passing through the passive element, the current creates a *voltage drop*, indicated as v in the circuit. This terminology signifies the fact that the current enters the passive element at the high-potential side (marked plus) and leaves the element at the low-potential side (marked minus). On the other hand, the current passes through the active element in the direction of a *voltage rise*, i.e., it enters the negative terminal and leaves the element from the positive terminal. Kirchhoff's voltage law (to be discused below) states that the total voltage supplied by the active elements in an electric circuit, i.e., the sum of the voltage rises, is equal to the sum of the voltage drops in the passive elements. Thus, in this circuit $e = v$.

A circuit with real elements is shown in Figure 7.3. Capital letters are conventionally used to denote voltages and currents when these quantities are constant. This is the case here since the active element is a 6 V battery; i.e., its terminal voltage is 6 V when no current is drawn from it. However, since it has internal resistance, the terminal voltage will decrease when the battery is supplying current because the 6 V generated by the battery is dropped partly across the internal resistance. The voltage remaining at the terminals, V_t, is dropped partly across the lead resistances (these are not perfect conductors in this case), and partly across the light bulb where we want to have it. The mathematical statement of this fact is

$$6V = V_1 + V_2 + V_3 + V_4 \qquad (7.9)$$

The voltage drops across the internal resistance and the lead resistance are undesirable effects that should be minimized. Equation (7.9) is derived from the application of Kirchhoff's voltage law (KVL) for this problem. This extremely important law states that *the algebraic sum of the voltages around a loop at any instant of time is zero.*

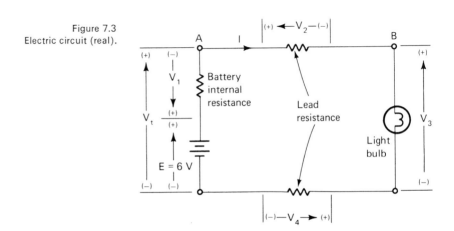

Figure 7.3
Electric circuit (real).

In applying this law, the following convention is used to determine the algebraic sign of the voltage: If the current passes through an element from the positive potential side to the negative potential side, the voltage is taken as positive, and vice versa. In Figure 7.3, for example, the algebraic sign of the voltages V_1, V_2, V_3, and V_4 is positive while that of the 6-V battery is negative. Kirchhoff's voltage law written for the circuit in Figure 7.3 is

$$-6 + V_1 + V_2 + V_3 + V_4 = 0 \qquad (7.10)$$

which is equivalent to (7.9).

The rate at which electrical energy is absorbed or supplied by a circuit element, i.e., the *power* associated with the element, is given by the product of the voltage across the element and the current through it. It is positive (power is absorbed by the element) if current enters the positive potential side and negative (power is supplied by the element) if current enters the negative potential side. For example, if (7.10) is multiplied by the current I, the result is

$$-6I + V_1 I + V_2 I + V_3 I + V_4 I = 0 \qquad (7.11)$$

The unit commonly used for electrical power is the watt (W). Thus, (7.11) states that the battery is supplying $6I$ W and that this power is being absorbed by the passive elements. It is, in essence, an example of the conservation of energy. Another way to write (7.11) is

$$P_{\text{supplied}} = P_{\text{int. res.}} + P_{\text{leads}} + P_{\text{bulb}} \qquad (7.12)$$

where

$$P_{\text{supplied}} = 6I, \qquad P_{\text{int. res.}} = V_1 I, \qquad P_{\text{leads}} = V_2 I + V_4 I$$

and

$$P_{\text{bulb}} = +V_3 I$$

Ohm's law relates the quantities electromotive force, current, and resistance. It states that *the current through a resistor is determined by the voltage across it divided by the resistance.* Applied to the circuit shown in Figure 7.4, this gives

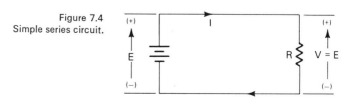

Figure 7.4
Simple series circuit.

$$I(\text{amps}) = \frac{V(\text{volts})}{R(\text{ohms})} \tag{7.13}$$

This seems intuitively reasonable since one would expect that current should increase as the voltage is increased and decrease if the resistance is increased.

Another form of Ohm's law, found by rearranging (7.13) to read

$$V = IR \tag{7.14}$$

is useful when one knows the current requirement and the resistance in a circuit and wants to determine the necessary voltage.

The resistor is classified as a *linear element,* because the voltage across it is related to the current through it by the linear, or straight-line, relationship, $V = IR$.

Still another form of (7.13) is useful when V and I are given and one must determine the resistance required. This form is

$$R = \frac{V}{I} \tag{7.15}$$

Note carefully that in our discussion of Ohm's law the voltage current relationship is for the resistor, not for the active element. The voltage across the active element, if it is ideal (no internal resistance), is independent of the current through it.

Example:

In Figure 7.4 let $E = 6$ V and $R = 6\,\Omega$. Find the current I. Equation (7.13) may be used directly to give

$$I = \frac{6\text{ V}}{6\,\Omega} = 1 \text{ A}$$

Some alternative expressions for power derived from Ohm's law are

$$P = VI = (IR)I = I^2 R = V \times \left(\frac{V}{R}\right) = \frac{V^2}{R} \tag{7.16}$$

Power may be calculated from any of these relationships. In the preceding example, $P = 6 \times 1 \text{ W} = 1^2 \times 6 \text{ W} = 6^2/6 \text{ W}$.

7.2 THE BASIC CIRCUIT FORMS

The three basic forms of electric circuits are (1) series circuits, (2) parallel circuits, and (3) series-parallel circuits. These forms are investigated below, and it is shown how Kirchhoff's voltage law, Ohm's law, and Kirchhoff's current law (yet to be introduced), are used in their analysis.

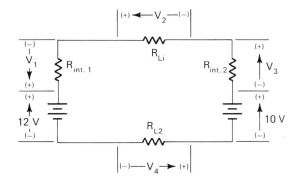

Figure 7.5
Series circuit with two sources.

A series circuit is so named because all elements in the circuit are connected in series. The circuit in Figure 7.3 is a series circuit as is the one with two batteries shown in Figure 7.5. This circuit is one that might be used to charge the 10 V battery from the 12 V battery. The lead resistances, R_{L1} and R_{L2}, must be kept reasonably small if the charging operation is to work. Perhaps some readers have had the experience of connecting two sets of battery-jumper cables together in order to reach from one battery to another and found that the charging operation would not work. This occurred because there was so much voltage drop in the cables that the low-voltage battery could not be charged up to full voltage. The loop equation found from KVL for this circuit is

7.2.1
Series Circuits

$$-12\,\text{V} + V_1 + V_2 + V_3 + 10\,\text{V} + V_4 = 0 \tag{7.17}$$

or

$$2\,\text{V} = V_1 + V_2 + V_3 + V_4 \tag{7.17a}$$

A fundamental fact in series circuits is that the same current must flow through every element. If the electric current is considered to be analogous to a water current in a hydraulic circuit, one can appreciate that this must be the case.

It is often convenient to combine the resistances in a series circuit into a single equivalent resistance. In the series circuit shown in Figure 7.6 we

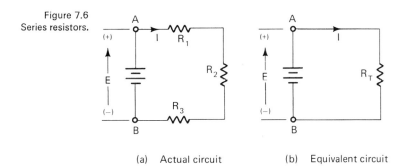

Figure 7.6
Series resistors.

(a) Actual circuit

(b) Equivalent circuit

would like to find a single value of resistance to insert between terminals A and B in place of R_1, R_2, and R_3 and that would be equivalent in the sense that it results in the same current, I. To find the equivalent resistance, write the loop equation for the circuit in Figure 7.6:

$$E = IR_1 + IR_2 + IR_3 = I(R_1 + R_2 + R_3) \tag{7.18}$$

From (7.18) it is seen that a suitable value for R_T is (subscript T stands for total)

$$R_T = R_1 + R_2 + R_3 \tag{7.19}$$

If this R_T replaces the three resistances, then the current I remains the same since

$$I = \frac{E}{R_1 + R_2 + R_3} = \frac{E}{R_T} \tag{7.20}$$

Thus, R_T is equivalent in the sense desired. In general, if n resistors are connected in series, the equivalent resistance is simply the sum

$$R_T = R_1 + R_2 + R_3 + \cdots + R_n \tag{7.21}$$

7.2.2 Parallel Circuits

A simple parallel circuit is shown in Figure 7.7. The resistances R_1 and R_2 are said to be connected in parallel; i.e., each has one terminal connected to node A and one connected to node B. The current I is seen to divide at node A, part going through R_1 and part through R_2, and recombine at node B. Kirchhoff's current law is very helpful in analyzing parallel circuits. Therefore a discussion of this law precedes further discussion of parallel circuits.

Kirchhoff's current law (KCL) states that the algebraic sum of the currents into a node at any instant is zero. In using this law, we assign a positive sign to each current entering a node and a negative sign to each current leaving a node. Thus, a positive current leaving a node is interpreted as a negative current entering the node. For example, in Figure 7.7(a), $+I$ comes into

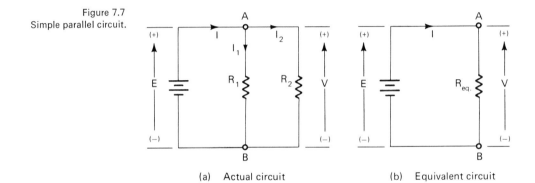

Figure 7.7 Simple parallel circuit.

(a) Actual circuit   (b) Equivalent circuit

node A, $+I_1$ and $+I_2$ come out of node A, or, alternatively, $-I_1$ and $-I_2$ go into node A. Therefore, KCL is

$$+I - I_1 - I_2 = 0 \quad \text{or} \quad I = I_1 + I_2 \tag{7.22}$$

In solving a problem it is not necessary to know whether or not a current enters or leaves a node. Simply assume current directions to start. A negative curent in the results means that the direction is opposite to what had been assumed.

Returning to our discussion of parallel circuits, we see from Figure 7.7(a) that I_1 and I_2 may be calculated from Ohm's law to be

$$I_1 = \frac{E}{R_1} \quad \text{and} \quad I_2 = \frac{E}{R_2} \tag{7.23}$$

Then KCL gives us

$$I = I_1 + I_2 = E\left(\frac{1}{R_1} + \frac{1}{R_2}\right) \tag{7.24}$$

Our goal now is to find the single equivalent resistance in Figure 7.7(b) to place between the nodes A and B that takes the place of the parallel combination R_1 and R_2 but that leaves the current I unchanged. We can find this equivalent resistance by noting from (7.24) that

$$\frac{E}{I} = \frac{1}{(1/R_1) + (1/R_2)} = \frac{R_1 R_2}{R_1 + R_2} = R_{eq} \tag{7.25}$$

It is interesting and important to note that R_{eq} in a parallel connection is always less than the smallest resistance in the parallel combination.

Now let's consider any number of parallel resistors. Thus, in Figure 7.8 there are n resistors in parallel, $R_1, R_2, \ldots, R_n$. What is the equivalent resistance between terminals A and B looking into the network from the source? To find this resistance, we make use of Ohm's law n times as follows:

$$I_1 = \frac{E}{R_1}, \quad I_2 = \frac{E}{R_2}, \quad \cdots, \quad I_n = \frac{E}{R_n} \tag{7.26}$$

Figure 7.8
Parallel resistors.

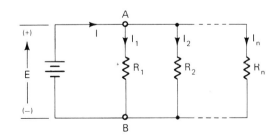

Hence, the total current by Kirchhoff's current law is

$$I = \frac{E}{R_1} + \frac{E}{R_2} + \cdots + \frac{E}{R_n} = E\left(\frac{1}{R_1} + \frac{1}{R_2} + \cdots + \frac{1}{R_n}\right) \quad (7.27)$$

The equivalent resistance must be such that

$$\frac{E}{I} = R_{eq} = \frac{1}{(1/R_1) + (1/R_2) + \cdots + (1/R_n)} \quad (7.28)$$

Or, equivalently,

$$G_{eq} = \frac{1}{R_{eq}} = \frac{1}{R_1} + \frac{1}{R_2} + \cdots + \frac{1}{R_n} = G_1 + G_2 + \cdots + G_n \quad (7.29)$$

Note that conductances in parallel add, as do resistances in series. When all resistances are equal, i.e., $R_1 = R_2 = R_3 = \cdots = R_n = R$, then R_{eq} is simply calculated by $R_{eq} = R/n$ since

$$G_{eq} = \frac{1}{R_{eq}} = \frac{1}{R} + \frac{1}{R} + \cdots + \frac{1}{R} = \frac{n}{R} \quad \text{or} \quad R_{eq} = \frac{R}{n} \quad (7.30)$$

Thus, there are two special cases where R_{eq} is easily calculated: (1) When there are only two resistances in parallel (7.25) may be used, and (2) when all resistors in parallel are equal, (7.30) may be used.

7.2.3 Series-Parallel Circuits

A simple series-parallel circuit is shown in Figure 7.9, where R_1 is in series with the parallel combination R_2 and R_3. To analyze this type of circuit, one first reduces the parallel combination, R_2 and R_3, to its equivalent resistance using (7.25). This allows the circuit in Figure 7.9(a) to be represented by its

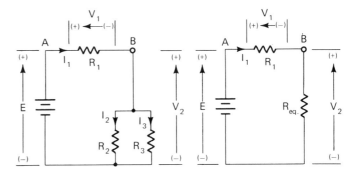

Figure 7.9
Simple series-parallel circuit and its series equivalent.

(a) Actual circuit   (b) Series equivalent

equivalent representation shown in Figure 7.9(b). The latter representation is a series circuit that has a total resistance

$$R_T = R_1 + R_{eq} = R_1 + \frac{R_2 R_3}{R_2 + R_3}$$

Having found R_T, we now apply Ohm's law to find I_1:

$$I_1 = \frac{E}{R_T}$$

The next thing of interest here is to find the currents in each branch of the parallel part of the circuit, i.e., I_2 and I_3. These are easily calculated once V_2 is known, and V_2 may be found by the relation $V_2 = I_1 R_{eq}$. Thus, $I_2 = V_2/R_2 = IR_{eq}/R_2 = IR_3/(R_2 + R_3)$ and $I_3 = V_2/R_3 = IR_2/(R_2 + R_3)$. Note that the ratio of the currents is inversely proportional to the resistances through which they flow, i.e.,

$$I_2/I_3 = R_3/R_2.$$

Example:

In the circuit shown in Figure 7.9(a), let $E = 10$ V, $R_1 = 5\,\Omega$, and $R_2 = R_3 = 10\,\Omega$. We find $R_{eq} = (10)(10)/20 = 5\,\Omega$ and $R_T = 5\,\Omega + 5\,\Omega = 10\,\Omega$. Thus, $I_1 = (10\text{ V})/(10\,\Omega) = 1$ A. $V_2 = 1\text{ A} \times R_{eq} = 1 \times 5 = 5$ V. $I_2 = V_2/10 = 5\text{ V}/10\,\Omega = .5$ A, and $I_3 = V_2/10\,\Omega = 5\text{ V}/10\,\Omega = .5$ A. Since $R_2 = R_3$, the current I_1 divides evenly between these two resistances at node B.

7.2.4 Two-Loop Network and the Superposition Theorem

The circuit shown in Figure 7.10 is a two-loop network containing a voltage source in each loop. The voltage source in loop 2 makes this network different from the series-parallel circuit studied previously. We shall examine methods for analyzing this type of network, and determine what value of current flows in each part of the network, given the resistance and emf values.

Kirchhoff's current law and Kirchhoff's voltage law are applied to write the necessary equations to solve this problem. KCL written at node a says that

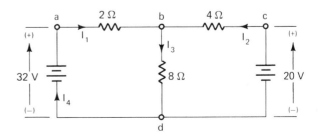

Figure 7.10
Two loop network with branch currents.

$I_4 - I_1 = 0$. Therefore, $I_1 = I_4$, so we do not have to carry I_4 around any further. At node b, KCL gives

$$+I_1 + I_2 - I_3 = 0 \tag{7.31}$$

This equation has three unknowns. If we had two additional independent equations in I_1, I_2, and I_3, then we could solve the system of three linear algebraic equations in three unknowns to find the currents. The two additional equations required are obtained by writing KVL for each loop.

Before writing these, consider the voltage current relations for each resistor:

$$V_{ab} = +R_1 I_1 = +2I_1 \tag{7.32}$$

The notation V_{ab} means the potential at node a (the first subscript) is positive relative to the b node (the second subscript). This notation and the plus sign in (7.32) are consistent with the convention that the current enters a passive element at the positive potential terminal. Similarly,

$$V_{bd} = +R_3 I_3 = +8I_3 \tag{7.33a}$$

and

$$V_{cb} = +R_2 I_2 = +4I_2 \tag{7.33b}$$

The potential differences for the active elements may be written as $V_{ad} = +32$ V and $V_{cd} = +20$ V. The subscript notation would then be used to write the potential at d relative to a as $V_{da} = -32$ V and that at d relative to c as $V_{dc} = -20$ V. KVL applied to loop 1 gives the loop equation

$$V_{da} + V_{ab} + V_{bd} = 0 \tag{7.34}$$

Note that we started at node d, proceeded around the loop in a clockwise sense, and returned to node d. Equation (7.34) may be written as

$$-32 + 2I_1 + 8I_3 = 0 \tag{7.35}$$

The loop equation for the second loop is

$$V_{dc} + V_{cb} + V_{bd} = 0 \tag{7.36}$$

or

$$-20 + 4I_2 + 8I_3 = 0 \tag{7.37}$$

Equations (7.31), (7.35), and (7.37) are three simultaneous linear, algebraic equations. The solution of these for I_1, I_2, and I_3 is $I_1 = +4$ A, $I_2 = -1$ A, and $I_3 = +3$ A. The negative sign associated with I_2 indicates that it flows in a direction opposite to that implied by the arrow direction.

A second method for solving this problem is to assume only two *loop currents* are flowing, as shown in Figure 7.11. Since there are only two unknowns in this formulation, two equations, the KVL equations for each loop, are sufficient to solve the problem. For loop 1, KVL gives

$$32 = 2I_1 + 8I_1 + 8I_2 \tag{7.38}$$

and for loop 2 it is

$$20 = 4I_2 + 8I_1 + 8I_2 \tag{7.39}$$

These two equations may be solved simultaneously for I_1 and I_2. Note that the terms $8I_2$ in (7.38) for loop 1 and $8I_1$ in (7.39) for loop 2 represent *coupling* or interaction terms between the loops.

There is a third technique for solving this problem that employs the principle of superposition. Let's state this theorem and see how it applies to this problem.

The superposition theorem states that if cause and effect are linearly related, then the total effect of several causes acting simultaneously is equal to the sum of the effects of the individual causes acting one at a time.

In relation to this problem the theorem tells us that we may obtain a solution by first finding the currents (effect 1) caused by voltage source 1 (cause 1) acting alone, then find the currents (effect 2) caused by source 2 (cause 2), and then find the actual current (total effect) by adding these two sets of currents together. Thus, the problem with two sources is reduced to solving a problem with one source two times.

The term "linearly related" in the theorem means that cause and effect are related by a constant. Since this is true of voltages and currents for resistors the superposition theorem may be applied to solve this problem.

First let the 32-V source act in the circuit and set the 20-V source equal to zero. This is equivalent to putting a short circuit across the 20-V source and leads to the series-parallel network of Figure 7.12. The currents when the 32-V source is acting alone are

$$I_1' = \tfrac{96}{14} \text{ A} \qquad I_2' = \tfrac{32}{14} \text{ A} \qquad I_3' = \tfrac{64}{14} \text{ A}$$

Repeating this procedure with the 32 V source set equal to zero leads to the

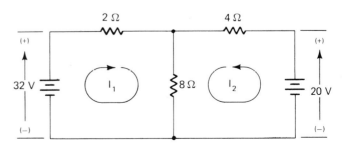

Figure 7.11
Two loop network with loop currents.

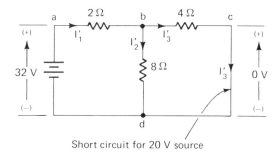

Figure 7.12
Two loop network with source #2 short circuited.

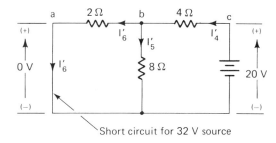

Figure 7.13
Two loop network with source #1 short circuited.

network shown in Figure 7.13. This circuit has $I'_4 = 50/14$ A, $I'_5 = 10/14$ A, and $I'_6 = 40/14$ A. The solution to the original problem then is found simply by noting that $I_1 = I'_1 - I'_6 = (96 - 40)/14 = +4$ A, $I_2 = I'_4 - I'_3 = (50 - 64)/14 = -1$ A, and $I_3 = I'_2 + I'_5 = (32 + 10)/14 = +3$ A.

Our discussion of circuits in this section has been on a rather abstract level. Real circuits take a great variety of forms. In Figure 7.14, for example, an electronic amplifier board constructed using printed circuit techniques and discrete components is shown. The conductors between components

Figure 7.14
Electronic amplifier constructed using printed circuit techniques and discrete components.

250

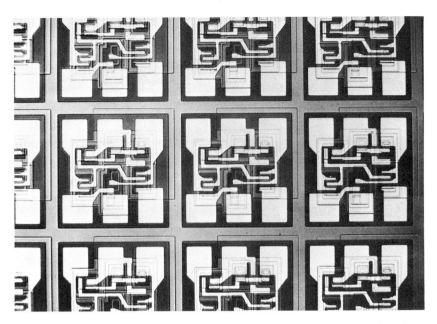

Figure 7.15
Many identical integrated circuits fabricated simultaneously on a single wafer of silicon.
(From *Electronic Principles* by P. E. Gray and C. L. Searle © 1969,
published by John Wiley & Sons, Inc.)

and the nodes are clearly visible. Integrated circuits, shown in Figure 7.15, represent the most recent development in circuit technology. They are constructed from semiconductor materials such as silicon and do not have discrete components. Instead, active and passive elements are both constructed on a single silicon wafer of extremely small dimensions. Integrated circuits generally are less expensive than an equivalent circuit of discrete components and considerably smaller, containing the equivalent of 100 to 200 discrete components within a volume of only one discrete component. They are also more reliable, an extremely important engineering consideration.

7.3
TIME-VARYING SIGNALS

Discussion so far has been concerned with constant voltages and currents, termed dc (direct current) voltages and currents. This terminology is somewhat cumbersome and inexact, but it is commonly used. Many voltages and currents are not constant but vary with time. These are sometimes called ac (alternating current) signals. One of the most common and important of these is the sinusoidally varying voltage used extensively in power generation and distribution. The potential difference between the two terminals of a wall socket is varying sinusoidally at 60 hertz (Hz, cycles per second). The voltage *waveform* (plot of potential difference versus time) is as shown in Figure 7.16.

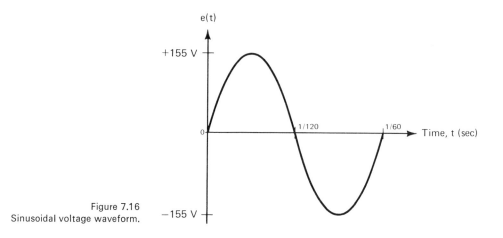

Figure 7.16
Sinusoidal voltage waveform.

It is characterized by its peak amplitude (155 V in this case) and its frequency (here, 60 Hz). Waveforms such as this may actually be observed on the face of a cathode-ray tube (CRT) of an oscilloscope, a measuring instrument for displaying electrical signals visually (see Figure 7.17). The CRT (probably more familiar to you in the form of a television picture tube) picture is obtained by having an electron beam strike a fluorescent screen. The beam is moved horizontally by a sawtooth voltage (Figure 7.18) and vertically by the voltage being observed. In this way, the picture of Figure 7.16 is traced out on the face of the CRT.

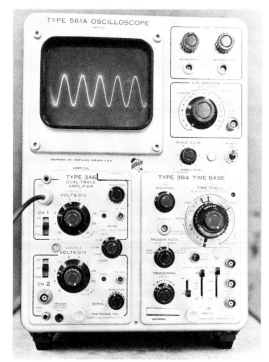

Figure 7.17
The Tektronic Type 503 oscilloscope.
(From *Electronic Components and Measurements* by B. D. Wedlock and J. K. Roberge © 1969, published by Prentice-Hall, Inc.)

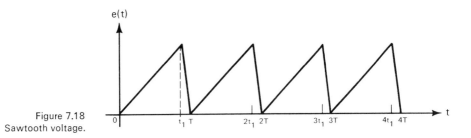

Figure 7.18
Sawtooth voltage.

The sawtooth voltage waveform shown in Figure 7.18 is of considerable interest in connection with TV receivers and oscilloscopes. This type of voltage waveform, commonly called a sweep voltage, is the type used to move (or "sweep") the electron beam in a TV picture tube horizontally and vertically. The picture is produced by simultaneously moving the beam across the screen horizontally and down vertically, as shown in Figure 7.19.

The voltages used to produce the magnetic field for moving the electron beam are as shown in Figure 7.20, where it is seen that there are 525 horizontal "sawteeth" occurring during one vertical sawtooth; i.e., the frequency of the horizontal sweep is 525 times as great as the vertical sweep frequency.

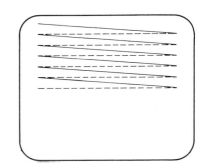

Figure 7.19
Beam scanning pattern for T.V.

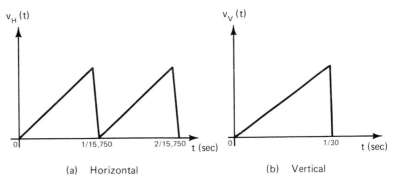

(a) Horizontal   (b) Vertical

Figure 7.20
Horizontal and vertical sweep signals.

253

During one vertical sweep, there are 525 horizontal lines drawn to produce one frame of the picture; 30 such frames are produced each second. Note that the so-called horizontal lines are not quite horizontal because of the simultaneous vertical motion.

Voltages with exponential waveforms (discussed in section 7.5) are of great importance because they typify the behavior of currents and voltages in many circuits during the transient state, i.e., immediately following a sudden change in the state of the circuit.

Time-varying voltages and currents may be generated in a variety of ways. One device commonly used for this purpose is the carbon microphone, shown in Figure 7.21. The important feature of this device is the variable resistance offered by the carbon granules. The microphone acts as a transducer that converts sound energy to electrical energy. It functions by having the sound pressure waves transmitted through the diaphragm to the pack of carbon granules. This varying pressure on carbon granules causes its resistance to vary in accordance with the incoming sound wave. With no sound wave entering the microphone the packet has a particular fixed value of resistance corresponding to atmospheric pressure. If a sound wave then is "heard" by the microphone, the pressure on the granules varies above and below atmospheric pressure, causing the resistance to decrease or increase, respectively. For example, if a single frequency tone enters the microphone, as shown in Figure 7.22(a), the resistance of the packet varies as shown in Figure 7.22(b). Since the source in the microphone circuit is dc (a battery), as the resistance varies sinusoidally with time, the current $i(t)$ will also vary sinusoidally. The voltage $v(t)$ across the fixed resistor must have the same time variations as the current through it. Thus, a sinusoidally time-varying current and voltage are produced in the microphone circuit. In a good quality (high-fidelity) microphone, the time variations of the current and voltage will be nearly exact replicas of the time variations of the sound waveform. If the current and voltage waveforms are not like the sound waveform, we say that "distortion" has been introduced into the sound system of which the microphone is a part. The voltage waveforms for speech are more complex than the single frequency tone. For example, for the word *alters*, the waveform is shown in Figure 7.23. The waveform of a 500-Hz pure tone is shown for comparison.

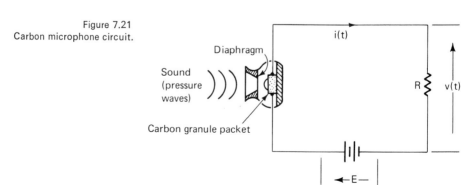

Figure 7.21
Carbon microphone circuit.

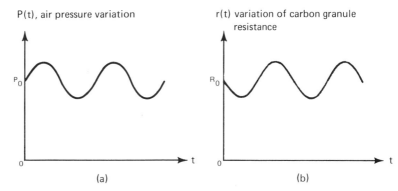

Figure 7.22
(a) Pressure variation about atmospheric pressure, P_0.
(b) Resistance variation about normal resistance, R_0.

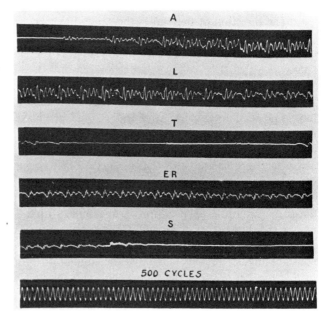

Figure 7.23
Waveform of the word *alters*.
(From *Speech and Hearing in Communication* by Harvey Fletcher © 1953 by Litton Educational Publishing, Inc., published by Krieger Publishing)

An ac generator that provides our 60-Hz sinusoidal house voltage is a machine arranged to move conductors in a magnetic field. Although the details of its construction are complex, the operating principle of the generator is relatively easy to understand and may be explained by Faraday's law, which states that there is an electromotive force generated in a conductor that moves in a magnetic field.

Assume that there is a conductor in a magnetic field as shown in Figure 7.24. The magnetic field has constant strength B and is pointed as shown. The velocity of the conductor is v, pointed at angle θ relative to the magnetic field. Faraday's law specifies the magnitude and direction of the emf generated between the ends of the conductor. It states that the strength of the emf is

$$e = lBv \times \sin\theta \qquad (7.40)$$

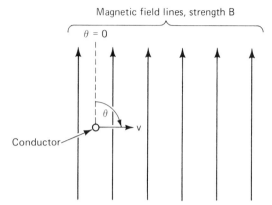

Figure 7.24
Conductor in a magnetic field.

The direction of the emf is the same direction in which a right-hand screw moves as the screw is rotated from the direction of the magnetic field toward that of velocity v. Thus, in Figure 7.24 the emf is pointed into the page.

If the velocity is parallel to the magnetic field, then emf $= 0$ because $\theta = 0°$ and $\sin \theta = 0$. The end of the conductor facing out of the page will have a positive potential relative to the opposite end if the velocity is to the right. If the conductor moves to the left, then the end of the conductor facing out of the page is at a negative potential with respect to the opposite end. These facts are basically all that are necessary to understand the operation of an ac generator.

In an ac generator, the stator (fixed) windings carry a current that produces the magentic field within the machine. Rotor windings move in the magnetic field to generate an emf. Consider a single coil of the rotor winding moving in the magnetic field as shown in Figure 7.25. The rotor is driven at a constant angular velocity ω, so that the angular position of the rotor is given by $\theta = \omega t$. Let $t = 0$ be the time when the plane in which the rotor winding lies is horizontal, as shown in Figure 7.25(a). At this time both sides of the

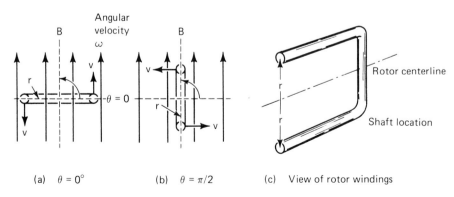

Figure 7.25
Single coil moving in a magnetic field.

256

rotor winding are moving parallel to the magnetic field; consequently, no voltage is being generated [$\theta = 0$ in (7.40)]. When $\theta = \pi/2$, the plane of the rotor winding is vertical, as shown in Figure 7.25(b), and maximum voltage is being generated in each side of the rotor winding. These voltages add so that a voltage twice that in either side appears between the ends of the winding. Since $\theta = \omega t$ and the voltage is proportional to $\sin \theta$, the voltage generated between the ends of the winding during one complete revolution of the rotor then is sinusoidal (Figure 7.16). This simple basic theory essentially explains the operation of the highly complex hydroelectric generator shown in Figure 7.26.

7.3.1 Average Values and Effective Values of Periodic Signals

It is often useful to know what value of direct current in a resistor would dissipate the same power as the average power dissipated by a sinusoidal current of a particular peak value. This value may be found and is called the *effective value* of the sinusoidal current (effective value in the sense that it is effectively generating the same heat as a direct current of the same magnitude). The instantaneous power in a resistor of value R carrying current $i(t)$ is

$$p(t) = i^2(t)R \tag{7.41}$$

Now the average power is written as

$$P_{\text{av}} = \{\text{average of } [i^2(t)]\} R = I_{\text{eff}}^2 R \tag{7.42}$$

Thus, the effective value of current would be

$$I_{\text{eff}} = \sqrt{\text{average of } i^2(t)} \tag{7.43}$$

Figure 7.26
Sectional view of large hydroelectric generator.
(*Courtesy General Electric Co.*)

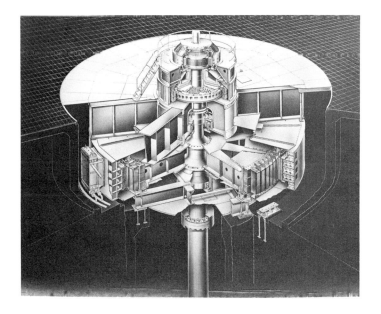

This is getting ahead of our story, however, since we need first to discuss the concept of average value that is employed in the definition of effective value.

The average value of any function $f(t)$ in the time 0 to T sec is defined by

$$F_{av} = \frac{1}{T} \int_0^T f(t)\, dt \tag{7.44}$$

The integral of $f(t)$ over the interval 0 to T is just the net area under the $f(t)$ versus t curve evaluated between $t = 0$ and $t = T$. Let's apply this definition to the function shown in Figure 7.27. The integral is easily found for this function. It is

$$\int_0^T f(t)\, dt = \int_0^{3T/4} A\, dt + \int_{3T/4}^T (-A)\, dt$$
$$= \frac{3T}{4}A - \frac{T}{4}A = \frac{A}{2}T$$

Therefore, $F_{av} = +A/2$, or, *on the average*, the value of $f(t)$ is $+A/2$.

The effective value of a periodic function is just the square root of the average of the squared function. Another name for effective value is the root mean square (rms) value. The abbreviation stands for the square *root* of the *mean* (average) value of the *square* of a function. For $f(t)$ the average value of the squared function, called F_{rms}^2, is

$$F_{rms}^2 = \frac{1}{T} \int_0^T f^2(t)\, dt \tag{7.45}$$

Then

$$F_{rms} = \sqrt{F_{rms}^2} = \sqrt{\frac{1}{T} \int_0^T f^2(t)\, dt} \tag{7.46}$$

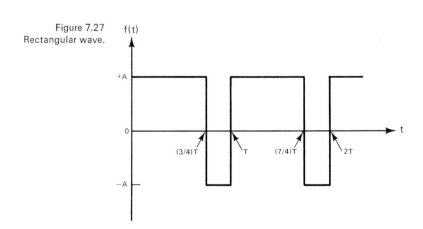

Figure 7.27 Rectangular wave.

How is this definition used? Consider the power in a resistor driven by a voltage $v(t) = E_0 \sin \omega t$. The average power cannot be expressed in terms of average values of voltage and current because these are zero. Therefore, another approach is taken. The instantaneous power is

$$p(t) = i^2(t)R = \left(\frac{E_0}{R} \sin \omega t\right)^2 R \qquad (7.47)$$

Now, obviously, the average value of $\sin^2 \omega t$ cannot be zero since it is always positive. The average value of $p(t)$ is

$$P_{av} = \frac{R}{T} \int_0^T i^2(t)\, dt = RI_{rms}^2 \qquad (7.48)$$

where $T = 2\pi/\omega$. Note that

$$I_{rms}^2 = \frac{(E_0/R)^2}{T} \int_0^T \sin^2 \omega t\, dt \qquad (7.49)$$

The trigonometric identity $\sin^2 \omega t = .5(1 - \cos 2\omega t)$ may be used to put (7.49) in the form

$$I_{rms}^2 = \frac{(E_0/R)^2}{T} \int_0^T \frac{1}{2}(1 - \cos 2\omega t)\, dt \qquad (7.50)$$

The average of $\cos 2\omega t$ is zero, leaving

$$I_{rms}^2 = \frac{(E_0/R)^2}{2} \qquad (7.51)$$

Let $E_0/R = I_0$, the peak value of current. Then

$$I_{rms} = \frac{I_0}{\sqrt{2}} = .707\, I_0 \qquad (7.52)$$

This relationship is true for any sinusoidal signal; i.e., the rms or effective value is .707 times the peak value. A direct current of this value passing through the resistor would dissipate the same power as the sinusoidal current does. Hence, house voltage of 110-V rms has a peak amplitude of $(\sqrt{2})(110)v = 155v$ (Figure 7.16).

7.3.2 Phase Angle and Power Relationships

The extremely important concept of phase angle is used to describe the relationship in time between two sinusoidal waveforms of the same frequency. For example, if $v_1(t) = V_1 \sin \omega t$, and $v_2(t) = V_2 \sin(\omega t - \theta)$, these two voltages are related in time as shown in Figure 7.28. We say that $v_1(t)$ "leads" $v_2(t)$ by θ radians because it reaches its peak value, V_1 at time $t_1 = \pi/2\omega$, which is before the time $t_2 = (\pi/2 + \theta)/\omega$, when $v_2(t)$ reaches its peak value

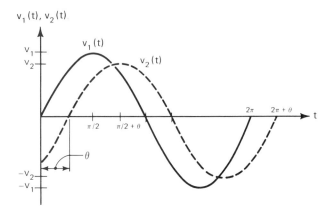

Figure 7.28
Phase relationships.

V_2. Alternatively, we say that $v_2(t)$ "lags" $v_1(t)$ by θ radians. The phase angle is θ in this case. The concept of phase angle only has meaning when the time relationship of one signal relative to another is being discussed.

The average power dissipated by a passive element may be found in terms of the rms voltage across it, the rms current through it, and the phase angle between the voltage and current (assuming sinusoidal signals). The equation for average power in this case is

$$P_{av} = V_{rms} I_{rms} \cos \theta \qquad (7.53)$$

Thus, if the voltage and current are 90° out of phase, the average power dissipation is zero. This is the situation in ideal inductors and capacitors, which are discussed in Section 7.4. In resistors, voltage and current are in phase. Therefore $\theta = 0$, $\cos \theta = 1$, and $P_{av} = V_{rms} I_{rms}$.

7.4 NONDISSIPATIVE PASSIVE ELEMENTS—INDUCTORS AND CAPACITORS

Inductors and capacitors are energy *storage* elements, whereas the resistor is an element that dissipates energy. An inductor stores electrical energy in the magnetic field surrounding its winding. A capacitor stores the energy in the electric field between its plates. Inductors are constructed by winding a coil of wire on a core. If the core is a magnetic material such as iron, the value of inductance is higher than it is if the core is of a nonmagnetic material. The construction of two typical inductors is shown in Figure 7.29. A current passing through the coil creates a magnetic field. The energy in the magnetic field is given by the expression

$$w = \tfrac{1}{2} L i^2 \qquad (7.54)$$

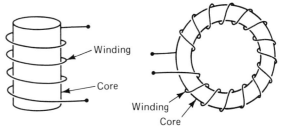

Figure 7.29
Two typical inductors. (a) Cylindrical inductor (b) Toroidal inductor

where L is the value of inductance in henries, which is proportional to the square of the number of turns and also depends on the dimensions, type of material, and shape of the core.

The symbol for an inductor and the equation relating the voltage across it and the current through it are shown in Figure 7.30. Why is the voltage across the inductor proportional to the time derivative of the current rather than to the current itself as in a resistor? As a voltage is applied a current tries to develop, which in turn creates a growing magnetic field. The growing magnetic field "cuts" the conductors and thereby causes a voltage to be induced in them. This induced voltage has a polarity that opposes the applied voltage (Lenz's law). Consequently, it tends to delay the buildup of the current, causing it to lag the applied voltage. It also accounts for the very important fact that the current through an inductor will not change instantaneously if the voltage across it undergoes a finite instantaneous change in amplitude. Compare this to the current through a resistor under the same circumstances. If the switch in Figure 7.31(a) is closed at $t = 0$, the current in the resistor instantaneously jumps to the value E/R. In Figure 7.31(b), however, the current is zero before the switch is closed and also just following its closing. This must be true for the inductor since $v(t) = L\, di(t)/dt$; then by integrating this equation we obtain

$$i(t) = \frac{1}{L} \int_0^t v(x)\, dx = \frac{1}{L} \int_0^t E\, dx \qquad (7.55)$$

and since E is finite, the value of the integral must be zero if $t = 0$.

The voltage-current relation for the inductor implies a 90° phase shift between current and voltage for sinusoidal signals. Suppose that in Figure 7.30 $i(t) = I_0 \sin \omega t$. Then $v_L = L\omega I_0 \cos \omega t = L\omega I_0 \sin(\omega t + \pi/2)$. Another

Figure 7.30
Voltage-current relation for an inductor.

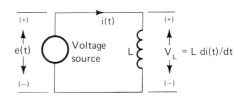

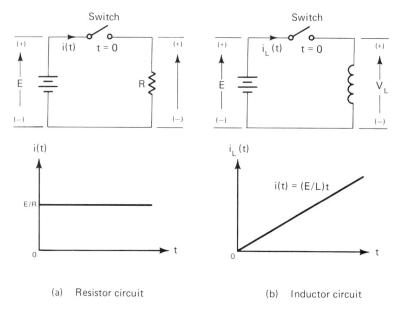

Figure 7.31
Comparing initial currents in R and L circuits.

important property of the inductor is brought out by this relationship; i.e., the *inductive reactance* (opposition to current flow) increases linearly with frequency. This means that the voltage v_L required to maintain $i(t) = I_0 \sin \omega t$ is proportional to ω. The inductor offers more opposition to current flow at higher frequencies because a faster changing magnetic field induces a larger opposing voltage in the inductor winding.

When the inductance is driven by a sinusoidal voltage source, the resulting current is also sinusoidal and lags the voltage by 90°. The average power dissipated is given by $P_{av} = I_{rms} E_{rms} \cos \theta$ and is zero because $\theta = 90°$ and $\cos \theta = 0$. Power is not dissipated on the average because the voltage and current are 90° out of phase, and therefore the device is a storage element. The magnetic field varies in phase with the current. The electrical energy is put into the magnetic field as the current builds up and then is put back into the circuit from the field as the current decays.

A capacitor is constructed by placing a dielectric material, i.e., an insulator, between two conducting plates, as shown in Figure 7.32. The value of the capacitance, C, in farads, is directly proportional to the area of the conducting plates and inversely proportional to the distance separating them. The capacitor, like the inductor, is an energy storage element. When a voltage v is applied across the plates, energy is stored in the electric field that is created between them. The energy is given by

$$w = \tfrac{1}{2} C v^2 \tag{7.56}$$

The symbol for a capacitor and the relationship between the voltage across it and the current through it are shown in Figure 7.33, where the lower limit of integration, 0, is the time when current starts to flow. It is assumed that there is no voltage across the capacitor at this starting time. The voltage is proportional to the integral of the current because it results from an accumulation of charge [this may be seen by substituting (4), $i(t) = dq/dt$, into the integral for i_C].

Since this is this case, the voltage across the capacitor will not change instantaneously when the current through it undergoes a finite instantaneous change. This important fact is illustrated by letting $i_C(t)$ in Figure 7.33 be I_0, a constant. Then, as shown in Figure 7.34, the voltage is

$$v_C(t) = \frac{1}{C} \int_0^t I_0 \, dt = \frac{1}{C} I_0 t \tag{7.57}$$

Since I_0 is finite, $v_C(0+)$ (the voltage across the capacitor just after closing the switch) is zero.

If in Figure 7.33 $i_C(t) = I_0 \sin \omega t$ and $v_C(0) = 0$, then $v_C(t) = -(I_0/\omega C) \cos \omega t = (I_0/\omega C) \sin(\omega t - \pi/2)$. Hence, it is seen that the current leads the voltage by 90° (the voltage lags the current by 90°). In contrast to inductive

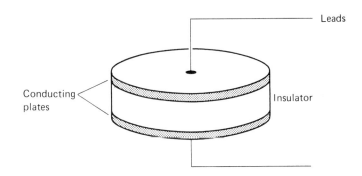

Figure 7.32
Capacitor construction.

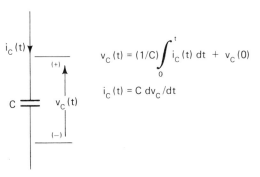

Figure 7.33
Voltage-current relation for a capacitor.

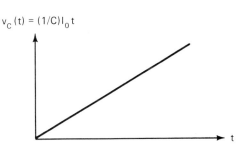

Figure 7.34
Capacitor voltage for constant current.

reactance, capacitive reactance decreases inversely proportionally to frequency; i.e., to maintain a peak current amplitude equal to I_0, the peak voltage amplitude required is $I_0/\omega C$. Also, as in the inductor, no power is dissipated since the voltage and current are 90° out of phase. Here, the energy goes into the electric field between the plates according to (7.56) as the voltage is increasing. When the voltage is decreasing, energy is put back into the circuit from the field. The net result is that no power is dissipated but simply transferred back and forth between the electric field and the circuit.

7.5 TRANSIENT BEHAVIOR IN SERIES *RL* AND *RC* CIRCUITS

Before discussing transient behavior in series *RL* and *RC* circuits, it will be helpful to examine the exponential function that arises naturally in the analysis of the transient problem.

The exponential function may be written as

$$f(t) = Ae^{-t/\tau} \tag{7.58}$$

where $f(t)$ is the value of the function at time t; A is a constant; e, the base of natural logarithms, is equal to 2.718...; and τ, called the time constant has units of seconds. Let's normalize (7.63) with respect to A and plot the result as a function of time in Figure 7.35. The function is plotted only for positive time since $t = 0$ may be arbitrarily chosen to signify the starting time of the problem. When $t = \tau$, one time constant, the exponential function has the value $e^{-1} = .368$; i.e., the function has decayed to approximately 37% of its initial value. In a period of five time constants, the function decays to less than 1% of its initial value.

The derivative and integral of the exponential function are both essential to the analysis of the transient problem. The time derivative, plotted in Figure 7.36, is seen to be

$$\frac{df(t)}{dt} = -\frac{A}{\tau}e^{-t/\tau} \tag{7.59}$$

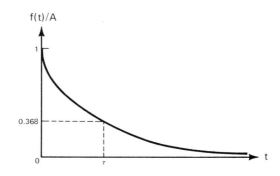

Figure 7.35
Universal exponential curve (decaying exponential function).

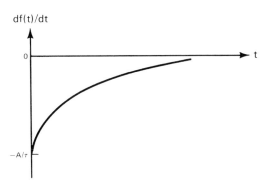

Figure 7.36
Derivative of the exponential function.

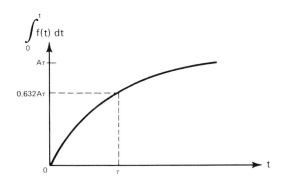

Figure 7.37
Integral of exponential function.

It is clear that the derivative is also an exponential function with the constant multiplier now equal to $-A/\tau$. Note the effect of the time constant on the magnitude of the derivative. If the time constant is small, the magnitude of the derivative is large and vice versa. The time constant is commonly used as a measure of the rate of change of an exponentially varying physical variable by noting that it has a large time constant if the rate of change is slow or a small time constant if the rate is fast.

The integral of the exponential function, plotted in Figure 7.37 is given by

$$\int_0^t f(x)\,dx = \int_0^t A e^{-x/\tau}\,dx = A\tau(1 - e^{-t/\tau}) \qquad (7.60)$$

It too involves the exponential function itself, as does the derivative. This function represents the area under the exponential curve from time 0 to any other time t. It reaches approximately 63% of its final value (100% − 37%) in one time constant.

Consider the series resistor-inductor circuit shown in Figure 7.38. Before closing the switch at $t = 0$, the current is zero and the circuit is said to

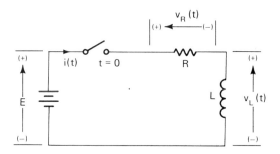

Figure 7.38
Series RL circuit.

be at rest. The equation that satisfies KVL following the closing of the switch at $t = 0$ is the differential equation

$$E = Ri(t) + L\frac{di}{dt} \quad \text{for } t \geq 0 \tag{7.61}$$

Just after $t = 0$, $i(0) = 0$ because current through the inductor cannot change instantaneously. The presence of a resistor and a constant, finite voltage source makes it reasonable to argue that a constant, finite final value of current should result after the transient period. An increasing exponential current would satisfy both of these requirements. Let us assume then that the current is of the form

$$i(t) = I_0(1 - e^{-at}) \tag{7.62}$$

and find out if this current can be made to satisfy (7.61) for all times $t \geq 0$. I_0 and a are unknown constants that will later be determined in a way that allows (7.62) to be a valid solution. Substituting (7.62) into (7.61) gives

$$E = RI_0(1 - e^{-at}) + aLI_0 e^{-at} \tag{7.63}$$

Now since the left-hand side of (7.63) is constant, the right-hand side must also be constant if the equation is to be satisfied at all times. The right-hand side will be constant (independent of t) if we set

$$RI_0 = aLI_0 \tag{7.64}$$

Then (7.63) becomes

$$E = RI_0 \tag{7.65}$$

Thus, (7.64) and (7.65) give conditions on a and I_0 that must be satisfied if (7.62) is to be, in fact, a solution to (7.61) for all time. These conditions are that

$$a = \frac{R}{L} \quad \text{and} \quad I_0 = \frac{E}{R} \tag{7.66}$$

The current that does satisfy the differential equation (7.61) is then

$$i(t) = \frac{E}{R}(1 - e^{-Rt/L}) \tag{7.67}$$

Note that $i(0) = 0$ and $i(\infty) = E/R$.

The quantity $1/a = L/R$ has a special significance. It is the time constant of the circuit and is given the symbol $\tau = L/R$. The current will increase more slowly if L is increased (or R decreased) and vice versa. The time constant is related in a special way to the initial value of the derivative, $(E/R)(1/\tau)$. If the current continued to increase linearly at this rate, i.e., if the current were given by

$$i(t) = \left(\frac{E}{R}\right)\left(\frac{t}{\tau}\right) \tag{7.68}$$

then it would reach its steady-state value, E/R, in one time constant (when $t = \tau$).

Consider next the series resistor-capacitor network, shown in Figure 7.39. The equation that satisfies KVL after closing the switch at $t = 0$ is

$$E = Ri(t) + \left(\frac{1}{C}\right)\int_0^t i(x)\, dx \tag{7.69}$$

assuming that there is no initial voltage across the capacitor. Again we shall argue physically that an exponential form of current can satisfy (7.69). Immediately after closing the switch the capacitor voltage is exactly what it was before closing the switch, i.e., $v(0+) = v(0-)$. Now if the voltage across the capacitor is zero at $t = 0+$, then the full voltage E must appear across the resistance to satisfy KVL. Thus, the initial current must be $i(0+) = E/R$. Eventually, the voltage across the capacitor builds up to equal the applied voltage E. At that time, there can be no voltage across the resistor. Therefore, the current must go to zero in the steady state. Consistent with these observations, let us assume a current of the form

$$i(t) = I_0 e^{-t/\tau} \tag{7.70}$$

Figure 7.39
Series RC circuit.

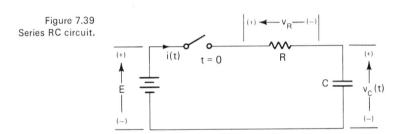

As in the *RL* circuit case, we have two constants to determine, I_0 and τ. We have already argued physically that I_0 must be E/R. The requirement on τ may by determined by substituting (7.70) into (7.69) to yield

$$E = RI_0 e^{-t/\tau} - \left(\frac{\tau}{C}\right) I_0 (e^{-t/\tau} - 1) \tag{7.71}$$

To make the right-hand side independent of time, it is necessary that

$$\tau = RC \tag{7.72}$$

With this value of τ, (7.71) becomes

$$I_0 = \frac{E}{R} \quad \text{or} \quad E = RI_0 \tag{7.73}$$

Therefore, (7.72) and (7.73) specify values for τ and I_0 that make (7.70) a valid solution for (7.69). This solution is

$$i(t) = \left(\frac{E}{R}\right) e^{-t/RC} \tag{7.74}$$

Note that the current decays more slowly as the product *RC* is increased and vice versa.

7.6 SUMMARY

Some of the most important ideas used by electrical engineers have been introduced and discussed. The analysis of elementary dc circuits was used to show that Kirchhoff's voltage and current laws and Ohm's law are essential in finding the voltage, current, and power associated with each circuit element. The importance of time-varying signals in network analysis was also considered. Voltages and currents with sinusoidal, exponential, or sawtooth waveforms are common. Some of the ways these signals are generated were presented. Resistors, although classified as passive elements along with inductors and capacitors, are fundamentally different in nature from the last two devices. They dissipate electrical energy, whereas inductors and capacitors store it. Laws relating the voltage and current for all three of these devices were introduced, and the transient behavior of simple *RL* and *RC* circuits was studied.

REFERENCES

1. DELTORE, V., *Principles of Electrical Engineering.* Englewood Cliffs, N.J.: Prentice-Hall, Inc., 1965.
2. SMITH, R. J., *Circuits Devices and Systems.* New York: John Wiley & Sons, Inc., 1971.
3. WEDLOCK, B. K., and J. K. ROBERGE, *Electronic Components and Measurements.* Englewood Cliffs, N.J.: Prentice-Hall, Inc., 1969.

1. The resistivity, ρ, of copper at 20°C is 1.7×10^{-8} Ω-m. At 20°C, what is the resistance of a 10-ft piece of copper wire if it has a diameter of .010 in.?

EXERCISES

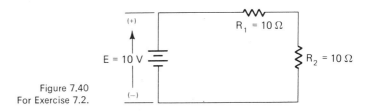

Figure 7.40
For Exercise 7.2.

2. In the circuit shown in Figure 7.40, what is the voltage drop across resistor R_1? Across resistor R_2? How much power is dissipated in each of these resistors?

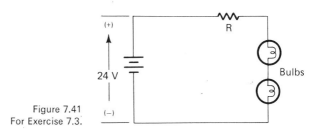

Figure 7.41
For Exercise 7.3.

3. In the circuit shown in Figure 7.41, what value of resistor R is required to cause a 6-V drop across each bulb if the resistance of each bulb is 8 Ω?
4. Three resistances having values of 10, 20, and 30 Ω are connected in series in a circuit. What is the total resistance?
5. The resistors in Exercise 4 are placed in parallel. What is the equivalent resistance?
6. Resistors of values 50×10^3 Ω, 250 kΩ, 1 MΩ, and 500 kΩ are connected in parallel. Compute the equivalent parallel resistance.

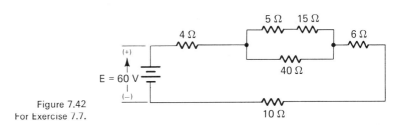

Figure 7.42
For Exercise 7.7.

7. In the circuit shown in Figure 7.42, determine the voltage drop across each circuit element as well as the power dissipated in each element.

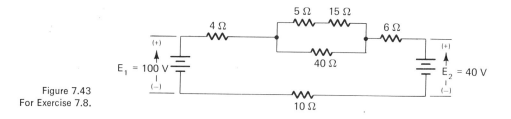

Figure 7.43
For Exercise 7.8.

8. In the circuit shown in Figure 7.43, determine the voltage drop across each circuit element as well as the power dissipated in each element. What is the total power dissipated?

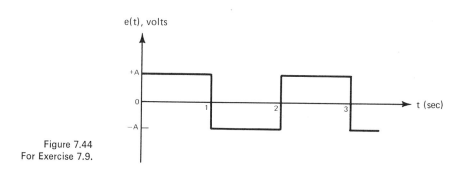

Figure 7.44
For Exercise 7.9.

9. A conductor is moving with velocity $v(t)$. The motion is in a magnetic field, B, and is perpendicular to the direction of the magnetic flux. The voltage generated between the ends of the conductor is periodic, as shown in Figure 7.44. Plot the position of the conductor as a function of time.

10. What is the average value of the voltage waveform shown in Exercise 9? What is the mean square value? The rms value?

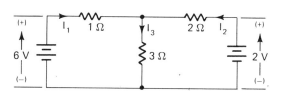

Figure 7.45
For Exercise 7.11.

11. Find I_1, I_2, and I_3 in the circuit shown in Figure 7.45.

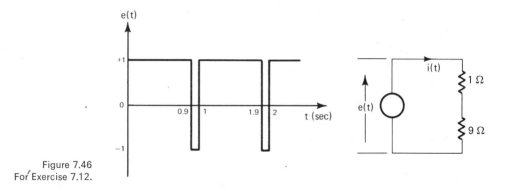

Figure 7.46
For Exercise 7.12.

12. The periodic voltage, $e(t)$, is applied to the circuit with two resistors in series (see Figure 7.46).
    a. Sketch the voltage waveform across each of the resistors.
    b. Find the average power dissipated in each resistor.
13. The voltage across part of an electric network is $v(t) = 10 \sin 2\pi 60 t$ and the current through it is $i(t) = \sin(2\pi 60 t - \pi/3)$. What is the average power being dissipated in this part of the network?

8

computer engineering

R. M. Glorioso

Computer engineering is one of the newest disciplines to emerge as a separate field of engineering. Generally, computer engineering is associated with electrical engineering, because computers were originally designed, constructed, and used by electrical engineers. Thus, the computer engineer is concerned with all aspects of the computer including the details of the internal structure and logic, the overall system design, the system instructions and software, programming, the interface to the ultimate human user (the so-called man/machine interface), as well as applications of computers in many areas. Graduate computer engineers do not work only for computer manufacturers. The use of computers in control, communications, and manufacturing systems has opened several opportunities for computer engineers in all areas of industry and government. For example, the ever-growing need to increase productivity will be met by incorporating computers of all kinds into the control and automation of production lines. In this chapter we shall examine the fundamental nature of the general-purpose digital computer and the exciting discipline of computer engineering.

8.1
HISTORICAL PERSPECTIVE

Although the electronic digital computer has been in existence only since 1946, man's involvement with ideas fundamental to the creation of the modern computer predate this development by at least 200 years. In fact, one can argue that the development of the concepts of numbers and symbols and indeed all man's history has contributed to the development of the modern digital computer. However, here we shall be concerned only with the developments beginning with Charles Babbage (1792–1871), who is generally acknowledged to have conceived many of the principles of automatic computing devices or analytic engines, as he called them.

In 1833 Babbage proposed the construction of a device that would receive its instructions and data from a set of punched cards of the type devised for Jacquard looms. This machine would add, perform logical operations, store intermediate results on fresh cards for later use, ring a bell when the next batch of cards were needed, and ring an even louder bell if the wrong card were placed in the machine. Alas, the analytic engine would never be constructed, and Babbage's genius would not be demonstrated in his lifetime. The reasons for this are many. First, Babbage had already spent £22,000 of government money on a never-to-be-completed difference engine-calculator and was generally recognized as an impractical dreamer. The impracticality of these ideas at that time was demonstrated by the mechanical nightmare of wheels, cogs, and gears that Babbage proposed to use in his engines. Besides the fact that the state of the machine tool industry was not up to satisfying the close tolerances of these parts, the power required to run the engine was formidable and not easily obtained in those preelectric power years.

In retrospect it is clear that Babbage was a century ahead of his time, for in 1937 Howard Aiken proposed and in 1944 completed his automatic sequence-controlled calculator or Harvard Mark I. This electromechanical machine, which was considerably more modest than Babbage's engine, used punched cards for data, instruction, and output and was actively used until 1959. Following the Mark I, Aiken at Harvard and Stibitz at Bell Telephone Laboratories built faster electromechanical computers using relays to store intermediate results.

The introduction of electronic techniques into the logic and memory of digital computers marks the real turning point in the eventual design of the general-purpose digital computer. J. Presper Eckert and John W. Mauchly at the Moore School of Electrical Engineering of the University of Pennsylvania, Philadelphia, began work on such a machine under a contract with the U.S. Army in the early 1940s. This machine was designed primarily for computation of ballistic tables and was called the electronic numerical integrator and calculator (ENIAC). The ENIAC used 18,000 vacuum tubes, was housed in a room 100 ft long, and needed 100 kW of electric power to run it. Programming of this machine was accomplished by changing the actual wiring by rearranging plugs on a patch panel, a rather inconvenient arrangement at best.

The final step in design, which also describes most contemporary computers, was proposed in the mid-1940s by John von Neumann.[1] He suggested that both the instructions and data be stored in a common memory within the machine. The instructions would be stored in consecutive memory locations and an external pointer would be used to indicate the location from which the next instruction was to be taken. This feature endowed the computer with the limitless flexibility and staggering potential, which has yet to be fully exploited.

From the 1940s to the late 1960s technology caught up with the ideas of Babbage and von Neumann. In the mid-1950s the first solid-state mobile military digital computer was built for the U.S. Army by Sylvania. This computer, designated MOBIDIC, utilized temperature-sensitive germanium transistors and needed extensive cooling for reliable operation. Less sensitive and more reliable silicon transistors were introduced into the now burgeoning computer industry in the early 1960s. Integrated circuits (ICs), which again increased the reliability and improved the performance of the computers, followed closely, and by the late 1960s computers without ICs were considered "old." The present trends in computer technology are toward faster, even more reliable, and physically smaller computers that use complex ICs called LSI (large-scale-integrated). Devices that promise to perform more computations in 1 sec than the ENIAC could compute in several hours are now available (and these take up 1 in.[3] and consume less than 1 W of power). The new pocket calculators which use these devices are new areas of application for computer technology. An LSI computer and a pocket calculator are illustrated in Figure 8.1. Indeed, computers and the associated technology have come a long way from the mechanical monstrosities of Babbage.

The impact of computers is not confined to technology alone. Computers and their sometimes insidious idiosyncrasies have pervaded many aspects of all our lives. In business, computers keep track of our bank accounts and charge cards, pay bills, control inventory, and even make some business decisions. In industry, machines are controlled by computers, manufacturing schedules are made up by computers, and in many companies crude robots that can be programmed to perform several different tasks whir about the production lines. Today's high-speed transportation systems on both the ground and in the air depend as much on the controlling computer as on the man at the controls.

Perhaps the best indicator of the influence of the computer is its effect on the economy. Presently the annual sales of computers and computer services accounts for nearly 10% of the gross national product (G.N.P.)[2] and it has been predicted[3] that by the year 2000 the computer industry will surpass

8.1 historical perspective

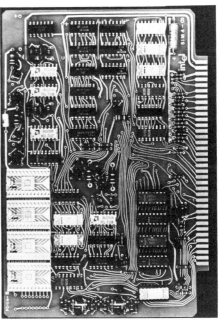

Figure 8.1
Pocket calculator and LSI computer.
(*Courtesy Hewlett-Packard Corp. and Intel Corp.*)

the automotive industry in dollar volume. Presently, the automotive industry accounts for approximately 25% of the G.N.P. The computer industry is also making its contribution to the U.S. balance of payments, since $12 billion worth of goods and technical services were exported in 1970 by this $100-billion-a-year industry.

The future of the computer industry seems assured by two basic factors. On the one hand, technological advances are continually lowering the costs associated with the manufacturing of computers and logical hardware, which in turn opens previously untapped markets, while, on the other hand, new applications for this technology are being discovered daily. It seems a certainty, for example, that computers or computer services will soon be used in every home as a calculating device, for rapid access to large stores of information, and even for placing orders from the Sears catalog.[4] The encyclopedia salesman of the future may only sell access to some particular computer in which an encyclopedia hundreds or even thousands of times larger than the conventional home set is available at the push of a button. The applications one can think of seem endless. However, before one can really appreciate the intracacies as well as the beauty of the general-purpose digital computer it is necessary to study its structure and operation.

8.2
GENERAL-PURPOSE DIGITAL COMPUTER

The block diagram in Figure 8.2 illustrates the basic elements of a general-purpose digital computer system. It is important to recognize that any system, no matter how complicated, must be concerned with the eventual user of the system—people. The user may not even be directly in contact with the machine; for example, the user may be the recipient of a computer-generated bill or a paycheck for $1 million. Thus, the utility and indeed the salability of a computer system centers about the so-called man/machine

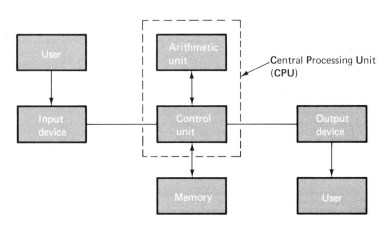

Figure 8.2
Block diagram of general-purpose digital computer.

interface. In a large computer system, there are often several interfaces between man and machine. In addition to the ultimate user mentioned above, there are the interfaces to the person who programs the computer, the service personnel who must run diagnostic programs and trace machine failures, and the managers of an organization using computers who must see computer-generated data and remain convinced that the expensive system is really needed. In the small or minicomputer systems that are becoming popular today the user often performs several of these functions.

For the moment, however, let us assume that a new machine has been delivered, that it is operational, and that we must get a problem solved. First, we must specify the problem in a logical step-by-step manner. The desired output and the form of the input data must also be specified. Here, we may be faced with many alternatives, as there are many input devices that convert data from an external machine form to an internal machine form. Some examples of input devices are the following: (1) *Punched card and punched tape readers* convert input data from coded holes in paper to electrical signals. (2) *Teletypes*, which are often used for remote computer terminals, provide a keyboard input and a teleprinter output at a relatively low cost. (3) *Graphic systems*, which generally use a large cathode-ray tube as an output device and a light pen, allow the user to communicate with the computer by virtually writing on the face of the tube. (4) *Analog-to-digital (A/D) converters* change inputs that vary continuously to signals that can change only in discrete steps. The conversion of a voice signal from a microphone to a form that can be handled by a computer requires an A/D converter. (5) *Discrete input devices*, such as measuring instruments and data-gathering equipment, are now being made with digital outputs that can be coupled directly into a general-purpose digital computer. The simplest discrete input device is a two-position switch that indicates a 0 or 1, the numbers with which the machine normally operates. Examples of some of these input devices are shown in Figure 8.3.

The control unit is the heart of the computer, as it is this device that controls and coordinates the flow of data depending on the instructions placed in the memory by the user. Thus, the flow of information within the central processing unit as well as to the outside world is under the direct control of this device.

The arithmetic unit, as its name suggests, is the portion of the computer that has the ability to carry out all arithmetic operations such as addition, subtraction, and multiplication much like a desk calculator. However, unlike most desk calculators, the arithmetic unit has the capability of making logical decisions: Is one number greater than, less than, or equal to some other number? The result of this decision can then be used to modify the course of the computer's operation. It is this feature, decisions that modify execution, that endows the computer with much of its power, since most complex decisions can be made using a sequence of the simple decisions easily made by a machine.

The arithmetic and control units are the fastest sections of the general-purpose digital computer. For example, the simple decisions described above

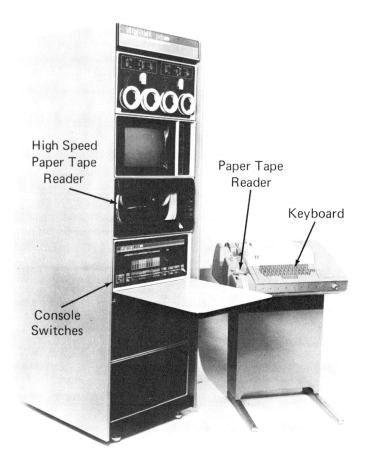

Figure 8.3
General-purpose computer system with, from the top down, magnetic tape, graphic display, paper tape reader/punch, CPU, and teletype on the right.
(*Courtesy Digital Equipment Corp.*)

can be carried out within 10 nsec (1 nsec = 10^{-9} sec). Additions are often carried out in 2 μsec and multiplications in 4 μsec (1 μsec = 10^{-6} sec).

The memory unit is one of the primary determinants of the "size" of a computer system because more instructions and more data can be handled without operator intervention on a computer with a large memory than on the same computer with less memory. Memory size is usually specified by the size and number of binary words that the computer can store. For example, a machine with a 12 × 4,096 memory has 4,096, 12-bit words. The exact definition of a word will be given later. Computer memory systems can take on many forms, but the mainstay of most computers is the magnetic core, shown in Figure 8.4. Its advantages are low cost, small size, high speed, and the ability to be accessed in a random manner. The speed of a memory is determined by the time required to change the magnetization from one direction to the other. This direction change is accomplished by reversing the directions of the currents I_x and I_y flowing in the X and Y planes. Each current is set to be one-half of that required to magnetize the core in either direction. In this way, we can now define a plane of cores in which current flowing in one x wire and one y wire will influence only one core. This so-called coincident current random-access memory is illustrated in Figure 8.5 with a 9-bit memory where the core designated $x = 2, y = 2$ is being addressed since current is flowing in only these two wires. Therefore, we can select any core at "random" in one step by simultaneously applying current to the appropriate x and y select lines. To read out the contents of any one core in the memory, however, we must use the third wire shown in Figure 8.5 as the sense lead. A voltage is "induced" on this wire when there is a change in the direc-

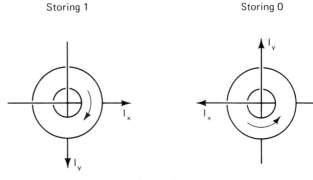

Figure 8.4
Magnetic cores storing 1 and 0.

$$||I_x| = ||I_y| = I_M/2$$

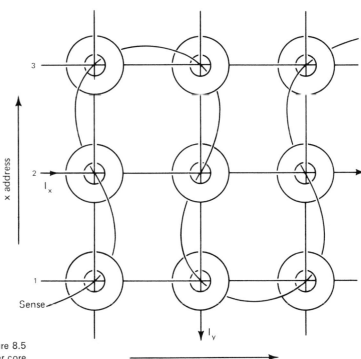

Figure 8.5
A 3 by 3 core plane where the center core is being addressed.

tion of magnetization. Thus, the read cycle proceeds as follows. The core to be read is addressed by applying X and Y current, which will set the appropriate core to 0. If no voltage appears on the sense wire, we know that a 0 was stored. On the other hand, if a voltage pulse appears on the sense lead, then we know that the direction of magnetization of that core has switched from the 1 direction to the 0 direction, thus inducing the voltage pulse on the sense lead. However, if a 1 was initially stored in memory, the read cycle has changed it to 0 and we must restore the 1 if we wish to continue to remember the contents of that location. Thus, we store the contents of the cores being read in some intermediate storage medium and immediately write that information back into that location. These gyrations are necessary to preserve the integrity of information stored in a core memory and nearly all computers today use this so-called read-write cycle. As involved as this process may appear, it is not unusual for it to take place in less than 800 nsec.

Although the magnetic core memory has enjoyed widespread use, it is now being challenged by the high-speed random-access solid-state electronic LSI memory. This device, which uses electronic rather than magnetic storage elements, is faster than core memories (often less than 200 nsec), consumes less power than cores [several milliwatts (1 mW = 10^{-3} W) versus several watts], and is considerably smaller than core memories of the same bit size. These very attractive features are achieved by using the photographic and fabrication technologies developed by the solid-state industry in the manufacture of transistors and ICs. The photograph in Figure 8.6 of a 1,024-bit LSI memory taken through a microscope graphically illustrates the high packing density achieved with this technology. The primary factor presently against the solid-state memory is its cost. However, the high speed of these

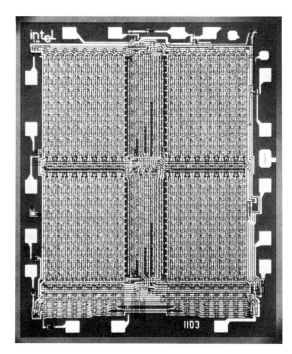

Figure 8.6
1024-bit LSI memory taken through a high-power microscope.
(*Courtesy Intel Corp.*)

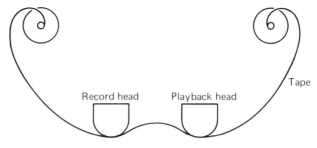

(a) Magnetic tape unit

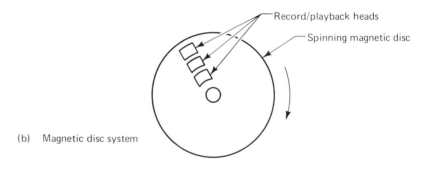

(b) Magnetic disc system

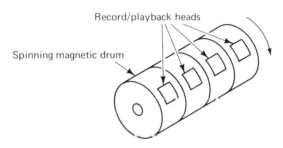

Figure 8.7 (c) Magnetic drum system

devices is being used to advantage in several computers as "scratch pad" memories. A "scratch pad" is used to store intermediate results much as one uses a tablet while solving a complex problem with a small desk calculator. Also, the low power requirement, light weight, and small size of these devices make them ideal for application in an aerospace environment. In these applications their weight, speed, and size outweigh their relatively high cost.

While cores and solid-state memories are fast, they are still relatively expensive devices for storing large volumes of data such as the data for many scientific experiments or the status of many bank accounts. Thus, if speed is not of the essence, other less expensive storage media are used. These other media use serial rather than random access. For example, the magnetic tape system illustrated in Figure 8.7(a) requires that the beginning of the tape pass the playback head before one gains access to the middle or the end of the tape. Thus, the information is stored sequentially (or serially), and access to a particular piece of information (a datum) may require that we pass through a series of data before that datum reaches the playback head. Two other types of memory that are commonly used in computer systems are the magnetic disc and drum. These are similar to magnetic tape in operation. In one case the magnetic material is deposited on a disc, and the record and playback heads are spaced over the disc. In the other case the magnetic material is deposited on a drum and the heads are placed over tracks, as shown in Figure 8.7(b) and (c). The information stored on a disc or drum is available within one revolution, and access times are shorter than tape (several milliseconds as opposed to several seconds). In all three of these systems the medium must be moved mechanically in order to write into or read from the memory. Thus, the reliability of these storage media is generally less than that of the electronic or electromagnetic storage media. For example, it is not uncommon for a mechanical failure to cause a head to "crash" into a magnetic disc, thereby destroying the disc as well as the information stored therein.

The final element of the block diagram of Figure 8.2 is the output device, which is the electronic or electromechanical means by which the computers' electrical signals are translated into a form fit for human consumption. Since the input and output devices are often used together, many of these devices were discussed previously. Some common output devices are teletypes, paper tape and card punches, digital-to-analog (D/A) converters, graphical output devices such as CRT displays and electromechanical plotting devices, line printers, and a light bulb, which is used to indicate the status of major elements of a computer system on its control console by lighting to indicate a 1 and remaining off to indicate a 0. Figure 8.8 illustrates some of these output devices and their respective outputs.

Figure 8.8
(a) Cathode ray tube graphic I/O unit. (b) Line printer and keyboard I/O system.

The operation of the general-purpose digital computer then proceeds as follows. First, the instructions and data are entered into the memory by the user via the input device under the control of the control unit. The control unit then sequences through the instructions step by step by performing the indicated instructions using the arithmetic unit, obtaining additional data and instructions from the input devices, and presenting results or error statements to the user via the output devices. The processing of instructions will be considered in more detail in a later section.

8.3 COMPUTER CODING AND NUMERICAL REPRESENTATIONS

The instructions and data within a computer system must eventually be coded in a form that can be directly interpreted and manipulated by the machine hardware. Thus, no matter what language one uses to write his program (FORTRAN, BASIC, APL, etc.) or what format the data are in, the program eventually must be reduced to this machine language. The internal machine language used in general-purpose digital computers is a *binary code*, and the numbers in the machine use the *binary number system*. In this system, each discrete position of the code or each place in a number can take on only one of two different values, namely 0 or 1. This two-state system is used because it readily lends itself to the use of simple, reliable, and inexpensive components that are easily mass-produced. Because we must represent both numbers and operational instructions using the same binary system, we must consider two types of internal machine coding: symbolic and numeric representations.

A number in any number system can be represented by a string of digits,

$$d_{n-1}\cdots d_2 d_1 d_0 \underset{\text{digit point}}{\cdot} d_{-1} d_{-2} \cdots d_{-m}$$

for example, 34.6 or 10.11. Since one cannot necessarily tell what number system a given number is in, one frequently uses a subscript after the number to indicate the system being used, for example, 985.0_{10}, which is nine hundred eighty-five in the decimal number system, or 110_2, which is the binary number representation for 6_{10}. The decimal representation for a number in a number system which has r symbols (the decimal system has 10 symbols, 0 through 9) is

$$N_{10} = \sum_{i=-m}^{i=n-1} d_i r^i \tag{8.1}$$

where r is the radix or base of the number system being used. For example, the base of the binary system is 2, and of the decimal system, 10. The decimal

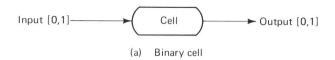

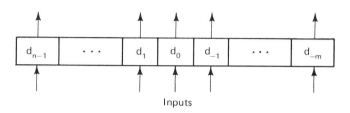

Figure 8.9

(a) Binary cell

(b) Binary register containing the binary number $d_{n-1} \ldots d_1 d_0.d_{-1} \ldots d_{-1}$

number 536.4_{10} can thus be given by

$$N_{10} = 5 \times 10^2 + 3 \times 10^1 + 6 \times 10^0 + 4 \times 10^{-1} = 536.4_{10}$$

and the binary number 110_2 is given by

$$N_{10} = 1 \times 2^2 + 1 \times 2^1 + 0 \times 2^0 = 6_{10}$$

See the Exercises for problems on number systems.

In digital systems binary numbers are organized in systematic arrays called registers. Each register consists of a set of basic elements called cells. A cell is a binary storage element with an input and an output into which a 1 or a zero can be stored. A basic cell is illustrated in Figure 8.9(a).

A single memory core is an example of a binary cell. Thus, the output of a cell is generally considered to remain at the previous value entered into it until the output is directed to change. Each cell can then be used to store one binary digit, or *bit*. A collection of these cells in a systematic array forms a register, as illustrated in Figure 8.9(b). The registers in the arithmetic and control units of a computer usually have the same number of cells as the words of the memory. Therefore, a computer with an 8×256 core memory has at least eight cells in each of its registers. The central registers in a computer, called accumulators, usually have at least one extra cell to enable the machine to handle carries (overflows as a result of addition of two numbers) of one additional bit. The register in Figure 8.10 has eight cells, and the

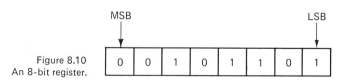

Figure 8.10
An 8-bit register.

decimal representation of its contents is given by

$$N_{10} = 0 \times 2^7 + 0 \times 2^6 + 1 \times 2^5 + 0 \times 2^4 + 1 \times 2^3 + 1 \times 2^2 \\ + 0 \times 2^1 + 1 \times 2^0 = 45_{10}$$

Note that the leftmost bit is the most significant bit (MSB) and the rightmost bit is the least significant bit (LSB). It is interesting to observe the MSB and LSB as a register counts from 0 to its maximum value. For example, consider the 4-bit register in Figure 8.11. The LSB changes every count, while the next bit changes every other count and the MSB changes after 8 counts. If one tries to count one more than 15, the register returns to 0. Thus, the register has 16 different states.

Decimal Value	2^3 MSB	2^2	2^1	2^0 LSB
0	0	0	0	0
1	0	0	0	1
2	0	0	1	0
3	0	0	1	1
4	0	1	0	0
5	0	1	0	1
6	0	1	1	0
7	0	1	1	1
8	1	0	0	0
9	1	0	0	1
10	1	0	1	0
11	1	0	1	1
12	1	1	0	0
13	1	1	0	1
14	1	1	1	0
15	1	1	1	1

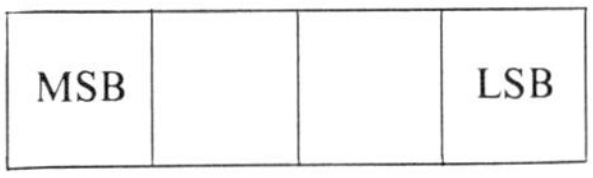

Figure 8.11.
Counting with a 4 bit register.

There are other number systems that are used in computer coding. Among them are the octal (base 8), hexadecimal (base 16), and the binary coded decimal systems. The octal and hexadecimal systems are natural extensions of the binary systems where 3 bits make an octal system and 4 bits make a hexadecimal system, as illustrated in Figure 8.12. Note that the last six symbols in the hexadecimal system may be chosen arbitrarily and can be any symbols other than 0–9. The octal number system is by far the most popular, as it uses only numbers rather than both numbers and letters in the representation. A binary register can be converted to an octal register by considering groups of 3 bits or to a hexadecimal system by considering groups of 4 bits.

Decimal	Octal	Binary Representation of Octal Numbers (3 bits)	Hexadecimal	Binary Representation of Hexadecimal Numbers (4 bits)
0	0	000	0	0000
1	1	001	1	0001
2	2	010	2	0010
3	3	011	3	0011
4	4	100	4	0100
5	5	101	5	0101
6	6	110	6	0110
7	7	111	7	0111
8			8	1000
9			9	1001
10			A	1010
11			B	1011
12			C	1100
13			D	1101
14			E	1110
15			F	1111

Figure 8.12.
Octal and hexadecimal number systems.

Example 1

Give the octal and hexadecimal representations of the following 8-bit word:

10101110

For the octal system, divide the register into groups of 3 bits starting with the LSB and use the table in Figure 8.10 for each group:

$10/101/110 = 256_8$

For the hexadecimal system, divide the register into groups of 4 bits starting with the LSB and use the table in Figure 8.10 for each group:

$1010/1110 = AE_{16}$

Example 2

Find the decimal representations of the following numbers:

173_8
CAD_{16}

$173_8 = 1 \times 8^2 + 7 \times 8^1 + 3 \times 8^0 = 123_{10}$
$CAD_{16} = C \times 16^2 + A \times 16^1 + D \times 16^0$
$= 12 \times 16^2 + 10 \times 16^1 + 13 \times 16^0 = 3{,}245_{10}$

It is now of interest to examine the coding of symbols in a digital computer and specifically the coding of instructions in machine language. To illustrate this language, we shall use the UMALAB (UMASS Laboratory) computer, which has 32, 8-bit words, labeled 0_8 to 37_8 consecutively, with 8-bit registers as an example. From the block diagram given in Figure 8.2, we know that we must be able to code certain instructions into the machine. For example, that we must have instructions that operate input-output equipment, that read from and write into the memory, and that cause arithmetic operations to occur are obvious from the diagram. These instructions are, in fact, the three major classes of instructions used in all general-purpose digital computers and are called I/O, memory reference, and operate-instructions, respectively.

Since there are 32 different memory locations, it is necessary to be able to generate $2^5 = 32$ different numbers to address the memory. Thus, when the memory is to be addressed, the last 5 bits of the 8-bit instruction word are used for the address. This leaves the first 3 bits of the word to uniquely specify the major instructions in the machine. These first 3 bits of the word then contain our *operations code* or OP code. Also, the OP code is used to differentiate between the different memory reference instructions.

The UMALAB computer has only one central register—accumulator—in which arithmetic operations take place. Thus, most memory reference instructions involve the accumulator. The organization of the memory reference instructions and the corresponding 3-bit OP codes are given in Figure 8.13. The first two instructions enable us to read from and write into the

8.3 computer coding and numerical representations

Octal Representation	OP Code	Instruction
0	000	*Transfer* contents of location given by the last 5 bits from *memory* into the *accumulator*
1	001	*Transfer* contents of *accumulator* into *memory address* given by the last 5 bits and clear accumulator (all bits 0)
2	010	*Add* contents from *memory* location given by last 5 bits to the contents of the *accumulator*
3	011	*Unconditional Jump* — Take as the next instruction the contents of the word in the address given by the last 5 bits

Figure 8.13
Memory reference instructions and their OP codes.

memory. The third code initiates a memory reference arithmetic instruction, add. Having only the ability to add may at first appear to be a drawback in this computer's instruction set. However, this is all that is needed because subtraction, multiplication, and division can all be achieved by combining the add with operate-instructions. The last memory reference instruction is the unconditional jump. As stated previously, a computer examines successive memory locations from some starting location, interpreting each as an instruction to be carried out. A special register called a program counter (PC) is used to indicate the address in memory of the next instruction. The jump instruction merely changes the contents of the PC to that indicated in the jump instruction.

There are still four OP codes that have not yet been described here. The OP code 4_8 is used to activate a particular input device, while 5_8 activates a particular output device. OP codes 6_8 and 7_8 are used to denote the operate-instructions. These instructions are summarized in Figure 8.14. Since

	OP Code	Octal Representation	Remaining 5 Bits	Instruction
Input-output	100	4_8	00000	Read from teletype key board into the accumulator
	100	4_8	00001	Read from paper tape reader into accumulator
	101	5_8	00000	Print character stored in accumulator on teleprinter
	101	5_8	00001	Punch character stored in accumulator with paper tape punch
Operate-instructions	110	6_8	00000	Shift bits in accumulator 1 bit right
	110	6_8	00001	Shift bits in accumulator 1 bit left
	110	6_8	01101	Complement accumulator—change 1s to 0s and 0s to 1s in accumulator
	110	6_8	01110	Increment accumulator—add 1 to the accumulator
	111	7_8	00000	Skip the next instruction if all bits in the accumulator are 0
	111	7_8	00001	Skip the next instruction if MSB of accumulator is 0
	111	7_8	00010	Skip the next instruction if MSB of accumulator is 1
	111	7_8	00101	Clear the accumulator—set all bits = 0.
	111	7_8	11111	Halt—stop machine

Figure 8.14.
Coding of I/O and some operate instructions for UMALAB.

UMALAB has only two input and two output devices, namely a teletype keyboard/printer and a paper tape reader/punch, physically attached to the computer, just two of the possible 32 input and 32 output instructions are used. It is clear, then, that we can accommodate 30 additional input devices and 30 additional output devices in our instruction set. Also, the instruction length will allow 64 micro instructions. However, only a selected few are shown here for illustration, as consideration of the complete UMALAB repertory is beyond the scope of this text. The shift, complement, and increment operate-instructions are used in the process of subtraction, multiplication, and division. For example, given the following word and the same word shifted left 1 bit,

$$00011101 = 29_{10}$$
$$00111010 = 58_{10}$$

a multiplication by 2 has occurred. Similarly, a shift right by 1 bit is a division by 2. It is clear that multiplication by 2 of numbers in which the MSB is 1 or divisions by 2 where the LSB is 1 cannot be accomplished within the 8-bit word. This requires the use of additional storage to create virtual double-length or 16-bit words.

The skip instructions are the simple decision-making operations mentioned earlier. It is these instructions that allow one to carry out the decisions that change the course of the program being executed. For example, assume that we wish to read in five successive words from the paper tape reader or perform some other operation five times. We merely set some location in memory to 4_{10} less than the largest value that the register can have, in this case 373_8 (since the largest number an 8-bit register can hold is $11111111_2 = 511_{10} = 377_8$). Then every time the operation occurs we add 1 to the contents of that location and check for a 0 in the accumulator before placing it back into memory. If the accumulator is not 0, we know that the operation has not been performed five times and the next instruction jumps back to perform the operation again. On the other hand, when our check indicates the accumulator contains a 0, then the operation has occurred the given number of times, the jump instruction is skipped, and the computer proceeds to the next statement. This process is illustrated in Figure 8.15. The last three statements in this program are as follows, where the read statement is in location 3_8 and the octal numbers indicate the locations where these instructions are stored:

    6_8 11100000   skip next instruction if accumulator is 0
    7_8 01100011   jump to 3_8
    10_8 11111111   halt

Next, let us assume that we have data stored in locations 32_8 through 37_8 and that we wish to add the numbers in locations 32_8 and 36_8, store the

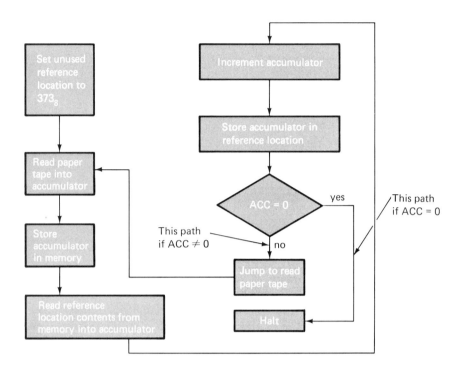

Figure 8.15
Flow chart to read paper tape five times and store in memory.

results in location 31_8, and jump to location 5_8 where some other program is stored. This program starts in location 24_8:

24_8	11100101	clear accumulator
25_8	00011010	transfer from 32_8 into accumulator
26_8	01011110	add contents location 36_8 to accumulator
27_8	00111001	store contents of accumulator in location 31_8
30_8	01100101	jump to location 5_8 for next instruction

The programming of a computer at its most basic level, then, must be concerned with the machine coded sequences of 0s and 1s. It should be apparent that a program in terms of sequences of binary numbers would place a burden on the programmer, especially for machines with word lengths of 64 bits. It is for this reason that higher-level languages such as FORTRAN, COBOL, ALGOL, and APL have been developed. These languages are much more people-oriented and therefore are easier for us to understand. Some statements in these languages, for example, may generate 50 or more machine language instructions, thereby eliminating much of the drudgery associated with machine language programming. The program that adds two numbers given above may be written as $C = A + B$ in one of these languages, which certainly is easier to understand than the machine sequence given above. The description of these languages seems to contradict the statement made earlier that a machine can understand only binary numbers. However, special programs called compilers have been written for computers that translate the people-oriented symbols of these higher-level languages to the machine-oriented sequences of 0s and 1s.

It is now apparent that the internal structure of a digital computer depends on the decoding and manipulation of sequences of 1s and 0s. Each instruction to be executed must create a unique sequence of binary operations. For example, the operations code for each instruction must be decoded in order to determine if that instruction requires a memory reference or not. Thus, the sequences 000, 001, 010, and 011 must all activate the computer's memory cycle. Also, the skip instructions, part of the OP code 7_8 instructions, require that logical decisions concerning the value of the accumulator be made in order to determine which instruction will be executed. The following sections are concerned with the representation and manipulation of the logical signals in a digital computer.

8.4 SIGNALS IN LOGICAL SYSTEMS

The logical operations that take place in a computer generally make use of electrical signals to represent the numbers associated with cells and registers. These electrical signals travel through the computer causing one or more parts of the machine to become active. It is interesting to note that although the electrical power associated with each signal in a computer is low, it is not zero, and the transfer of information in any system requires the expenditure of energy. Hence, it costs money to transfer information.

The interesting feature of signals in digital systems is their all-or-nothing nature. Since we need only to associate a 1 or a 0 with each cell in a computer, it is not necessary to determine the relative magnitude of all possible levels of the signals in these systems. It is necessary only to decide if a signal is at one level or the other level. Thus, we can define one voltage level to represent the logical 1 and the other level to represent the logical 0.

One way of creating these levels, illustrated in Figure 8.16, uses simple

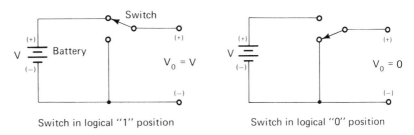

Figure 8.16
Switches used to generate binary voltage levels with positive logic convention.

Switch in logical "1" position · Switch in logical "0" position

switches. Thus, when the switch is on, a positive voltage appears across the terminals, and when the switch is off, no voltage is present. If we use the so-called positive logic convention, when the most positive voltage is present, we have a binary or logical 1 and, conversely, when the most negative voltage is present (in this case 0 voltage), we have a logical 0. The logical 1 is often referred to as logically *truth*, and the logical 0 is called logically *false*.

8.5 DIGITAL LOGIC

Every computational function in a computer or digital system requires the use of logic. The most interesting thing about this logic is that it can be broken down into three rather simple logical operations sometimes called *fundamental logic operations*.

The simplest of the three fundamental logic operations is called *inversion* or *complementation* and is represented symbolically in Figure 8.17. The operation of an inverter merely reverses the logical sense of the input. That is, if the input is a logical 1, the output is a logical 0 and vice versa. There are several physical devices that are used to carry out the inversion process. These devices, called inverters, can be electromechanical relays, transistors, integrated circuits, or mechanical gears and levers. No matter what the physical implementation is, the input-output characteristics of an inverter can be represented in a tabular form called a truth table as given in Figure 8.18. The truth table is a complete representation of the behavior of a logical system, as all possible input combinations are examined. The output of the inverter is the complement of the input, or, since we are using the binary system, we sometimes say that the value of the output is *not* the value of the input. This relationship can also be represented symbolically by

$$Y = \bar{X} \qquad (8.2)$$

where Equation (8.2) is read "Y is NOT X". Thus, the inverter is sometimes called a NOT circuit.

Input X ——▷o—— Output Y

Figure 8.17
Logical representation of inversion.

Input X	Output Y
0	1
1	0

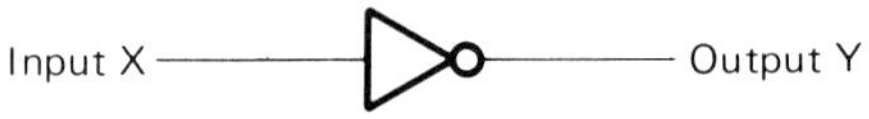

Figure 8.18
Truth table representation of inverter operation.

Figure 8.19
Inclusive OR (2-input) logic symbol and truth table.

Input		Output
X_1	X_2	Y
0	0	0
0	1	1
1	0	1
1	1	1

Although the inverter is an important logical element, it alone is not sufficient for computer logic. The ability to accommodate more than one input variable is needed to design computer logic systems. Thus, let us consider the logical connective between two or more variables, called *union*. Here we say that the output is logically true if one or the other or both of the inputs are logically true. This operation is also called the inclusive OR or merely OR and is illustrated in Figure 8.19. The expression for the OR operation in Figure 8.19 is

$$Y = X_1 + X_2 \tag{8.3}$$

where plus sign is used to indicate the relation between the input variables. The equation is read "Y is true if X_1 is true OR X_2 is true." Although we have considered only the two-input OR gate in our example, the operation can easily be extended to any number of input variables, and devices can be built to do this:

$$Y = X_1 + X_2 + \cdots + X_r \tag{8.4}$$

It can be shown mathematically that any logical function of any number of variables can be constructed using only OR gates and inverters. Thus, we have now defined the basic components needed to construct any digital logic circuit. It does not end here, however; and the constraints and advantages of particular technologies have created the need to define other logical functions. As it happens, the inverter circuit is simply an amplifier and as such is quite capable of driving several other logic elements. On the other hand, the simplest and therefore least expensive OR circuit is not capable of driving many other logic elements. The number of other logic elements that a particular logic element can drive is defined as the *fan-out* capability of that element. Thus, inverters have a high fan-out capability, while simple OR gates have a low fan-out capability. The solution to this problem is then to combine the logic of the OR function with the fan-out capability of the inverter, as shown in Figure 8.20. This combined device is called a NOR gate, which retains the universal logic capability of the individual logic gates.

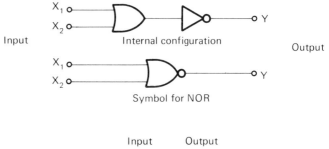

Symbol for NOR

Input X_1 X_2	Output Y
0 0	1
0 1	0
1 0	0
1 1	0

Figure 8.20
NOR gate symbology and truth table (2-input).

The notation that describes the operation of the NOR is, of course, the combination of the OR followed by the inversion:

$$Y = \overline{X_1 + X_2} \tag{8.5}$$

Now, let us examine the last of the three fundamental logical operations used in computer design. The logical connective associated with this device is called *intersection*, and the output is true if and only if all of the inputs to the device are also true. Thus, for a two-input gate (X_1, X_2) both X_1 and X_2 must be true for the output to also be true. Thus, we call this an AND gate, and its logic symbol and truth table are given in Figure 8.21. The expression for the AND operation is

$$Y = X_1 X_2 \tag{8.6}$$

The AND gate in its simplest form also has poor fan-out capability and is therefore teamed up with an inverter to create another universal logic ele-

Figure 8.21
AND gate (2-input) logic symbol and truth table.

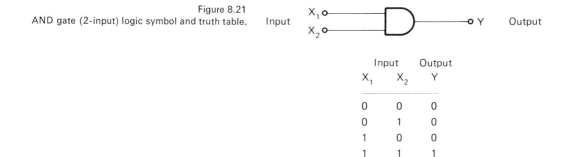

Input X_1 X_2	Output Y
0 0	0
0 1	0
1 0	0
1 1	1

ment called a NAND gate. Since this gate is a combination of the AND and the inverter, the expression that describes the NAND indicates that

$$Y = \overline{X_1 X_2} \tag{8.7}$$

The symbol and truth table for the NAND are given in Figure 8.22. The extension of any of these functions to any number of input variables is straightforward.

We have now examined the three basic logic functions (AND, OR, and NOT) needed to design digital logic. Any logic system with which one can realize these functions is called a universal logic system. This appears to contradict the statements made above concerning the universality of the NAND and NOR as each of these gates were defined as universal. It can be shown, however, that each of these elements can in fact be configured to form these three basic operations, AND, OR, and NOT (see the Exercises).

8.6 DE MORGAN'S THEOREM

De Morgan's theorem is one of the most important theorems associated with digital system design. It is easily used to convert the forms of the expressions associated with NAND and NOR logic. Simply stated, De Morgan's theorem says that the inverse of an AND is an OR with each variable inverted,

$$Y = \overline{X_1 X_2} = \bar{X}_1 + \bar{X}_2 \tag{8.8}$$

and the inverse of an OR function is an AND with each variable inverted,

$$Y = \overline{X_1 + X_2} = \bar{X}_1 \bar{X}_2 \tag{8.9}$$

The proof of (8.8) the first part of De Morgan's theorem, can easily be

Figure 8.22
NAND gate (2-input) logic symbol and truth table.

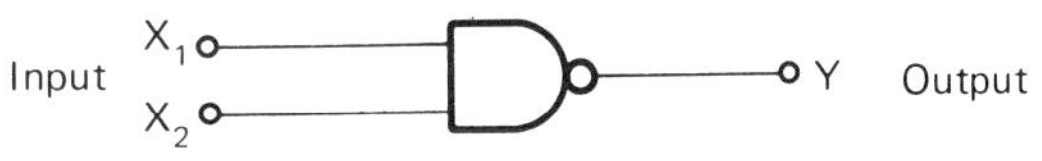

Input	Output
X_1 X_2	Y
0 0	1
0 1	1
1 0	1
1 1	0

carried out using a truth table:

Input		A	B	C	D	E
X_1	X_2	$X_1 \cdot X_2$	$\overline{X_1 \cdot X_2}$	$\bar{X}_1$	$\bar{X}_2$	$\bar{X}_1 + \bar{X}_2$
0	0	0	1	1	1	1
0	1	0	1	1	0	1
1	0	0	1	0	1	1
1	1	1	0	0	0	0

Columns B and E represent the two statements associated with the first part of De Morgan's theorem, and, since these columns are identical, the first part of the theorem is proved.

8.7 BINARY FUNCTIONS

Combinations of the logic devices described above are used to form the various logic functions used in computers and digital systems. For example, consider the truth table given below:

Input		Output	
X_1	X_2	S	C
0	0	0	0
0	1	1	0
1	0	1	0
1	1	0	1

The logic expression for the first output function, which is called the EXCLUSIVE OR, is

$$S = \bar{X}_1 X_2 + X_1 \bar{X}_2 \tag{8.10}$$

and the second output function is

$$C = X_1 X_2 \tag{8.11}$$

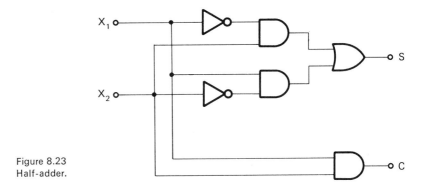

Figure 8.23
Half-adder.

The realizations of these two functions are given in Figure 8.23. The function S may be recognized as the binary sum of the inputs X_1 and X_2 and C as the associated carry function. Thus, the circuit in Figure 8.23 is called a half-adder. A full adder also has the capability of handling a carry input from an an adjacent sum bit.

Finally, let us consider the logic circuit given in Figure 8.24 and examine the procedure for analyzing it. First, label the intermediate logical outputs, say, z_1, z_2 and write logical expressions for them. Note that the two inputs on the first level NAND gates are tied together, thereby forming inverters. Thus,

$$z_1 = \bar{X}_1$$
$$z_2 = \bar{X}_2$$
(8.12)

Now, write the logic expression for the next level logic:

$$Y = \overline{z_1 z_2}$$
(8.13)

Applying De Morgan's theorem to (8.13),

$$Y = \bar{z}_1 + \bar{z}_2$$
(8.14)

and the final expression is the logical OR

$$Y = X_1 + X_2$$
(8.15)

We have now thus shown that NAND gates can be configured to perform the OR function.

Figure 8.24
Logic network.

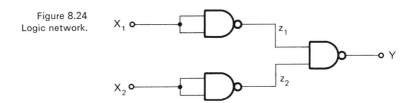

8.8
THE FLIP-FLOP AND ITS APPLICATION

The registers in the arithmetic and control units of a computer are usually made from integrated circuits in each of which one or more *flip-flops* or electronic cells are located. The basic flip-flop illustrated in Figure 8.25 has two logical inputs J and K and two logical outputs Q and $\bar{Q}$. The truth table for the cell shows the *next* output at Q given the present inputs as J and K. Note that the input is applied to the input first, next the clock goes from 0-to-1, and a short time later the output changes. Thus, the truth table assumes that the input appears at time n and the output changes at time $n + 1$, which is denoted by Q_{n+1}. Thus, logical 1s on both inputs cause the output to change states. The third input, labeled C (clock) in Figure 8.25, is used to control the changes in the state of the flip-flop from one time interval to the next. Thus, changes of the inputs are not reflected at the outputs until the clock input goes from 0 to 1. This device is also called a *gated* or *clocked J-K* flip-flop.

A collection of flip-flops can obviously be configured to form a register as shown in Figure 8.26. Here the clocks are all connected to the same line.

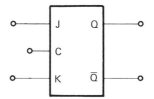

J	K	Q_{n+1}	$\bar{Q}_{n+1}$
0	0	Q_n	$\bar{Q}_n$
0	1	0	1
1	0	1	0
1	1	$\bar{Q}_n$	Q_n

Figure 8.25
Electronic cell—the J-K flip-flop.

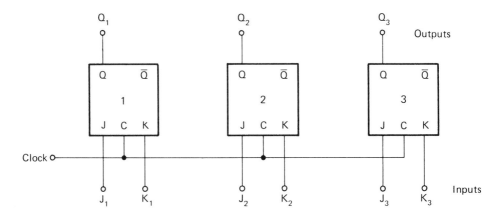

Figure 8.26
A 3-bit register using J-K flip-flops.

Thus, when the clock line goes from 0 to 1, the information present on the inputs appears on the outputs and the new information is stored in the register.

Flip-flops can also be connected as counters. A single *J-K* flip-flop with the *J* and *K* lines connected to a 1 becomes a one-bit counter. Every time the clock goes from 0 to 1 the flip-flop changes state. If the output of this flip-flop is connected to the clock of a second flip-flop, as shown in Figure 8.27, we have a 2-bit counter. This is called a ripple counter since changes take place in only one flip-flop at a time; the changes ripple from the LSB to the MSB. The addition of more bits to the counter in Figure 8.27 is straight-forward.

Registers made of flip-flops are used to store data, instructions, and addresses. Counters made of flip-flops are used to keep track of the machine's operation, for example, whether it should be fetching an instruction or data or should be executing an instruction. The location of the next instruction to be fetched from memory is stored in a counter called the *program counter*. Flip-flops, registers and counters are also found in a myriad of other applications such as electronic musical instruments, FM stereo tuners, burglar alarms, and electronic timers.

8.9
COMPUTER SYSTEMS

Let us now look at the combination of the machine language codes with the logic design developed thus far. As stated previously, the machine language codes must be interpreted properly for each machine instruction. Thus, it is reasonable to place an instruction that has been fetched from memory in an

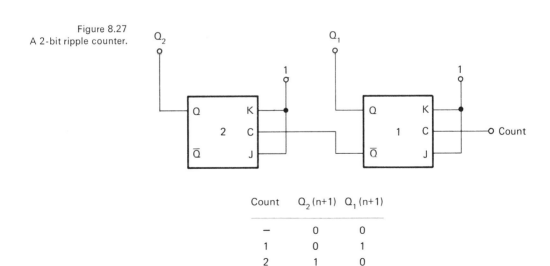

Figure 8.27
A 2-bit ripple counter.

Count	$Q_2(n+1)$	$Q_1(n+1)$
—	0	0
1	0	1
2	1	0
3	1	1
4	0	0

electronic register called the instruction register (IR). The output of this register must then be decoded by a logic network to determine the sequence of events needed to execute the instruction. A simplified diagram of the IR and decode logic for UMALAB is given in Figure 8.28. The decode logic places a 1 on the lines, which enables the execution of a given instruction to occur. Not all outputs will necessarily be logically 1 for any instruction. It is possible to have one, two, or more outputs at 1 for any arbitrary instruction. A specific example of this problem is given in the Exercises.

The operation of this system proceeds as follows. An 8-bit instruction word is fetched from the location in memory that is indicated by the program counter register. The contents of that location are stored in the instruction register. The instruction is then decoded, and the output of the decode logic enables the appropriate instructions to occur.

UMALAB is a very small computer with only one accumulator and a truly minimal memory. Other computers have many other features such as multiple accumulators, intricate addressing hardware, and the ability to perform multiplications directly with hardware. However, their basic principles of operation are identical to UMALAB.

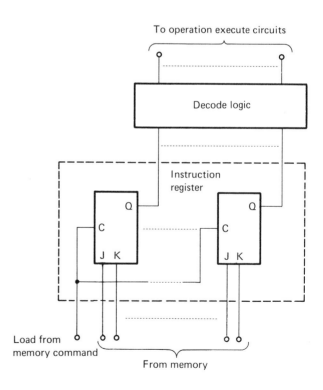

Figure 8.28
Instruction register and decode logic for UMALAB digital computer.

8.10 SUMMARY

This chapter has introduced some of the interesting aspects of computer engineering. The basic operation and machine language programming of a simple general-purpose digital computer has been outlined. Finally, the fundamentals of logic and logical design were introduced.

Perhaps the significance of this chapter lies in the material that is not included. For example, the basic configuration of the general-purpose digital computer has several interesting ramifications that allow computations to take place more rapidly. The intricacies of system programming that allow the efficient integration of several different storage media into one computer system is another area not discussed here. The logical design of computer elements requires the incorporation and trade-off between parameters such as economy, size, and power requirements as well as the fundamentals presented here. Also, many interesting applications of computers in areas such as data acquisition and processing, linguistics, pattern recognition, and artificial intelligence have not been discussed here. This material is included in many of the advanced courses in computer engineering and science.

REFERENCES

1. LEFF, A. L., "Franchised Computer Power," *Modern Data Systems,* April 1969.
2. "News Briefs," *Datamation*, June 1967.
3. FOSTER, C. C., "When the Chips Are Down," No. TN/CS/0028. Amherst: University of Massachusetts, Computer and Information Science Program Publication, Feb. 1972.
4. PYLYSHYN, Z. W., ed. *Perspectives on the Computer Revolution.* Englewood Cliffs, N. J.: Prentice-Hall, Inc., 1970. Contains several papers by contributors to the development of the computer including Babbage and von Neumann.

EXERCISES

1. Find the decimal values of the following:
    a. 1101_2.
    b. 101.101_2.
    c. 376_8.
    d. $C8A_{16}$.
2. Convert the following decimal numbers to hexadecimal, octal, and binary numbers:
    a. 10.
    b. 72.
    c. 128.

3. Describe the operations performed by the following machine language program. Assume that the input data are 10_8 and 17_8, respectively.
   a. 11100101.
   b. 10000000.
   c. 00100111.
   d. 10000000.
   e. 01000111.
   f. 11001101.
   g. 11001110.
   h. 10000111.
4. Find a method for subtracting using the UMALAB computer with the given instruction set.
5. Using a truth table, prove the second part of De Morgan's theorem.
6. Provide a logic symbol, truth table, and logic expression for three input NAND and NOR gates.
7. Show that the AND and NOT functions can be realized using NAND gates.
8. Show that AND, OR, and NOT functions can be realized using NOR gates.
9. Design a logic circuit to realize the following truth table: where Y is the output and X_1, X_2, and X_3 are the inputs:

X_1	X_2	X_3	Y
0	0	0	0
0	0	1	1
0	1	0	0
0	1	1	0
1	0	0	1
1	0	1	0
1	1	0	0
1	1	1	1

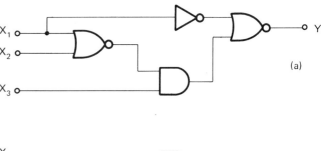
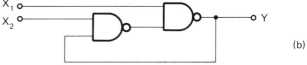

Figure 8.29
For Exercise 8.10.

10. Describe the operation of the logic circuits shown in Figure 8.29.
11. Design 3-bit pulse counters using *J-K* flip-flops with the features indicated successively in each part of the problem:
    a. It counts down from 111 to 000.
    b. It counts down and provides an output indicating that the state 011 has been reached.
    c. It counts up and stops counting when 011 has been reached.
12. Design an IR decoder for UMALAB that provides a 1 output for each memory reference instruction.

9

communications in the modern world

F. S. Hill, Jr.

9.1 INTRODUCTION

It is hard to imagine how different the world would be if all communications equipment suddenly disappeared. The ability to communicate easily over long distances has become so commonplace in our society that we take it completely for granted. In fact we are outraged when something goes wrong. The three most common communications systems, telephone, radio, and television, are used daily by almost everyone, and this has kindled an "information explosion," which is having powerful consequences in our culture. The world seems much smaller when one can sit at home and watch events as they occur on another continent. As video telephones and computer access become standard in the home, communications will be even more important in our lives.

Things were not always this way, of course: The incredible feats now performed by communication networks have only recently become possible. For centuries communications were very limited. Messengers delivered information by hand, or crude communication systems were devised for short distances using drums, smoke signals, or semaphore. A great leap forward came with the telegraph and Morse code in 1838, as information could rapidly be sent in coded form over much greater distances. Alexander Graham Bell invented the telephone in 1875, permitting speech itself to be transmitted, thus avoiding the need for translating the message first into letters and then into dots and dashes. Around 1890 Marconi and Popov demonstrated through radiotelegraph experiments that wireless communication was indeed feasible over long distances, and with the invention of the triode amplifier by de Forest in 1906 the age of electronic communication was firmly established. Since that time there have been dramatic leaps forward both in theory and practice, until today all modern countries are densely criss-crossed with communication networks. Now telephone, radio, and television make up only a small part of the communications traffic. A great deal of information is sent directly from one machine to another, with no human intervention. Computers can share data with one another, as well as monitor and control automated manufacturing processes. The field of communications has developed from an initial effort to enable people to converse at a distance into a gigantic area of study where information is treated in a very general way and manipulated freely to suit many different needs.[1]

In this chapter we shall discuss some of the fundamental ideas of communications and see how they are related to systems actually in use. Since the basic job of a communications engineer is to design and construct systems that do a job better than existing systems, we must see what it means to do the job "better," and what it is about the world that prevents a system from doing the job perfectly.

9.1.1 Signals and Waveforms

The basic communications commodities are *information* and *signals*. We all have a notion of what information is, although we shall see that to the engineer *information* has a very specific meaning. The concept of *signal*, too, is more or less familiar, only in communications we mean something more general than just a turn signal on a car or a traffic light. Signals *carry* the desired information from one point to another, and they take many forms. The handwriting on a postcard is a signal, as are the positions of the hands on a clock, but in the following sections we shall see that to the engineer a signal is most often a *waveform*. In fact we shall frequently use the two words interchangeably. What is a waveform? Most commonly it is a fluctuation in time of some quantity such as voltage, current, pressure, height, or brightness. To exhibit the *wave shape* of a waveform, one draws its *graph* as in the examples of Figure 9.1, plotting the fluctuating quantity on one axis versus time on the other. The scales on the axes mark off the units of the quantity and those of time. In some cases the actual time at which an event occurs is of importance, as in recording an earthquake tremor. For these situations the time axis is labeled with the correct time of day, as in Figure 9.1(a). However, one normally is interested only in the signal wave shape itself, rather than its time of occurrence, and so some arbitrary reference time is selected and denoted $t = 0$. Thus in Figure 9.1(b) and (c) no actual reference is made to a time of day; only the elapsed time from $t = 0$ is shown. In many figures that follow the axes are not even carefully scaled, because only the basic *shape* of the signal is important to the discussion, not its detailed values at each instant.

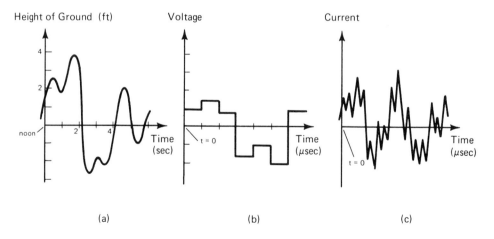

Figure 9.1
Some examples of signals.

There are many ways of actually obtaining a plot of a given signal. The oscilloscope is one such example, being very flexible and convenient to use. Figure 9.2 shows some typical oscilloscope displays (plots of voltage versus time).

There are other kinds of devices for recording signals. The oscillograph is very similar to the oscilloscope, only here a moving strip of paper passes under a pen. The pen moves back and forth in accordance with the voltage signal, and the shape of the waveform is traced out on the moving strip. As

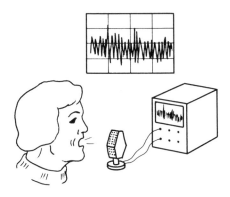

(a) Voice signals

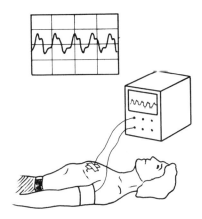

(b) Electrocardiagram (EKG)

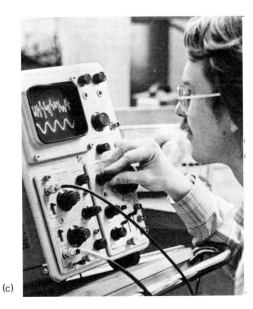

(c)

Figure 9.2
Typical signals viewed on an oscilloscope.

shown in Figure 9.3 a polygraph recorder can trace out several signals simultaneously, for such applications as lie detectors and brain-wave recordings. The oscillograph works only for relatively slowly varying signals, but it has the advantage of producing a permanent record of the waveform measured.

There are many other ways of measuring and plotting signals, some of which we shall discuss in detail below. *Storing* and *displaying* signals are now foundation stones of technology, and they play a key role in communications. To develop some of these ideas we shall turn to a rather general treatment of the notion of a communications system.

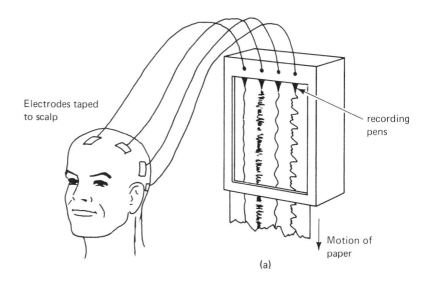

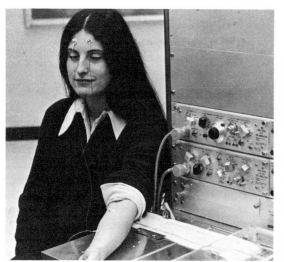

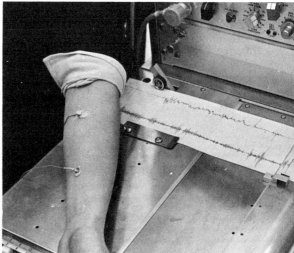

Figure 9.3
A polygraph recorder measuring brain waves.

9.2 A BASIC COMMUNICATIONS SYSTEM

We shall begin by considering a building-block form of a communications system, which shows the most basic components of all communications systems. Figure 9.4 shows the five basic blocks, each of which has a specific role to perform in generating, transforming, and carrying information. By viewing all communications systems in this single framework, we get a clear perspective of the way various parts of a specific communications network fit into the whole.

First we shall discuss what each block does in rather general terms, and then we shall consider some examples of actual communications systems, identifying their basic parts.

9.2.1 Information Source

The information source is the device from which the information originates as the *message*. It could be a person speaking, or a television camera tube face, or a computer that is to send data to another location. The term message will be used somewhat casually. Sometimes we mean an elemental message, which might consist of a single symbol such as 4 or B. In other discussions we could mean a long string of characters forming whole sentences, as in "COME TO DINNER TONIGHT." Clearly, this second message is composed of a succession of elemental messages, and whether we consider it as a single message or as a sequence of small messages is a matter of convenience. Some messages go on for a long time, as in a lecture or a telephone conversation. There we could consider the message as the waveform of a speaker's entire conversation, or it might be more convenient to consider the waveform for each spoken word as a message. The conversation would then be a sequence of messages. We shall see that there are several ways to send the same information and that information is a rather basic commodity that can be manipulated and transformed to suit a given need.

9.2.2 Transmitter

The job of the transmitter is to convert the original message into the *transmitted signal*. Whereas the message might be simply a sequence of characters or a

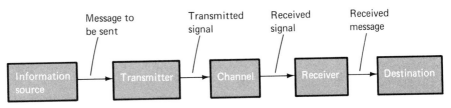

Figure 9.4
Basic communications system.

list of numbers, the transmitted signal is a physical quantity that can actually propagate over distance—that can be sent and carried by a transmission system. This signal must be in a form that permits efficient transmission, and the transmitter's job is to effect this. In the case of a telephone where the message is someone's spoken conversation, the transmitter converts the sound pressure variations into a varying electrical current, which can then travel along the wires to the listener. If the message is a sequence of 1s and 0s stored in a computer, the transmitter might convert the sequences into a succession of short electrical pulses for transmission along a wire to the receiving device. It might send a pulse of 5 V for a 1 and one of -5 V for a 0. This is one example of a process called *modulation*, which the transmitter performs. We shall discuss modulation in detail below.

9.2.3 Channel

The channel is basically the medium used to carry the signal from the transmitter to the receiver. In the case of a telephone circuit the channel consists of all the wires, amplifiers, switching circuits, etc., through which the signal passes on its way from one telephone set to the other. In a radio link from earth to a satellite the channel is the propagation path through space plus the antennas and associated circuitry between the transmitter and receiver. In a sonar system the channel is the water path or paths in the ocean from transmitter to receiver.

Many different ways of sending a signal are used today, and consequently there are many kinds of channels. One useful way of categorizing them rests on whether the signal is *guided* or *unguided* as it propagates from transmitter to receiver. A guided signal follows a rather well-defined path, say along a wire or through a pipe. For instance, telephone signals are typically carried on pairs of wire strung on telephone poles, either as open-wire pairs or twisted-wire pairs, as shown in Figure 9.5.

Figure 9.5
Telephone pole shown with two open-wire pairs and a bundle of twisted-wire pairs. An amplifier is shown that boosts the signals in the twisted-wire pairs.
(*Courtesy James Martin*)

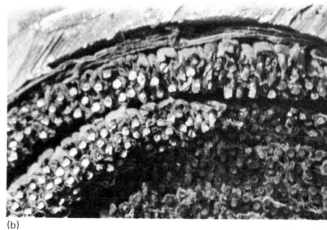

(b)

Figure 9.6
Sheathed cable containing many twisted-wire pairs for telephone signals.
(*Courtesy James Martin*)

Large bundles of twisted-wire pairs can be covered with a lead sheath to form a cable, as shown in Figure 9.6. Many hundreds of such pairs may be put in a single cable to carry large numbers of individual telephone signals.

For long-distance transmission it is more efficient to use *coaxial cable*, or bundles of them, as shown in Figure 9.7. A coaxial cable consists of a circular metal tube that acts as one conductor, enclosing a single center wire that acts as the other conductor. Insulating spacers are used to keep the inner conductor centered properly. Coaxial cables have the advantage that the signal is wholly contained inside the cable. There is no "leakage" of signal energy away from the cable as there can be with open-wire or twisted-wire pairs. This eliminates the problem of interference between two signals in

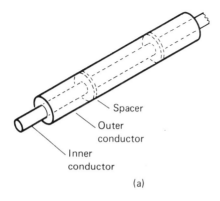
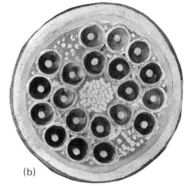

Figure 9.7
(a) Coaxial cable. (b) Bundle of coaxial cables.
(*Courtesy AT & T*)

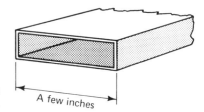

Figure 9.8
Section of rectangular wave guide.

neighboring cables. Another kind of guided signal channel is a wave guide, consisting of a hollow rectangular metal pipe, as shown in Figure 9.8. Electromagnetic waves of extremely high frequency called *microwaves* propagate through the wave guide for large distances and are immune to outside interference just as in the case of coaxial cable.

Examples of channels involving *unguided* signals are also numerous. The most familiar are radio and television broadcasting. Here the signal is fed to a large antenna, from which it travels out in all directions. This technique is, of course, appropriate when the desire is to reach many different receivers whose precise location is not known. The price that is paid is in signal power: Most of the signal is scattered and lost, only a small part being intercepted by radio or TV antennas.

There is also a large class of channels for which the signal is not guided but is *directed* to some extent. This occurs in transmission from a *directive* antenna, which transmits most of its signal power in a chosen direction. Figure 9.9 shows several examples of directive antennas. Directional antennas permit one to make much more efficient use of signal power and are very useful in communication to a receiver having a precisely known location. Sometimes the transmitting and receiving antennas are steerable, as in most radar and radio telescopy applications.

Laser communication promises both high directivity and the ability to carry an astonishingly high number of signals. As seen in Figure 9.10, the light emitted from a laser is already in an extremely narrow beam. Systems presently being developed will aim the beam through a long underground pipe to a receiver. The pipe may contain long optical fibers that guide the light around curves. The pipe also prevents birds, fog, and other opaque objects from getting in the way and cutting off the beam.

These examples have indicated the broad variety of physical channels one can encounter. There is also another sense for the term *channel*, a sense that illustrates some primary concerns of the communications engineer. Channel also refers to the *effects* a communications system can have on the signals passing through it. Some of these effects are undesirable, the most important being *distortion* and *interference*. Each of these types can be broken down into several categories, and we shall discuss their main features below.

(a) Distortion

When a signal is distorted by a channel its wave shape is altered. If the distortion is severe enough, the original signal will not even be recognizable

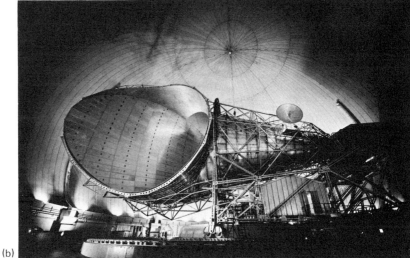

Figure 9.9
Typical directive antennas. (a) Microwave radio tower in a city. (*Courtesy AT & T*) (b) Early bird satellite system receiving antenna. (*Courtesy COMSAT*) (c) A large 97 ft. dish antenna (left) at the Andover, Maine, earth station transmits and receives all forms of communications to Atlantic INTELSAT communications satellites. Seen in the background of this photo is the "retired" horn antenna underneath the famous bubble. At the right of the photo is the "sugar scooped" antenna used to track and command INTELSAT communications satellites. (*Courtesy COMSAT*)

Figure 9.10
Laser transmitter.
(*Courtesy AT & T*)

at the receiver, and so engineers go to great pains to design systems with low distortion. The main types of distortion are *nonlinear distortion* and *dispersion*. A common example of the former type is *clipping*, as shown in Figure 9.11. Here the signal is so large that the channel cannot accommodate it, and the tops and bottoms of the waveform get lopped off. This is a common problem with high-fidelity music equipment: If the volume is turned up too high, the amplifiers clip the signal and distort it.

A communications engineer often describes this and other types of nonlinear distortion by means of an *input-output characteristic*, which plots the channel output voltage level resulting from each possible input voltage level. A characteristic for a channel with clipping is shown in Figure 9.12, where it is seen that signal amplitudes smaller than 3 V produce proportional outputs, while those greater than 3 V all produce the same output, causing distortion.

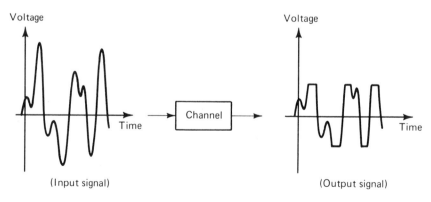

Figure 9.11
Clipping—a form of nonlinear distortion.

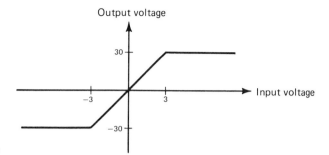

Figure 9.12
Characteristic for a clipper.

In the case of *dispersion*, the signal becomes "dispersed" or smeared out in time while passing through a channel, as shown in Figure 9.13. One important result of dispersion is that one cannot send pulses very close together over a dispersive channel, or their smeared out versions will overlap and obscure each other. This is significant in pulse transmission as we shall see below, for it reduces the rate at which information can be sent. There are two main causes of dispersion, which we shall loosely call *sluggishness* and *multipath transmission*.

Sluggishness. Some channels by their very nature take a certain amount of time to respond to a signal. For instance, in riding along in a car the wheels (the "transmitter") bounce up and down on every bump, but the passengers (the "destination") experience only a smooth up and down motion. The car's shock absorbers and springs (the "channel") introduce dispersion into the signal (the height of the bump) and transmit only a very smeared version to the passengers. Or consider the crude communications system pictured in Figure 9.14. Man A at the top of a deep well sends information to man B by positioning his end of the rope at "dash," "dot," or "space." Man B receives the message by noting the rope's position at the bottom. Both A and B agree to use the Morse code. This channel is clearly very sluggish, because any change in the rope position by A will cause the rope to swing back and forth and only slowly come to rest at the new position. If A changes the rope position too frequently, the swaying of the rope will be completely incomprehen-

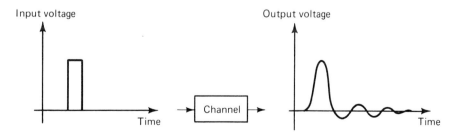

Figure 9.13
Pulselike signal before and after passing through the channel.

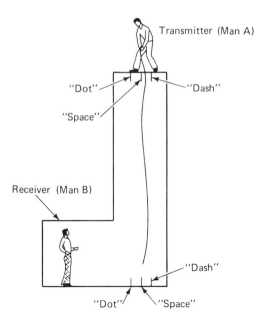

Figure 9.14
Sluggish channel for sending Morse code.

sible to B, and no information will get through the channel. We shall see later that a very similar phenomenon occurs in electrical systems and channels and that it is related to the channel's bandwidth.

Multipath Transmission. Multipath transmission is a kind of dispersion as suggested in Figure 9.15, where a signal not only follows the line-of-sight path between the microwave antennas but also follows several indirect paths. The line-of-sight path is shortest so the signal arriving along this path is received first, while the signals from the indirect paths arrive later. These "echoes" are usually much weaker than the main signal, but still they

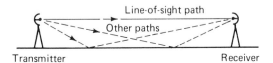

Figure 9.15
Multipath transmission between microwave antennas.

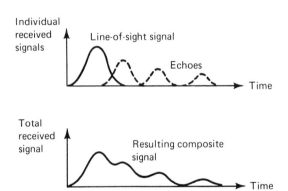

can severely distort it. If there are many indirect paths, then the resultant signal looks like a smeared-out version of the original just as in the case of a sluggish channel. Examples of multipath phenomena are very common. Television signals can follow multiple paths to an antenna, causing "ghosts" or faint displaced images on the screen. People shouting to each other in the Alps encounter so many echoes that they must shout their messages very slowly. Sound signals propagating underwater, as in the case of a sonar signal, often follow many paths, because the ocean surface and bottom reflect sound energy. This can make underwater communication very difficult indeed.

When channels have a lot of dispersion, the communications engineer must try to combat it, especially when the exact shape of the transmitted signal is important. In many situations an engineer can insert special devices called *equalizers*, which are designed to compensate for the distortion of the channel and which approximately reconstitute the original shape from the distorted version, as suggested in Figure 9.16.

(b) Interference

There is another kind of undesirable effect a channel can have on a signal besides distortion; it is called interference. The latter type can be much harder to combat, and it imposes one of the fundamental limitations on how well a communications system can convey information.

There are basically two kinds of interference: that arising from other signals, and *noise*. The first kind occurs when several signals share the same or neighboring facilities. If two microwave antenna links are placed too near each other, part of the signal from one will be picked up by the other antenna and will degrade the quality of the intended signal. In a later discussion we shall discuss *multiplexing* many signals together and sending the composite signal through a single channel. In this case one must carefully guard against "cross-talk" between the different signals. Another familiar example of interference is the babble of conversation at a party. Here the cross-talk between groups of people can be severe indeed.

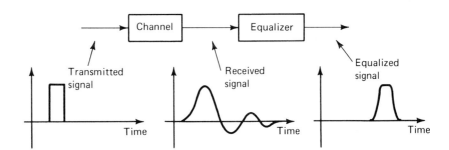

Figure 9.16
Effect of an equalizer to combat dispersion.

The other major type of interference is *noise*. This consists of an unpredictable fluctuation that gets added to the signal in the channel. The hiss of static that emanates from a poorly tuned radio is an example of noise. Although the term *noise* implies some kind of sound, the term is used to describe any kind of random interference—electrical, mechanical, etc.—that unavoidably perturbs the signal. Figure 9.17 shows a typical signal before and after being perturbed by noise.

Where does the noise come from? The most prevalent kind of noise is *thermal noise*, caused by the constant agitation of particles at the atomic level that is present in all matter. This agitation is caused in turn by the heat energy of the material, and indeed the amount of noise generated is proportional to the temperature of the material. This jiggling of atoms and random wandering of electrons generates small but unavoidable voltages across many circuit elements, and these voltages are the noise signals. For these reasons, noise is introduced whenever signals are amplified. Frequently this noise is very weak and causes little trouble, as in high-fidelity sound equipment. However, if the signal is also very weak, as it is when received from a satellite, both signal and noise are amplified together, and there is no way of separating them. So it is in situations where weak signals must be amplified up to usable strength that thermal noise is most troublesome, and indeed it imposes fundamental limitations on how rapidly and accurately information can be sent over a communications channel. It is so important to minimize the amount of this noise entering some systems that amplifiers are sometimes immersed in liquid nitrogen to cool them down, thus reducing the thermal noise strength.

To summarize, then, the channel is that part of a communications system that includes everything a signal encounters enroute from the transmitter to the receiver. In most systems one attempts to make the received signal as similar as possible to the transmitted signal. However, distortion and interference conspire to prevent this, and the two signals always differ to some

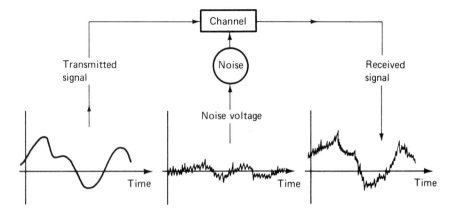

Figure 9.17
Effect of noise on a signal.

extent. Some types of signals are less sensitive than others to a particular kind of noise or distortion, and a major part of an engineer's task is to find ways of modulating signals (e.g., designing transmitters) so that the transmitted signal loses as little of its information as possible in passing through a given channel.

9.2.4 Receiver

The receiver operates on the received signal attempting to reproduce the original message. It essentially undoes what the transmitter did to the message, a process also known as *demodulation*. In the case of a telephone circuit, the receiver is the listener's set, which converts the fluctuating electrical current back into sound waves for reception by the listener's ear. A receiver may involve additional parts, however, as it may be designed to combat noise and distortion introduced by the channel. Because of the channel, the received message is almost never completely identical to the transmitted message, and the question arises, when will this make a difference? In a telephone system the channel interference may not be too serious, since the human brain is extremely proficient at making sense of poorly reproduced speech, but in other communications systems the effect of the channel may be the single most important aspect. We shall consider ways of determining how destructive noise and interference are in certain cases below.

9.2.5 Destination

The last basic block in a communications system is the destination, which accepts the *received message* and uses it in some way. In the case of a telephone the destination is the listener's ear (or brain, depending on one's point of view), and the overall goal of the telephone system is to make the received message a perfectly recognizable version of the transmitted message. In a computer-to-computer system the destination is, of course, a computer, and the system must be designed so that the received message, a sequence of 1s and 0s, is almost always identical to the transmitted message.

Now that the blocks in Figure 9.4 have been introduced in general terms, we shall give examples of communications systems and point out which elements correspond to which blocks. We shall see that the association of each block to a specific system component is sometimes a matter of one's point of view. There is no single correspondence that everyone would agree with. The whole point of making the correspondence at all is to set up a convenient framework within which one can work.

Example 1

Underwater Sound Communication. As man penetrates further into the vast underwater world that promises so much in the way of food and knowledge, the need to communicate underwater becomes more and more urgent. Neither light nor radio waves propagate very efficiently through water, and stringing wires between ships and frogmen will not do, so people have turned to sound as a means of communicating. A simple acoustic system is shown in Figure 9.18. A man on board the surface vessel speaks into a microphone, generating an electrical signal. This signal is amplified, suitably

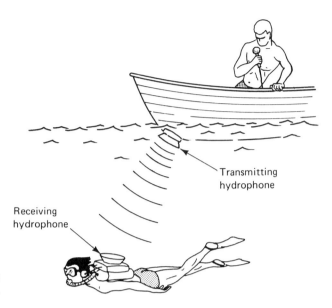

Figure 9.18
Underwater sound communications.

translated to high frequency (see Section 9.3.3), and used to drive the transmitting *hydrophone*. The hydrophone, which is essentially a loudspeaker designed for underwater operation, emits the signal as a traveling acoustic wave (a mechanical vibration rather than an electrical entity), and this wave propagates to the frogman at the speed of sound (about five times as fast as in air). The frogman carries a receiving hydrophone, which detects the sound. The original voice information is extracted from the received signal and carried to the frogman's earphone. There is, of course, a similar circuit for the frogman to talk back to the surface vessel.

In this system the man on the vessel is the *source*, and the frogman is the *destination*. There are two conversions of the man's voice, one to an electrical variation in the microphone and one to a strong acoustic signal at the hydrophone. Thus, there are two ways of viewing the circuit. An engineer concerned with the whole system would probably consider the microphone and earphone as the *transmitter* and *receiver*, respectively, and everything else would be the "channel." On the other hand, a specialist in acoustics would consider the transmitting and receiving hydrophones as the transmitter and receiver, while for him the channel would consist only of the water path between the vessel and the frogman. Both of these viewpoints are valid, and both are useful. It just depends on which parts of the system one wishes to concentrate on and which parts he is assuming he has no control or jurisdiction over. The channel is usually considered as being beyond one's control, and one designs a system around a given channel.

Example 2

Time-Sharing Computer System. In recent years time sharing has become a very powerful way of utilizing a large computer's abilities. Many people can interact with a central computer at the same time, and because people operate much more slowly than computers, each user has the impression that the entire machine is at his disposal.

A simplified time-sharing computer system is shown in Figure 9.19. Some of the users (there could be several dozen in an actual situation) are shown at teletype terminals, which could be situated a few feet or many miles from the computer. The teletype terminals are connected through coupling arrangements to standard telephone lines, and likewise the computer has couplers to a set of telephone lines. Information must go both ways in this system, so that the user can instruct the machine what to do and the computer can respond with the results of its computations.

To see this network as a communications system, let us consider only one of the users and assume that he wishes to send the message LIST to the computer. He is the source, then, and the transmitted message is LIST, which he sends simply by typing this word at the teletypewriter keyboard. The teletypewriter terminal converts LIST into a sequence of 0s and 1s according to a standard *code* that the computer is designed to interpret. For instance, the code for L turns out to be 00011001011. The *transmitter*, which is housed in the coupler, then converts this sequence of digits into a sequence of audible "beeps," which are fed into the telephone set as suggested in Figure 9.20. The beep for a 1 is at a lower pitch than that for a 0, so they can be distinguished at the receiver. Each beep lasts about .01 sec, so the message LIST would require about .4 sec to be sent.

The channel for this communications system is the entire telephone link

Figure 9.19
Time-sharing computer system.

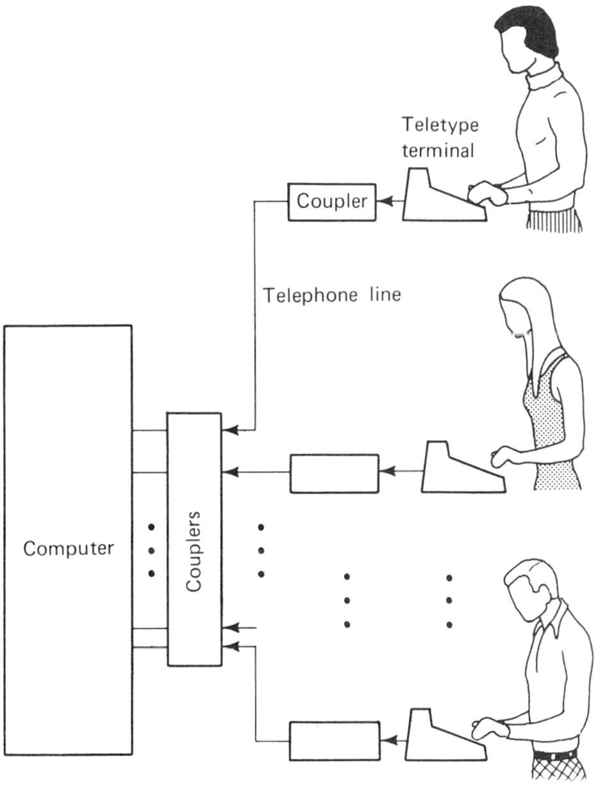

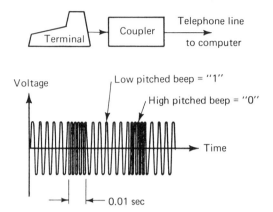

Figure 9.20
Sending information from a teletype terminal to a computer.

between the couplers at the teletype terminal and the computer. The received message is the sequence of beeps perturbed by any interference introduced by the channel. The receiver is the coupler at the computer, which is designed to recognize the two tones for 1 and 0 and to convert them back into the appropriate digits. Usually the received signal is recognized correctly at the computer so that the message LIST is indeed received. However, if the channel is poor, a somewhat different sequence may be received, resulting in LKST. An error has been made here, and the computer most likely will not be able to interpret this command. Most computer systems are designed to recognize certain kinds of errors, and the computer will immediately demand that the user retransmit the message for another try. The computer might send back the message HUH? or ERROR IN TRANSMISSION, RETYPE. This *error detection* capability in some kinds of communications systems is very important.

These examples have served to introduce several important concepts of modern communications systems, focusing on the building-block approach of Figure 9.4. In the next sections we shall consider in more detail some basic ideas associated with kinds of signals, modulation, and bandwidth in order to facilitate a closer look at modern communication systems.

9.3 SOME FUNDAMENTAL CONCEPTS IN COMMUNICATIONS

9.3.1 Frequency and Bandwidth

We shall begin with a discussion of *frequency*, which is one of the most important concepts used by communications engineers because many properties of communications systems are much better understood if one looks at the *frequency components* of a signal as they pass through a system. For instance, it would be difficult to say anything about electrical filters without the notion of frequency, and communications systems contain many filters for shaping signals, eliminating unwanted signals, etc.

The idea of frequency arises in connection with *sine waves*, which are perhaps the most important class of wave shapes in communications engineering. A sine wave (or *sinusoid*) is described mathematically using the trigonometric function sin(·) as:

$$s(t) = A \sin [2\pi(ft + \theta)] \qquad (9.1)$$

and it is completely described by only three numbers: the *amplitude A*, the *frequency f* (which is simply related to the *period P* by $f = 1/P$), and the *phase θ*, which gives the position of the beginning of a cycle relative to some given instant, say $t = 0$. Figure 9.21 depicts such a sine wave. Its shape is easily visualized in terms of a point p moving at uniform speed around a circle. As the point moves around the circle, its vertical height above the circle's center varies sinusoidally as shown. One can see that frequency is the number of cycles completed in 1 sec. The unit of measure for frequency is cycles per second, or Hertz. For example, the alternating current electricity delivered to our homes is in the form of a sine wave having a frequency of 60 Hz. An audible sine wave is heard as a pure "tone," and its frequency is what one means by the tone's pitch. Middle C on a piano has a frequency of 256 Hz.

Most signals are not simple sinusoids but instead have rather complicated shapes. However, an elegant branch of mathematics called Fourier analysis tells us that *any* signal is really just a sum of sine waves. That is, any signal can be decomposed into a number of sine waves of different frequencies. A particularly simple class of signals are *periodic waveforms*, which keep repeating a basic shape forever. The duration of the basic shape is the *period*. A sine wave is, of course, an example of a periodic signal. A more complicated periodic signal (which would sound more like a "buzz" than a tone) having a period of 1 msec is shown in Figure 9.22(a). Also shown in Figure 9.22 are the sine-wave components that add together to form this signal. There are three components in this simple case, having frequencies of 1,000 Hz (the fundamental), 3,000 Hz, and 4,000 Hz (the third and fourth harmonics). The amplitudes and phases are shown in Figure 9.22 for each component as well. In general a periodic signal of period T sec has a fundamental com-

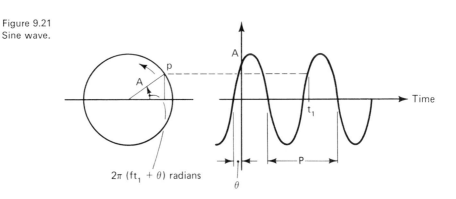

Figure 9.21
Sine wave.

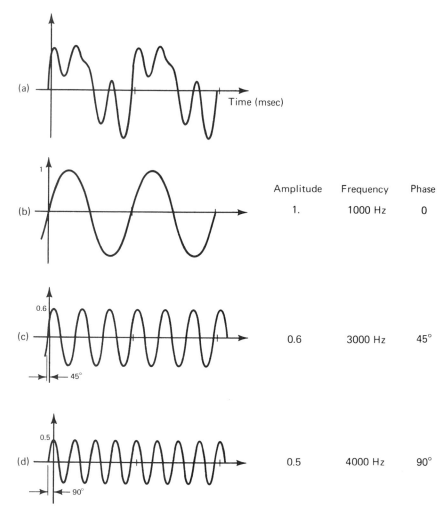

Figure 9.22
Decomposing a periodic signal into its components. (a) Periodic signal. (b) Its fundamental component. (c) Its third harmonic component. (d) Its fourth harmonic component.

ponent of $1/T$ Hz and several harmonic components at frequencies that are multiples of $1/T$ Hz.

The range of frequencies between the lowest- and highest-frequency components is called the signal's *bandwidth*, and for this signal the bandwidth is $4,000 - 1,000 = 3,000$ Hz. A useful display of the components present in a signal is provided by drawing its amplitude spectrum, as shown in Figure 9.23.

Roughly speaking, one plots the *amplitude* of each component in a signal versus the *frequency* of that component. The bandwidth is then easily spotted as the width of the band of frequencies occupied by the signal. As

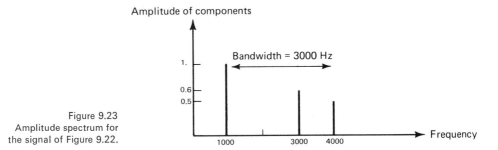

Figure 9.23
Amplitude spectrum for the signal of Figure 9.22.

additional examples, two pulse-train waveforms with periods of 1 msec are shown in Figure 9.24 along with their amplitude spectra. One can see that the bandwidths of the two signals are about 6 and 2 kHz, respectively (although here these are only approximations since there are an infinite number of harmonics, most of them of very small amplitude).

Most signals are not simply periodic, since most signals begin at some time and end at some time, whereas a periodic signal goes on forever. Nonperiodic signals have spectra also, but they are not simple lines. An example of a single pulse signal is shown in Figure 9.25 along with its spectrum. The spectrum again shows which frequencies are present in the signal and their relative strengths, but this signal has components at *every* frequency from 0 Hz up to about 3,000 Hz. Thus, the bandwidth of this signal is around 3,000 Hz.

A signal's bandwidth is of great importance to communications engineers because it indicates how large the bandwidth of a communications *channel* must be in order to carry the signal faithfully. If the bandwidth of the channel is not large enough, the signal can be severely distorted in passing through it. To get an intuitive feeling for the distortion a channel might cause,

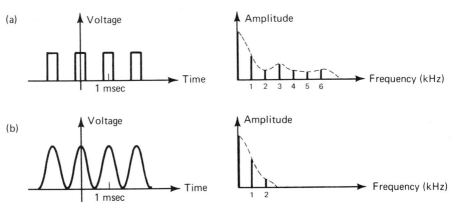

Figure 9.24
Typical pulse-train signals and their spectra.

we shall introduce the very useful notion of the *transfer function* of a system. The action of many channels (or *filters*) is completely described by their transfer functions. Filters of this class have the handy property that when a sine wave is fed into them, a sine wave of the same frequency will emerge at the output. The amplitude and phase of the sine wave may be changed as it passes through the filter, but the output waveform is still a sine wave. By contrast, an input signal of some other shape may emerge as a very different waveform, as we shall see below.

The transfer function tells us precisely how the filter *alters* the amplitude and phase of a sine wave at each frequency. There are two parts to the transfer function. The *amplitude response* gives for each frequency the *ratio* of the amplitudes of the output and input sine waves, while the *phase response* tells the *difference* in phase of the output and input sine waves:

$$\text{amplitude response} = \frac{\text{amplitude of output sine wave}}{\text{amplitude of input sine wave}}$$
$$\text{phase response} = (\text{phase of output sine wave}) - (\text{phase of input sine wave})$$
(9.2)

It is precisely because this ratio and difference are *not* the same for all frequencies that the filters have the ability to shape signals and eliminate unwanted interference. So some filter transfer functions are very desirable for producing properly shaped signals, whereas other functions can cause undesired distortion.

An example of a filter is shown in Figure 9.26, including a circuit diagram, and the transfer function. The circuit diagram, showing resistors ($\wedge\!\!\wedge\!\!\wedge$), capacitors ($\dashv\vdash$), and inductors ($\frown\!\!\frown\!\!\frown$) is included simply to indicate the complexity of a typical filter. Circuit analysis, learned quite early in an engineering curriculum, allows one to calculate the transfer function directly from the circuit diagram.

It is instructive to see how the shape of the periodic signal of Figure 9.22 is altered upon passing through this filter. To compute this, the response of each of the components of the signal is found using the transfer function, and

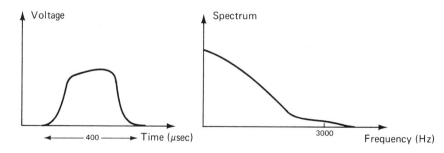

Figure 9.25
Single pulse waveform and its spectrum.

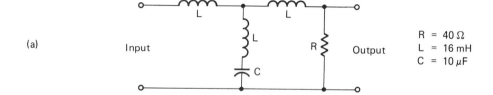
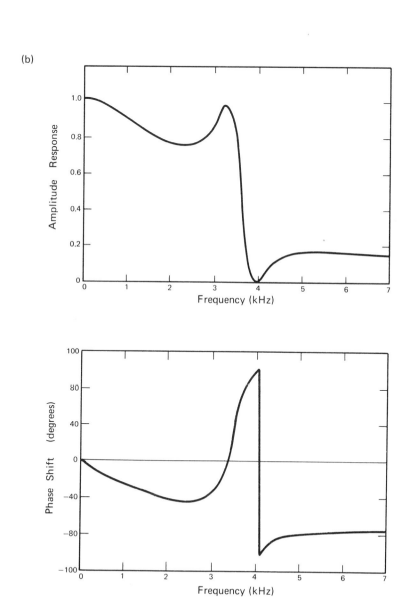

Figure 9.26
Typical filter.

then these are added to form the total output signal. Figure 9.27 indicates this process by showing how the amplitude and phase of each output component is found and by showing the resulting output shapes. It is seen that there is *no* component at 4,000 Hz at all: The filter has completely eliminated it. Clearly,

	Fundamental	Third Harmonic	Fourth Harmonic
Frequency:	1000	3000	4000
Input amplitude	1.	0.6	0.5
X Amplitude response	0.9	0.88	0.
Output amplitude	0.9	0.528	0.
Input phase	0.°	45°	90°
+ Phase response	−26°	−28°	0°
Output phase	−26°	17°	90°

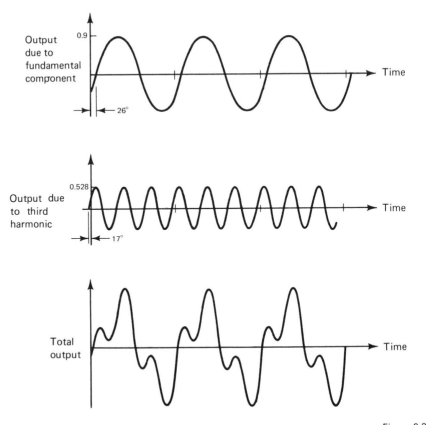

Figure 9.27
Output signal from the filter at Figure 9.26 when the input signal is that of Figure 9.22.

the output waveform is of a rather different shape than the input; some of the more rapid wiggles have been removed.

As another example, the responses of the same filter to a long and a short input pulse are shown in Figure 9.28. The long input pulse $s_1(t)$ gives rise to an output pulse $s_2(t)$ that is distorted but still recognizable as a nearly rectangular pulse, while the short input pulse $s_3(t)$ yields a very dispersed and almost unrecognizable output pulse $s_4(t)$. Clearly, a rapid succession of such short pulses would emerge from the filter in an indistinguishable muddle. This is precisely the kind of distortion that in some situations can limit the performance of a communications system.

To return to the notion of bandwidth for a *system*, we note that the bandwidth of the filter in Figure 9.26 is about 4,000 Hz, since frequencies above this are severely attenuated (reduced in amplitude). If this filter were an unavoidable part of a communications system, there would be little hope of sending signals with bandwidths greater than 4 kHz. In general, for a given signal to pass through a system undistorted it turns out that the amplitude response must be "flat" and that the phase response must be linear over all frequencies in the signal's spectrum. An "ideal" filter transfer function for signals having bandwidths less than 6,000 Hz is shown in Figure 9.29. High-fidelity sound equipment usually has amplifiers that are flat out to around 20,000 Hz. It does not matter that they are not flat beyond 20,000 Hz since the human ear cannot hear frequencies higher than this anyway. Some typical

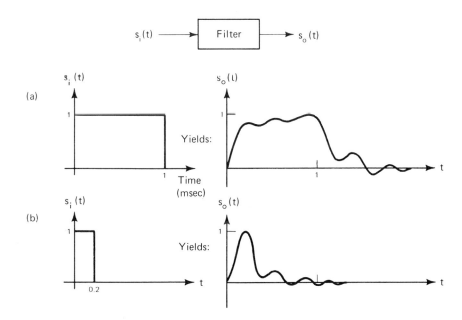

Figure 9.28
Long and short input pulse response of the filter of Figure 9.26.

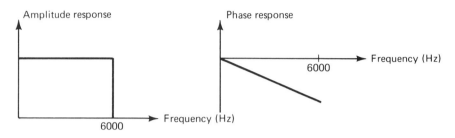

Figure 9.29
Filter that will not distort signals with bandwidths less than 6 kHz.

bandwidths required for various communication systems are listed below, showing a very wide range in system requirements:

Voice systems	4,000 Hz
Television broadcast	4 million Hz
Mariner space probe	12 Hz
Radar	1 million Hz
Telegraph	100 Hz
Electrocardiograph	200 Hz
High fidelity	20,000 Hz

There are basically two distinct kinds of signals, analog and digital. In this section we shall discuss their main properties and see how an analog signal can be changed into a digital one and vice versa. The fact that one can transform one type into the other with hardly any loss of information is very important in communications, for some applications require one kind of signal to be sent over a channel even though the source puts out the other kind. Consequently A/D (analog-to-digital) and D/A (digital-to-analog) converters are found in many modern communications systems.

In a communication system we think of the source as emitting information in either analog or digital form. Basically the analog form has continuously variable values, while the digital form can take on only a finite number of values. For example, consider a source that emits a number. If it can emit *any* number between, say, −5 and 5, it is an analog source, while if it can emit only discrete values, say the four numbers 5, 3.67, −1, and 2.736, it is a digital source. It is clear that an analog source can emit one of an *infinite* number of messages, although many of them differ only infinitesimally.

Most sources do not emit a single number and then stop. They emit a changing signal in which new information comes along from moment to moment. An analog source sends a varying waveform such as a voice signal or an electrocardiogram pattern, and a digital source emits a *sequence* of

9.3.2
Analog and
Digital Signals

symbols such as I AM HERE, 1384768, or 1001001010010. Each symbol is allotted a fixed amount of time, and then the signal changes abruptly to the next symbol's waveform. Figure 9.30 shows the basic difference between these two kinds of signals. The analog signal varies continuously, whereas the digital signal jumps abruptly to each new level. In this example the message is 149624. A pulse of 4 V represents a 4, a pulse of 6 V represents a 6, etc. The individual pulses here last $\frac{1}{10}$ sec.

One might think that analog waveforms are far superior to digital ones because they can carry so much more information. The digital signals seem rather crude by comparison, being limited to only a few levels. However, we cannot really make use of all the fine detail of analog signals, because of the ever-present *noise* in all communications systems. The noise in a channel perturbs the signal slightly, so that two analog signals that differ only in small details will no longer be distinguishable at the receiver. Hence, even though a source can emit an infinite number of different signals, only a finite number of different signals are *distinguishable* at the receiver. On the other hand, a digital signal is not so susceptible to noise as an analog signal. Looking at Figure 9.30(b), it would take a strong noise burst to offset one level so much that the receiver mistook it for a neighboring level. Digital signals are consequently "hardier" than analog signals, even though they cannot carry as much fine detail. There is no inherent advantage of one type of signal over the other until one specifies the nature of the system and channel in question.

(a) A/D Conversion (Sampling and Quantizing)

In those systems where the source emits analog signals yet the information is best sent digitally, one must convert the analog waveform to a sequence of symbols, a digital signal. The question naturally arises, How are we to do this, and how well can we do?

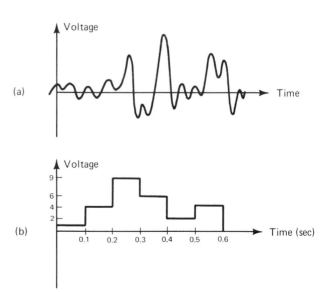

Figure 9.30
Typical analog and digital signals.

There are two basic operations to perform on an analog signal to create a digital signal that approximates it: *sampling* and *quantizing*. Each operation has many interesting properties, so we shall consider them separately at first.

Sampling. In sampling a signal, one examines it only at isolated instants, ignoring its behavior at all instants in between. Figure 9.31 shows a signal and its sampled version. Here a sample-and-hold technique is used, where the signal's value at each sampling instant is held until the next sampling instant. Clearly, this process throws away much of the fine structure of the signal, and one's first impression might be that precious information about the signal must be lost. Of course, taking more samples per second would reduce the amount lost, but one might suspect that unless you sample infinitely rapidly you must always lose some information. However, there is a remarkable fact, known generally as the *sampling theorem*, that shows that this is not true. It says that if one samples "often enough" there is *no* loss of information, because from the samples alone one can *perfectly* reconstruct the original analog signal. How often is "often enough?" The answer is that one must take samples at a rate twice as high as the highest frequency in the signal's spectrum. For instance, if a signal lies in a band of frequencies from 0 to 4,000 Hz (as speech signals do), one need take only 8,000 samples each second, and *no* information is lost. If a signal has no components at frequencies higher than 4,000 Hz, it simply cannot "wiggle" fast enough so that the samples will miss any information. This also says that *new* information is not coming along every instant, as there is some sluggishness to the signal.

Quantizing. We have seen above that the process of sampling a signal, if done properly, is *information-preserving*. This is not true of the other basic operation, *quantizing*. Here one deliberately throws away information in order to retain only a finite number of possible levels. An electrical quantizer has a number of built-in fixed voltage levels, and in the quantizing process it replaces the signal level at each instant by the nearest of these levels. The ac-

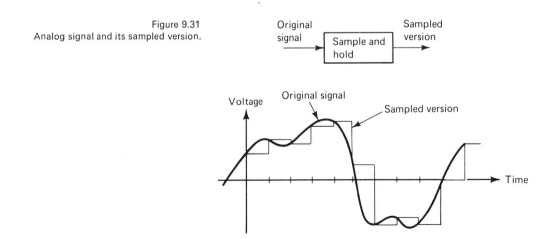

Figure 9.31
Analog signal and its sampled version.

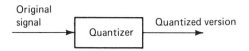

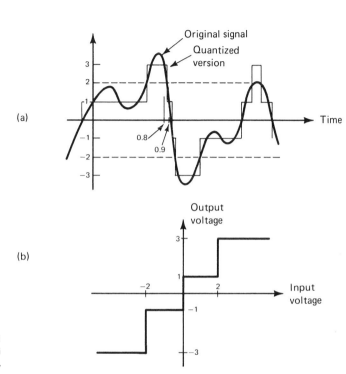

Figure 9.32
Four-level quantizer and its effect on an analog signal.

tion of a four-level quantizer is shown in Figure 9.32(a). The four levels here are −3, −1, 1, and 3 V. To see how it works, consider the instant $t = .8$. Here the signal shown has a level of about 2.1 V, which is nearer to 3 V than to any other quantizer level so the quantizer output is 3 V. However, at $t = .9$ the signal has diminished to 1.8 V, so the quantizer output is 1 V. This action is neatly summarized by the *quantizer characteristic* of Figure 9.32(b), which plots the function relating input and output levels. The levels −2, 0, and 2 V, being halfway between the quantizer levels, are the voltages at which the quantizer output abruptly jumps.

A four-level quantizer obviously throws away some information about a signal. To discard less information and make the quantization less *coarse*, more quantization levels must be used. In some telephone systems 128 levels are used, and here the quantization is so fine the ear cannot even tell the difference between the quantized and original signals. Figure 9.33 shows a speech waveform and its quantized version. It is clear that they are very similar, even though the original signal has an infinite number of levels, while the quantized signal takes at most 128 different levels.

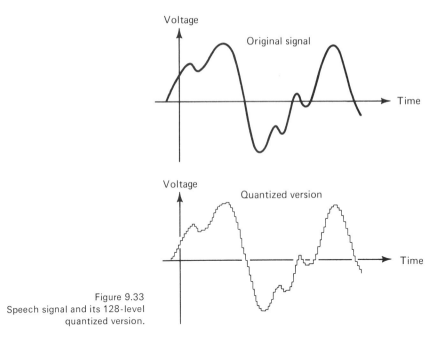

Figure 9.33
Speech signal and its 128-level quantized version.

Neither sampling alone nor quantizing alone produces a truly digital signal. With sampling alone the samples can still take on an infinite number of values, while with quantization alone the transitions from one level to the next can occur at an infinite number of instants. To convert an analog signal to a truly digital signal, one must perform both sampling and quantization. A complete A/D converter is shown in Figure 9.34(a), where a 16-level quan-

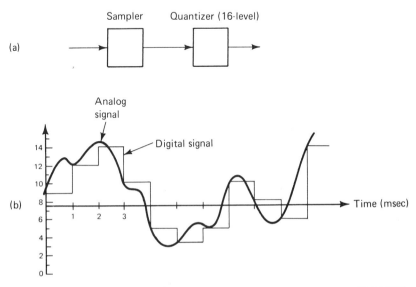

Figure 9.34
Analog signal and its digital counterpart.

tizer has been added to the sample-and-hold device of Figure 9.31. The effect of the converter on a typical signal is shown in Figure 9.34(b) where the sampling instants and the 16 output levels are specifically shown. Samples here are taken every millisecond. (This rate would not be fast enough for speech but is quite acceptable for signals of smaller bandwidth, such as EKG or oceanographic wave-height fluctuations.)

We shall discuss pulse code modulation more fully below, but it is useful at this point to note that if each of the 16 levels is assigned some *symbol*, say one of the numbers between 0 and 15, then this digital waveform could perfectly well be sent over a channel as a sequence of symbols. Using the assignment shown in Figure 9.34, the message for this particular waveform is 9, 12, 14, 10, 4, 4, 4, 10, ... In a very real sense the waveform (or at least its digital approximation) is equivalent to this sequence of numbers. Thus, the same information can be sent using either analog or digital signals. Information is a basic quantity, while the *form* it is transmitted in is frequently a matter of choice.

One can convert other kinds of analog signals such as two-dimensional images in a similar fashion. Figure 9.35 shows the result of quantizing a photograph into only two levels of brightness: black and white. (Some people recognize the image immediately, and others never. Can you see what is there?) This image is not sampled in the horizontal and vertical directions, however, since the black-white transitions can occur anywhere. Figure 9.36 shows a photograph along with a truly digital version: one that has been sampled and quantized. The picture has been divided into many small squares (sampled), and the brightness in each square has been made all black or all white (quantized). Here if we were to assign a 1 to black and a 0 to white, we

Figure 9.35
Mystery photo quantized to two brightness levels.

Figure 9.36
Original photograph with sampled and quantized version.

could transmit this digital image by scanning it (in the way one reads a page of print) and sending the sequence 00010010010100010010 . . . .

This is, in fact, precisely the way in which most photographic images are sent back from lunar and planetary explorer satellites. A photograph of the planet surface is taken, the photo is scanned slowly, and the image is converted to digital information by very fine sampling and quantizing (usually employing more than two brightness levels). Then the digital sequence is sent back to earth, and the image is reconstructed. In this way the power requirements of the satellite transmitter are greatly reduced. Figure 9.37 shows another digital image (a cluster of shaded squares) that is meaningless at close range, yet perfectly recognizable at a distance. It is interesting to speculate how the eye, brain, and memory make use of so little information to successfully construct the familiar picture.

(b) Digital-to-Analog Conversion

Normally, an analog signal that has been converted into a sequence of symbols by an A/D converter is eventually reconverted back into an analog waveform. This is a somewhat simpler procedure than A/D conversion, and it again consists of two steps. In the first step the D/A converter interprets each symbol in the sequence as it arrives and emits a pulse having the proper amplitude. For instance, Figure 9.38 shows the symbol sequence 4, 9, 6 being

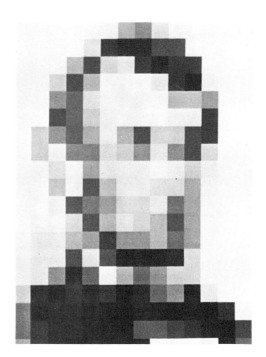

Figure 9.37
Familiar face in crude digital form.
(*Courtesy Leon D. Harmon and Bell Telephone Laboratories*)

converted into a sequence of pulses having amplitudes 4, 9, and 6 mV. In this case each pulse lasts the entire symbol period, although in other systems it might be very short. The second step consists of shaping the pulse stream into a smooth signal. If the pulse-shaping filter is chosen carefully, the recreated analog signal can closely resemble the original analog signal that entered the A/D converter.

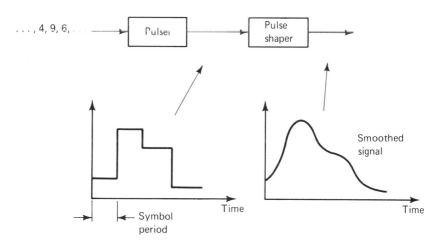

Figure 9.38
Digital-to-analog conversion process.

Before leaving the discussion of analog versus digital signals, there are two important applications of the techniques above to actual communications practice. The first shows how one can use sampling to send many different signals over the *same* wire or cable, while the second illustrates a powerful advantage of digital signals, their ability to be regenerated.

(c) Interleaving Samples:
Time Division Multiplexing

In Figure 9.31 the sampled version of the signal is a succession of constant levels, one for each sample taken. But it is wasteful to hold anything constant in communications: One can send more information per second by inserting samples of *other* signals in some of this available time. The technique of time division multiplexing is illustrated in Figure 9.39. Samples of four different signals are interleaved at the transmitter by the *multiplexer*, which is shown as a rotating switch, or *commutator*. This device (which would actually consist of electronic switches rather than a moving arm) successively "looks at" the sample levels of the four signals and combines them into a single stream of pulses. Of course, this new signal changes level four times more rapidly than each of the original sampled signals, so the transmission system must have a response four times faster (have four times as much bandwidth). In many cases, however, the extra bandwidth required is a small price to pay for being able to send four signals over a single channel. This technique is so successful that in some existing telephone systems several hundred separate signals are multiplexed together using this technique, and in fact we have all

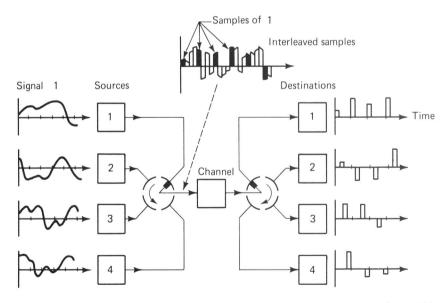

Figure 9.39
Time division multiplexing of four signals.

talked over such systems and been completely unaware of the processing our voice signals were undergoing. At the receiver shown in Figure 9.39 a demultiplexer separates the four signals again. Obviously the demultiplexer must turn in exact synchronism with the multiplexer in order to keep samples associated with their proper signals.

(d) Digital Signal Regeneration

Because analog signals can be converted into digital signals and back again with very little loss of information, the two kinds of signals are essentially equivalent. The speech signal for the spoken word *cat* is effectively the same as a long string of symbols, say 0s and 1s, if they are properly interpreted at the receiver. Much of modern communications centers around digital communications techniques, because digital signals have a very important advantage over their analog counterparts: their ability to be *regenerated.* Consider a telephone link from Boston to Los Angeles, for example. Any signal originating at one end of the link must be amplified many times before it reaches its destination, or else it will be uselessly weak and unintelligible. The usual method is to place amplifiers at many points along the way, each one making up for the attenuation that results from the portion of wire or cable between amplifiers. Thus, there can be several hundred amplifiers in the transcontinental link. Here is the problem. Each amplifier adds some noise to the signal and the noise is amplified along with the signal in all subsequent amplifiers. Once it is part of the signal it can never be removed, and so the noise contribution to the signal accumulates all along the way.

On the other hand, if the telephone signal is converted first to a digital signal, the situation is very different. Instead of building the system with simple amplifiers along the way, *regenerative repeaters* are used. A regenerative repeater receives the noisy digital signal and builds from it a "fresh" digital signal, one with virtually no noise. It can do this because it only has to decide which one of a small number of possible signals was sent. This new signal is then sent further along the communications link to the next repeater and so on, as shown in Figure 9.40. Finally, the signal is converted back to the original voice waveform in a D/A converter. Each repeater must correctly interpret the noisy digital signal it receives, or it will regenerate a digital signal that differs in some places from the original. As there is always some chance of noise perturbing the signal enough to produce an error at a repeater, errors can indeed accumulate en route. However, they can be kept to a minimum by careful engineering of the system.

9.3.3 Modulation

In the preceding sections we have mentioned in passing various modulation schemes but without giving a formal introduction to them. Now it is appropriate to look at modulation more closely. There are basically two senses in which the term *modulation* is used: (1) *impressing* information onto a physical quantity such as a voltage waveform, and (2) *altering* an already existing signal to make it more suitable for transmission or further processing.

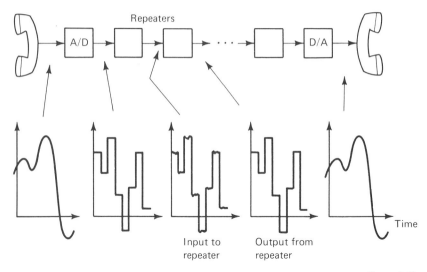

Figure 9.40
Digital telephone link with many regenerative repeaters.

(a) Case 1: *Pulse Modulation*

The first kind is related mainly to digital signals, and we have already seen some examples of it. Its role centers around the basic difference between a *symbol* and a physical *signal* that represents the symbol. For instance, in Figure 9.30(b) the sequence of symbols 149624 constituted the message. But one cannot send *numbers* over a piece of wire or out of an antenna. The numbers must be imprinted on a physical quantity that can travel over actual distance. A voltage fluctuation on a wire, for example, will do this, and so the fluctuations of the voltage source are made to correspond with the message using some predetermined scheme. In Figure 9.30(b) *pulse amplitude modulation* (PAM) was used, where it was decided to send a pulse of .1-sec duration for each symbol and where the amplitude of the pulse would equal in volts the numerical value of the symbol. There was nothing magic about this scheme; any method of *coding* the symbols into uniquely identifiable signals would do just as well. It is essential, of course, that the source and destination use the same scheme. Otherwise the received signal will be completely unintelligible at the destination. This is one of the problems of setting up communications with an intelligent life form in another solar system. How does one establish a common coding scheme and language? Even the sending of a photographic image (assuming they could make use of it) requires scanning the image and sending the data in some fashion, and this requires coding of one form or another. The same applies, of course, to signals we receive. Are some of the signals presently being received by radio telescope actually transmissions from intelligent life elsewhere in the universe? How could we tell? (See Reference 2 for an interesting discussion of this problem.)

There are many other ways of "modulating a signal with a message." One of the most popular is *pulse code modulation* (PCM). Here each of the

Table 9.1

Symbol	Binary Word
E	000
T	001
A	010
O	011
N	100
R	101
I	110
SPACE	111

9.3
some fundamental concepts in communications

symbols that one will ever want to send is assigned a binary code word. Then the binary words in the message are sent using two-level pulse amplitude modulation. For instance, if the symbols are the eight most frequently used symbols in English—E, T, A, O, N, R, I, and SPACE—one might assign the code in Table 9.1. Then the message NO RAIN would be equivalent to the binary message 100011111101010110100 and it would be sent over the channel using the waveform of Figure 9.41 (5 V for 1 and 0 V for 0). Note that since there are three binary digits (or bits) for each symbol, one must send these bits three times as fast in order to send the same number of symbols per second. Thus, one pays for the simplicity of two-level transmission with increased speed. It turns out that increasing the speed requires more bandwidth, so basically the trade-off is between simplicity and bandwidth.

A PCM transmission scheme is being used in some parts of the telephone system today, and we have all talked over a link using this scheme many times without even knowing it. In this scheme our voice waveform is sampled 8,000 times a second and then quantized into 128 possible levels. Each of the 128 levels has been assigned a 7-bit code word, and the appropriate stream of 1s and 0s is sent for each sample level. Thus, 56,000 bits are sent each second. This might seem rapid indeed until it is realized that the binary streams of 95 other speakers are commonly interleaved with our stream using time division multiplex, so that the rate of transmission is actually about 5.4 million digits/sec.

Figure 9.41
PCM signal for the message NO RAIN.

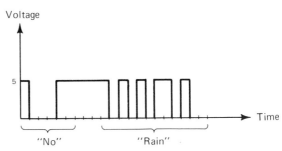

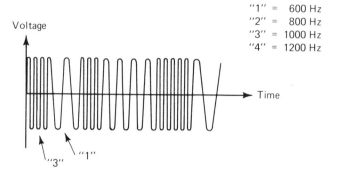

Figure 9.42
Example of pulse frequency modulation.

We shall now give examples of two other pulse modulation schemes to give some idea of the variety of methods available to the engineer. We assume that the alphabet consists of the four symbols 1, 2, 3, and 4 and that the message to be sent is 3142241.

Pulse Frequency Modulation (PFM). Here a pulse of different *frequency* is sent for each of the four symbols. The pulses are sine waves, lasting the same amount of time and having the same amplitude in each case. Figure 9.42 shows an example where the chosen frequencies are 600, 800, 1,000, and 1,200 Hz.

Pulse Width Modulation. In this scheme each symbol is allotted a fixed time interval, and the information is impressed on the signal according to the *duration* of a pulse within this interval, as suggested in Figure 9.43. The pulses

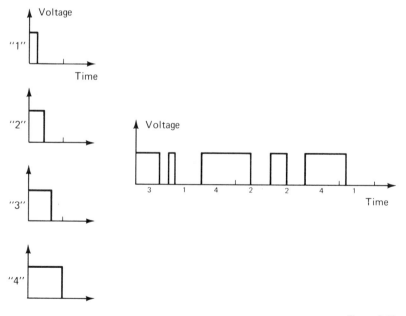

Figure 9.43
Example of pulse width modulation.

always begin at the start of their time interval, but they end at different instants within the interval.

We now turn to the other sense in which the term *modulation* is used.

(b) Case 2: *Carrier Modulation*

The other type of modulation involves altering a signal so that it is more suitable for transmission or processing. In many systems the method of transmission, or the medium itself, forces one to use only high-frequency signals. That is, while the signal carrying the message might have frequencies only up to 4,000 Hz, in order to transmit the signal we must use frequencies no lower than 10^6 Hz. Consequently some shifting of the signal's frequencies must be accomplished. The most familiar example is radio transmission. Here a signal is fed to an antenna for transmission "over the air" (actually through free space, of course) to one or more distant antennas. But antennas radiate energy efficiently only at high frequencies. Roughly speaking, a signal cannot "get out of" an antenna unless the antenna has a length at least one-quarter of the *wavelength* of the signal being sent. (A wavelength is the distance a sine wave propagates during one of its cycles. The wavelength L of a sine wave of frequency f traveling at speed c may be calculated as $L = c/f$.) In a vacuum waves travel at the speed of light, so a sine wave of frequency 4,000 Hz has a wavelength of $186,000/4,000 = 47$ miles. A quarter wavelength is then about 12 miles, and clearly this is too large a length for one to build an antenna conveniently. However, if the signal could be shifted up in frequency so that its frequencies were located around 10^6 Hz, a quarter wavelength is only 235 ft, and broadcast antennas are easily constructed this long.

The question is then, How do we shift up the frequencies in a signal? The answer is *carrier modulation*, and it basically involves impressing the signal onto a sine wave of high frequency. The sine wave is called the *carrier*, a reasonable name in light of the discussion above, and its basic frequency is the *carrier frequency*. The carrier has no information content of its own; the information in the signal to be sent is impressed on the carrier and then removed from the carrier at the receiver.

We now examine the two most important carrier modulation methods, *amplitude modulation* and *frequency modulation*. These techniques work well for both analog and digital signals, and so we will give examples of each.

Amplitude Modulation. (AM). This is the simplest type of carrier modulation. The amplitude of the carrier is varied in accordance with the signal, so that the signal fed to the antenna is a sine wave of varying amplitude. Amplitude-modulated signals are shown in Figure 9.44 along with the signals being modulated. Note that the sine wave's amplitude is large when the signal is positive and small when it is negative. The dotted line shows the *envelope* of the sine wave, which is a replica of the signal. Thus, the information is stored in the carrier's envelope.

It is interesting to see what happens to the signal's *spectrum* when amplitude modulation is performed. It turns out that the spectrum of the amplitude-modulated signal is centered about the carrier frequency and has

9.3 some fundamental concepts in communications

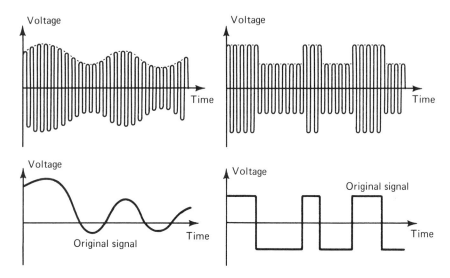

Figure 9.44
Amplitude-modulated signals.

the shape of the original spectrum plus its mirror image, as shown in Figure 9.45. We see that the amplitude-modulated signal has a bandwidth twice as large as that of the orginal signal.

When the amplitude-modulated signal has traveled through the channel and arrives at the receiver, it must be *demodulated* in order to recover the original signal. Since the information is contained in the carrier's envelope, this is sometimes referred to as *envelope detection*. The standard home AM radio receiver incorporates an envelope detector that discards the rapidly oscillating carrier and retains only the envelope variations. These are then amplified and fed to the speaker of the radio.

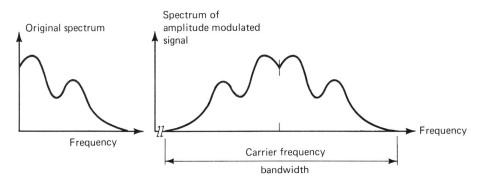

Figure 9.45
Spectrum of the original signal and its amplitude-modulated counterpart.

One can get a rough feeling for how this amplitude modulation might be accomplished by considering a slightly different scheme. Here, instead of using a sine wave as the carrier, a square wave is used. The amplitude of the square wave is made to vary with the signal by simply multiplying the two waveforms instant by instant, as shown in Figure 9.46. Thus, the carrier just alternately changes the sign of the signal at a rapid rate. The resulting signal is a high-frequency waveform because of its very rapid fluctuations. Now if this signal is sent over the channel and then simply multiplied by another identical square wave at the receiver, the sign reversals are just rereversed, and the original signal is recovered exactly. This is sketched in Figure 9.47, where it is clear that the carrier at the receiver must be carefully *synchronized* with the carrier at the transmitter. This scheme is not normally used in real transmission systems as it turns out that the sharp edges of the high-frequency signal are wasteful of bandwidth. However, it is a neat way of seeing how carrier modulation and demodulation work.

Frequency Modulation (FM). The other major type of carrier modulation is frequency modulation in which the carrier has a constant amplitude but a frequency that varies in accordance with the signal. Figure 9.48 shows typical frequency-modulated versions of some analog and digital signals. When the signal has level zero the frequency of the carrier is at its nominal center value. When the signal goes positive the carrier frequency increases, and when the signal is negative, the frequency decreases. Note that for the digital signal we essentially have a case of pulse frequency modulation. The total excursion in frequency of the carrier is normally a small fraction of the center value. However, a frequency-modulated signal usually has a rather

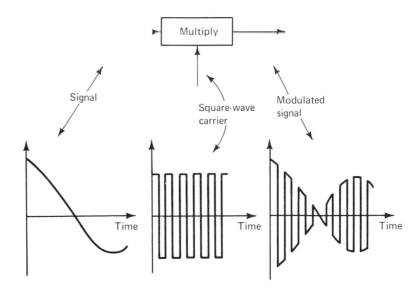

Figure 9.46
Amplitude modulation using a square-wave carrier.

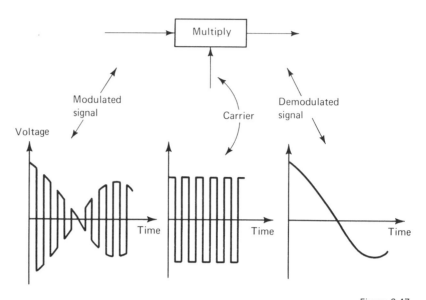

Figure 9.47
Demodulation using the same square-wave carrier.

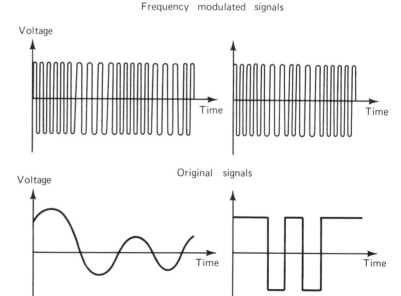

Figure 9.48
Basic signals of Figure 9.42 frequency-modulated.

large bandwidth, and its spectrum is very complicated mathematically. As suggested in Figure 9.49, it does not bear any simple relation to the spectrum of the original signal but usually has a much larger bandwidth. This seems wasteful of bandwidth, but this extra bandwidth can be shown to have powerful advantages. Specifically, an FM signal is much less sensitive to noise interference. In fact, unless the noise is very strong, there is actually a *noise suppression* action in an FM receiver such that one hears no "static" at all. This is why the audio portion of a TV signal is sent using frequency modulation, while the video part uses amplitude modulation. The ear is very sensitive to static, while the eye does not mind occasional "noisy" spots on the TV screen.

A frequency-modulated signal must be demodulated before it can be used by the destination. The receiver in this case is called a *frequency discriminator*, as it extracts the instantaneous frequency of the signal and generates a voltage proportional to it. FM receivers are somewhat more complicated than AM receivers and consequently more expensive, but for such things as high-fidelity music they are far superior, having lower noise and better sound reproduction.

There is one other kind of carrier modulation, called *phase modulation*, which is very similar to frequency modulation. Here the *phase* of the carrier is varied in accordance with the signal. A phase "lead" proportional to the signal is effected when the signal goes positive, and a phase lag when the signal goes negative. The spectrum of a phase-modulated signal is also very complicated and closely resembles that of a frequency-modulated signal. In fact, the reader familiar with calculus will be interested to note that phase-modulating a carrier with a signal is identical to frequency-modulating the carrier with the *derivative* of that signal.

9.3
some fundamental concepts in communications

9.3.4
Choice of Carrier Frequency: The Electromagnetic Spectrum

One very important point in both amplitude and frequency modulation techniques is that one can *choose* the value of the carrier frequency in the modulation process. As we have seen, this choice determines where in frequency the spectrum of the modulated signal lies. If one wants to send several signals over the same wire or cable, he need only modulate each signal up to a differ-

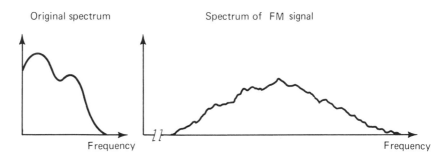

Figure 9.49
Spectrum of a frequency-modulated signal.

ent frequency region and then add them all together. Figure 9.50 indicates how this may be done. The process is called *frequency division multiplex*, and it is completely analogous in frequency to the action in *time* of time division multiplex. Here the frequency band to be used is divided into several parts, and each part is used to carry a different signal. For the three voice signals (having bandwidths of 4,000 Hz) shown in Figure 9.50, it is clear that the three carrier frequencies must be chosen at least 8,000 Hz apart to avoid overlap of the individual spectra. This composite signal is then sent through the cable to the receiver, where the three signals are separated again. As shown in Figure 9.51, three filters are used, each one passing only those frequencies relevant to one of the original signals. Each signal is then demodulated and routed to the desired destination.

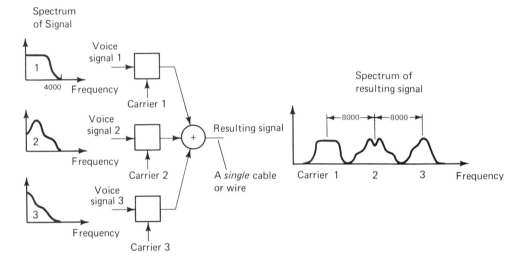

Figure 9.50
Frequency division multiplex.

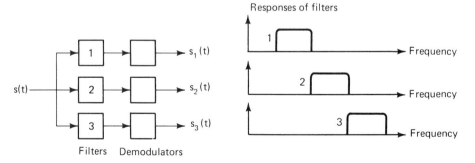

Figure 9.51
Demultiplexing the three signals.

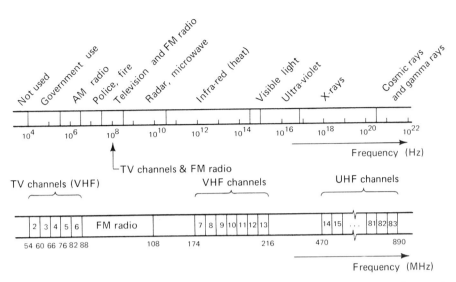

Figure 9.52
Allocation of the electromagnetic spectrum in the United States.

Frequency division multiplex is used very commonly today to carry hundreds or even thousands of voice signals over a single cable or over a microwave beam. Because so many signals share the same equipment, the cost of sending each signal has been dramatically reduced.

In broadcasting much the same is true, only here the equivalent of the cable is free space. That is, if several radio or TV stations wish to send signals in the same geographical area, they must use different frequency bands so that their signals will not overlap in frequency and become hopelessly jumbled. To regulate the use of this precious frequency space the Federal Communications Commission (FCC) has allocated many frequency bands to various types of broadcast and tightly controls how they are used. This usage is roughly shown in Figure 9.52, where frequencies from 10 kHz up into the X-ray region (10^{20} Hz) are depicted. The lower part of this electromagnetic spectrum is very crowded with users, because engineers have long been able to build equipment that works efficiently in these regions. However, the upper frequency regions are still for the most part unused, as appropriate electronic devices are not yet readily available. Masers and lasers are becoming more common and beginning to use small bands in this region, but there is a tremendous communications capacity just waiting to be used.

To get a feeling for how much "room" there is left in the spectrum, note that the entire band now used for all of radio, television, police, etc., extends only from 10 kHz to about 10,000 MHz, a bandwidth of about 10,000 MHz. Now consider the band used by the human eye, extending over a mere octave from 5×10^{14} to 10^{15} Hz*. But this octave has a bandwidth of 5×10^{14} Hz, so it has 50,000 times the communications capacity of all the bands presently

---

*Just as in music, one speaks here of an *octave* as a frequency ratio of 2 to 1.

being used. This band and all those around it simply await devices that can use them. This is one reason the laser has such potential for communications: A single beam could carry an immense number of signals (using frequency division multiplex, perhaps) and still occupy only a very small part of a one octave band.

9.4 CONCLUSION

We have examined several fundamental concepts and techniques of modern communications. These ideas are used every day by engineers to design and build practical systems. The concepts of frequency and bandwidth are central to all communications theory, for they permit one to discuss and analyze potential systems in a coherent manner. The whole field of electrical filters has traditionally focused on these frequency concepts.

The basic equivalence of analog and digital signals—in the sense that they can carry the same information—has led to a marriage between communications systems and computers. Now not only can computers "converse" with one another over communications links, but computers are also used as part of the communications system itself, processing and routing signals. This alliance between communications and computer science is a relatively recent development in engineering and promises great things in the future.

Modulation and multiplexing of signals have provided the keys to efficient use of channels. Signals can be moved around in frequency and packed one next to the other, so that a channel is used to its fullest ability. Hundreds of signals are sent over a single communication circuit, and this is only the beginning. With the development of laser channels the number of signals sent over a single circuit will grow to the hundreds of thousands. The next 50 years will see extraordinary new ideas and inventions applied to permitting man to communicate ever faster and more freely.

REFERENCES

1. MARTIN, J., *Telecommunications and the Computer*. Englewood Cliffs, N. J.: Prentice-Hall, Inc., 1969.
2. SHKLOVSKII, I. S., and C. SAGAN, *Intelligent Life in the Universe*. New York: Dell Publishing Company, Inc., 1966, Chapter 30.

EXERCISES

1. Consider two nonlinear voltage characteristics: (1) the clipper of Figure 9.12, and (2) the dead-band characteristic shown in Figure 9.53. The input signal is triangular, fluctuating back and forth between $-A$ V and $2A$ V every second. Sketch the output signal for each of the characteristics and for the case $A = 1$ and $A = 4$ (four cases in all).
2. Information can be converted to many forms. Some communications systems send only 0s and 1s, so we must frequently convert an English message into a sequence of 0s and 1s.

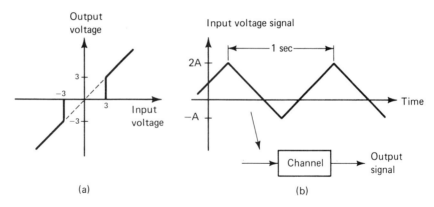

Figure 9.53
For Exercise 9.1. (a) Input-output characteristic.
(b) Input signal voltage.

a. Using the code given below, write the corresponding sequence of 0s and 1s for the message.

GOOD MEN LIKE CAIN AND ABEL

Code:

A = 0000	E = 0100	I = 1000	M = 1100
B = 0001	F = 0101	J = 1001	N = 1101
C = 0010	G = 0110	K = 1010	O = 1110
D = 0011	H = 0111	L = 1011	SPACE = 1111

b. If this message is sent over a channel and because of interference some of the 0s are received as 1s and vice versa, an incorrect message will be received. What message (in English letters) is received if the following digits (0 or 1) are received incorrectly: the third, ninth, twenty-fourth, forty-first, and seventy-third?

3. A certain periodic signal $s(t)$ consists of the sum of two sine waves. One sine wave has

   amplitude = 1 V
   frequency = 2,000 Hz
   phase = 0°

The other has

   amplitude = .5 V
   frequency = 4,000 Hz
   phase = 0°

Sketch each sine wave and the total signal s(t). Carefully label the time axes.

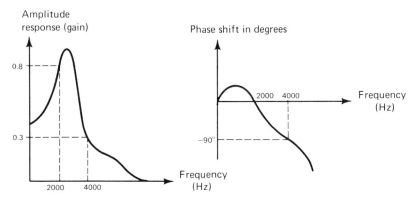

Figure 9.54
System transfer function for Exercise 9.4.

4. The signal s(t) of Exercise 3 is the input to a system with the transfer function shown in Figure 9.54. What are the amplitudes, frequencies, and phases of the output components as a result of each input sine wave in Exercise 3? Sketch each output component. Also sketch the total output signal (the sum of the two output components).

5. The coarsest quantization is performed by a two-level quantizer, sometimes called a *sign detector*. The output of such a device is $+1$ V whenever the input is positive, and -1 V whenever it is negative.
   a. Draw the input-output voltage characteristic of a sign detector.
   b. Draw a signal of your choice, and find the quantized version, using the characteristic of part a.

6. Does one get the *same* digital signal regardless of whether it is sampled first and then quantized or quantized first and then sampled? Experiment on some typical signals, doing it both ways.

7. Convert the analog signal shown in Figure 9.55 to a digital signal by sampling it every .01 sec (beginning at time 0) and using the quantizer characteristic shown in Figure 9.32. Sketch the digital signal. Write the equivalent sequence of binary digits for the digital signal, using the coding scheme

VOLTS	CODE
3	11
1	01
−1	10
−3	00

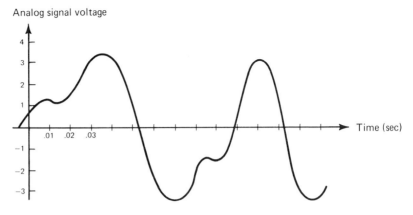

Figure 9.55
For Exercise 9.7. Analog signal to be converted to digital form.

8. One can assign code words to more complicated "symbols" as well. Using the code assignment below and the PCM scheme of 1-msec-long pulses, 5 V for a 1, and 0 V for a 0, draw the signal that carries the message "Happy birthday, the captain is aboard, we are returning home."

Symbol	Binary Word
Help	000
We do not read you	001
Is there a doctor aboard	010
Happy birthday	011
The captain is aboard	100
We are Under Attack	101
We are returning home	110
Identify yourself	111

9. Consider a source that sends only the symbols A, B, K, N, O, T, one every .1 sec. *Pulse position modulation* is used, wherein each .1-sec interval is divided into six equal parts, and a short pulse is sent in the first subinterval to represent A, in the second for B, etc. Sketch the signal that results for the message BANK BOOK NOT OK.

10. There is another type of pulse modulation that might be called *pulse shape modulation*. Here each symbol is represented by a signal that lasts the whole time interval allotted to it but has a unique wave shape. Devise six different shapes for the symbols in Exercise 9, and sketch the signal corresponding to the message BANK BOOK NOT OK.

11. For the square-wave modulation scheme of Figures 9.46 and 9.47, what is the relationship between the original signal and the demodulated signal when the square-wave carrier at the demodulator is out of synchronization with that at the modulator; i.e., it is -1 when it should be $+1$, and $+1$ when it should be -1?

10
introduction to biomedical engineering

D. E. Scott

10.1 INTRODUCTION

Within the last 10 or 15 years there has occurred a great increase in the number of engineers who are working with both physicians and scientists in the areas of biology, psychology, and the other life sciences. This type of engineering endeavor has been given the name bioengineering. If an engineer is working specifically in the area of human health care, he is termed a *biomedical* engineer, which is a subcategory of bioengineer.

Historically, engineers became involved in this field because of the needs of physicians and biologists for more sophisticated measuring instruments and more precise mechanisms of all sorts for use in operating rooms and in biological research laboratories. Originally, then, the engineer was essentially a high-level technician and a supplier of sophisticated hardware but not a member of the biological research team or operating room medical staff.

More recently the engineer has gradually become accepted as a professional "equal" by increasing numbers of medical and biological scientists, largely because of a spreading recognition that engineers have more to offer the life sciences than biologists and physicians had at first assumed. This additional value of the engineer is what may be called the "rigorous viewpoint." Until fairly recently, biology and medicine had a "taxonometric viewpoint." That is, a biologist studying a certain insect would tabulate all the known and observable facts about this particular little animal (it has six legs, is greenish in color, has an obvious large double neuron in its thorax, etc.), filling a volume or two with his detailed and complex observations. Then, when he had amassed all the facts he could about this insect, he would begin over again with another type of insect (or, if he were a physician, with another organ of the body or particular disease).

The engineer's rigorous viewpoint is somewhat different. It consists of an immediate desire to put numbers on things, to make accurate measurements, to use mathematics, to make simplifiying assumptions when dealing with extremely complicated systems (which may or may not be correct), and above all to make mathematical models. A mathematical model of some biological entity is not a model in the sense that a little ship model is a reproduction of the real ship. A mathematical model is a set of equations, or a computer simulation, sometimes but not always accompanied by a picture or schematic diagram, that will (more or less accurately) predict the actions of the real biological system.

The advantages of making mathematical models of biological systems are now becoming evident to enlightened biologists and medical personnel. A good model focuses the investigator's attention on the essentials of the system being studied. Also, when the actions of the real system are almost, but not quite, predicted by the model, the investigator knows that he understands the workings of the system in an incomplete way and must now refine the model, if he can, to more closely mimic the actions of the real biological system. This, then, is the real purpose of a mathematical model; it is not to be developed and placed on a shelf to be admired as the miniature sailing vessel is, but is rather to be a conceptual aid during the development of understanding and familiarity with a biological system. Once the system is adequately understood, the model has no further purpose and is discarded. It is also discarded if it cannot even after a first meager success in predicting the actions of the biological system be further refined to give more accurate predictions. The latter situation indicates that the basic ideas of the investigator about the biological functions involved are probably wrong, and new concepts, or at least a critical review of the original assumptions concerning this function, are needed in order to create a new model. It has been this new point of view that the bioengineer offers the life scientist that has been largely responsible for the rapid growth of bioengineering.

Now scientists are beginning to see the need for more than just professional cooperation between the professional engineer and the biological scientist. Warren S. McCulloch, a famous research scientist, put it this way, "The difficulty with every field like this is that it tends to require an increasing number of people who know two scientific disciplines.... One has to have a reasonable knowledge of both engineering and biology in his head, and there is no use in having in one room what should be in one head." As a result, many universities are offering a program of an interdisciplinary nature called Bioengineering or Biomedical Engineering, the purpose of which is to put sufficient amounts of engineering and biology into the student's mind to enable him to make a real scientific contribution to this growing field.

The following sections of this chapter indicate some of the areas where bioengineers have contributed their rigorous viewpoint as well as their ability to create necessary complex instruments. Owing to limitations of space, the topics discussed are not intended to be an all-inclusive list. Rather they are only a random sampling of the ways bioengineers are aiding in the quest for new knowledge and new ways to aid mankind achieve a longer and more healthy life-span.

10.2 NERVES

One of the first biological systems to be modeled was the neuron or nerve cell in conjunction with the muscle that such a neuron might control. Until the seventeenth century, the contraction of muscle was thought to be due to an increase in muscle volume. The model then accepted consisted of a long inflatable rubber tube whose ends came closer together as the tube was pumped up. The nerve, it was thought, had to be a pipe with stiff walls that carried a "subtle fluid" under hydraulic pressure from the brain to the muscle, thus enabling voluntary muscle control. Not surprisingly, because of the slow pace of scientific and medical progress in those days, it took many years before this model was discarded as being basically wrong. In about the 1790s nervous activity was discovered to be basically electrical in nature. Many false concepts were advanced and disproved. Perhaps no other field of biology is so filled with models of varying degrees of complexity, each one in turn being discarded as being incorrect even though it predicted neural activity slightly more accurately than its predecessor.

One model, put forth by a scientist named Lillie in about 1915, used a galvanized (zinc-coated) iron wire immersed in dilute sulfuric acid. When the zinc coating was scraped off at a point on the wire, an electrical current was observed flowing from the iron to the zinc. Here, it was then thought, was a direct analogy to the wall membrane of an actual neuron and to what happens when this membrane is injured or electrically stimulated. However, there were several important inconsistencies between this analog and real neurons, and in addition it was easier to make measurements on real neurons than on the model. These findings led to its being generally relegated to the category of "historical interest only" in about 1945.

Briefly, let us consider some aspects of real nerves that are more or less faithfully reproduced by present-day models (both transistorized and purely mathematical).

Nerves transmit information from sensory receptors (e.g., heat receptors in the skin, taste buds, and pressure receptors in the skin) to the higher levels of the central nervous system (CNS) and also from the CNS out to the muscles and other body organs. Nerves that carry information toward the upper CNS (or "brain") are termed *afferent nerves*, while those that carry information from the brain are termed *efferent nerves*. Information flow on a typical nerve is from left to right in Figure 10.1. If a dendrite is electrically stimulated, a voltage change will be observed at point *A* and very slightly later with undiminished magnitude at point *B*. The voltage is measured by inserting a microscopic-sized electrode into the axon and amplifying the voltage between this electrode and the fluid surrounding the nerve cell. A

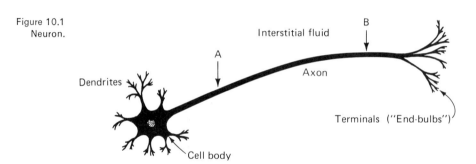

Figure 10.1
Neuron.

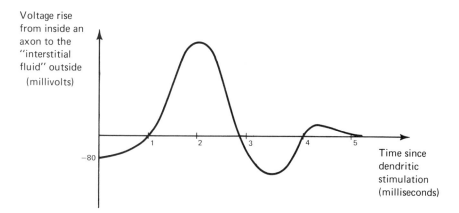

Figure 10.2
Typical action potential.

typical plot of this voltage versus time is shown in Figure 10.2 and is called an *action potential*. This voltage waveform travels down the length of the axon. This is the means by which nerves transmit messages. When the voltage wave arrives at the terminals or *end bulbs* of the neuron it is ready to stimulate one or more of the dendrites of a following nerve cell. The terminals are never quite in contact with the dendrites they are able to stimulate. There is a microscopic gap called a *synapse*. The neural message rides across this gap on a complex chemical substance called a *carrier* or *transmitter* that is secreted by the terminals. Every nerve has a stimulation threshold such that if a stimulus is applied to its dendrites of magnitude less than this threshold, there results either no effect at point *A* or, at most, an extremely small voltage shift (which is even less by the time it gets to point *B*). This phenomenon of a propagating subthreshold response being observed is called *electrotonus*. If the stimulation that causes electrotonus is a constant electrical current, then the distance along the axon (say between points *A* and *B*) necessary for the steady-state value of the observed voltages to decrease by a factor of 63% is called the *space constant* of the axon. The time it takes for this response at any fixed point to reach steady state is related to the *time constant* of the axon. There are many other well-known phenomena that may be observed such as *accommodation* and *postcathodal depression*. All these have been successfully modeled. We now know that the mechanism involved in neural propagation is electrochemical in nature. Following up the idea of electrotonus being a phenomenon that varies with distance down the axon and also that varies with time, A. L. Hodgkin and A. F. Huxley announced in 1952 that they had developed a partial differential equation that simulates most of the known phenomena observed in actual living neurons. For this work they later received the Nobel Prize.

Much work remains to be done in this field, especially in two main directions. First, the Hodgkin-Huxley equation predicts with extremely good accuracy most, but not all, observed phenomena; also some of the phenomena that can now be predicted accurately by their model are actually caused by biological mechanisms that are unknown or incompletely understood. Bioengineers can help in efforts to achieve this understanding, both through their ability to construct the complex measuring instruments that will be needed and through their relative familiarity with problems that require a rigorous quantitative attack. Second, models of neurons in the form of simple integrated circuits that can be mass produced are needed.

The schematic diagram shown in Figure 10.3 is an artificial nerve.[5] It contains five transistors and is capable of being stimulated by both exitatory and inhibitory inputs. These nerve models can be interconnected to form artificial neural networks. They help researchers to analyze how actual biological nervous systems work and even more importantly help researchers to make use of the ways nature does certain things in order to design more efficient electronic networks and communications systems. This imitation of nature by engineers has been given the name *bionics*. A notable example is the biological phenomenon of *lateral inhibition*, observed first in the eye of the horseshoe crab.

The effective result of lateral inhibition is to enhance and sharpen "edges." Consider a picture that is quite out of focus (as, for example, the way the horseshoe crab views things because of his very poor optical system). If just the edges of the various items in the picture can be made reasonably sharp, the picture will be of apparently acceptable quality. An artificial equivalent of this process has been used by engineers to sharpen edges and thus enhance picture fidelity in color television transmission.

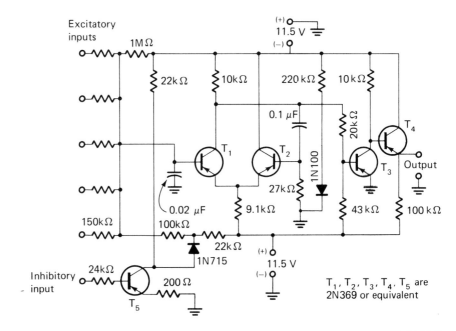

Figure 10.3
Electronic circuit for an artificial nerve model.

10.3 THE ELECTROENCEPHALOGRAM (EEG)

An electroencephalogram is a picture drawn on graph paper of the varying voltages that appear at the surface of the skin of the head. These voltage waves (sometimes called *brain waves*) result from the electrical nervous activity going on in the brain. There may be several "upstream" neurons that synapse on the dendrites of any single given "downstream" neuron. Any one of the upstream neurons, or any group of them, may fire, thus producing action potentials in the downstream neuron. It has been observed that sometimes certain of the upstream neurons tend to inhibit rather than excite the formation of action potentials in the downstream neuron. Any collection of neurons interconnected by such inhibitory and excitatory synapses is termed a *neural net*. The human *brain* is a neural net containing approximately 10^{12} nerve cells.

Consider the following "experiment." If one were able to attach a highly sensitive pickup onto the outside of the main telephone cable between New York and San Francisco, what useful information would one be able to extract from the conversations being transmitted over the cable? Obviously, the answer is none at all. Even if one were able to distinguish an individual word or syllable now and then, the exact thought being transmitted would remain obscure. Moreover, we would have no idea as to the identity or location of the speaker or the listener.

Such is the state of encephalography today. An electrode is placed on the scalp and connected to a high-gain amplifier. The amplified voltage is either recorded on a strip chart recorder or observed on an oscilloscope. Surprisingly, a rather large amount of useful information can be obtained from this process. Clearly what is being measured is the *amount of neural activity* in the region of the brain nearest the sensing electrode. From experience it has been determined that a second electrode located in a symmetrical location on the opposite side of the skull from the first electrode normally yields a signal extremely similar to the signal from the first electrode. A situation where one signal is either much larger than the other or fluctuates much more rapidly than the other would indicate a possible pathological situation to the physician.

One other technique that can be used to extract meaningful information from the EEG signal is to obtain its *power spectral density* (PSD). For example, if one were to count the number of waves or undulations in a given EEG recording taken from a patient whose eyes are closed, in any one second you would get results such as 10, 6, 12, 10, or 15. Suppose that you did this for each second of the day ($60 \times 60 \times 24 = 86{,}400$ sec) and then made a graph of the percentage of time between f and $f + 1$ waves/sec, and plot this for values of f between 0 and 100. If you then connect the points you have plotted with a smooth curve, you have the PSD curve for this EEG. It might look typically like Figure 10.4. Of course, only rarely would one

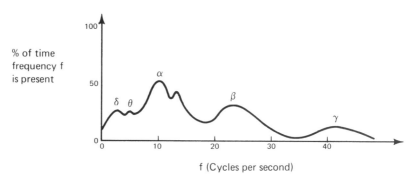

Figure 10.4
Typical day-long EEG PSD.

obtain an EEG recording 24 hr long, so it would be difficult to obtain such a plot; but, if one could obtain the EEG, the resulting PSD would look very much like Figure 10.4.

Note that the various peaks in Figure 10.4 are labeled. It turns out that EEG activity is normally concentrated near one of these peaks for any given psychological state of the patient:

EEG Wave Name	Dominant Frequency (Hz)	Psychological State
α, alpha	10	Awake, pleasantly relaxed
β, beta	20–25	Alert, problem solving
γ, gamma	40	Excitement, alertness
δ, delta	1–5	Light sleep
θ, theta	6	Normal ongoing activity
Spindles	13	Deep sleep

There are, of course, other neural activities that are observed, such as the extremely large undulations of grand mal epilepsy and the like. A great deal of work is presently being done by bioengineers in seeking better ways of analyzing EEG signals. Automatic analysis is being attempted utilizing the speed advantages of digital computers. A recent attempt to teach patients to get into the relaxed α (10-Hz EEG) state as an alternative to taking tranquilizing drugs has received much notice in the popular press. This area of biomedical engineering affords wide opportunities both for increasing the sophistication of the hardware that is needed and for developing better methods of analyzing EEGs.

An example of a technique familiar to engineers that is being used presently in efforts to extract more information from the human EEG waveform is the *time-autocorrelation function*. Very simply this function is obtained from a plot of the EEG as follows. Pick any two times τ sec apart and record the *product* of the amplitudes of the EEG at these two points. Now move to any two different times (still exactly the same time, τ, apart) and record the product of these two EEG amplitudes. Do this many times. You will then have a large column of numbers, each one being a product of the EEG amplitude measured τ sec apart. Now arithmetically average all these numbers. Your result will be an estimate of *the average value of the product* of the amplitudes of the EEG measured τ sec apart. It will only be an "estimate" unless you obtain literally an infinite number of products, but for most practical purposes, a large collection of such products will serve nicely. You have now found a single point on the autocorrelation function. If you repeat this process many times over, each time using a different value for τ, you will finally attain enough points to plot the autocorrelation function.

Fortunately, there is a simpler way to do this in the laboratory. A trick often used by engineers in finding the average value of a varying quantity is to remember that a geometrical interpretation of the "integral of a function $f(x)$ from $x = a$ to $x = b$" is nothing more than the area under the plot of $f(x)$ versus x from $x = a$ to $x = b$. (See Figure 10.5) So, if we divide this area by the base length, $b - a$, we obtain the *average height* of the curve $f(x)$ while it is inside the interval marked by points a and b. This trick can be used to find the autocorrelation function $K_{ff}(\tau)$ of a patient's EEG. Let us call this EEG, $f(t)$. Remembering that what we want is the average value of the product $f(t) \times f(t + \tau)$, we can write

$$K_{ff}(\tau) = \frac{1}{T} \int_{t=0}^{t=T} f(t) f(t + \tau) \, dt \tag{10.1}$$

where T is the length of time we have observed and plotted the EEG waveform. Essentially, then, this process consists of multiplying the EEG waveform times a copy of itself, which is delayed in time, and then integrating the resulting product waveform. This delay can easily be obtained by using a tape recorder with separate record and playback heads that are movable with respect to one another. The time delay, τ, can be varied by changing the

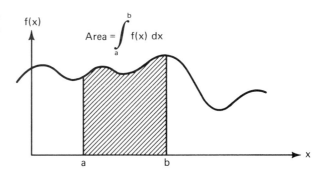

Figure 10.5
The integral as area under a curve.

distance between the heads. Figure 10.6 shows how the autocorrelation function may be obtained in the laboratory.

The autocorrelation function gives the physician a quantitative measure of the speed of the oscillations in the EEG and can be a very accurate diagnostic tool. Also the PSD curve shown in Figure 10.4 can automatically be obtained from a plot of $K_{ff}(\tau)$. It should be clear that engineers have much to offer the biologist and the physician in this area, because what we have been discussing here is almost pure engineering mathematics and yet it can be of lifesaving importance when applied in the area of biomedicine.

10.4 THE HEART

Many engineers have been interested in the cardiovascular system (the heart, arteries, and veins). Many biomedical engineers are working on investigations of blood chemistry and blood flow. For example, atherosclerosis, the disease wherein the arteries become more and more clogged resulting in a diminution of blood flow, is attracting a large amount of interest. Tentative results seem to show that the material that coats the main arterial walls begins to form in locations adjacent to where branches leave those main arteries. Engineers well versed in the science of hydrodynamics immediately realize that this is reminiscent of the effects that are observed in normal piping systems where smooth (*laminar*) flow breaks up into eddies and ripples (*turbulant flow*).

Such analogies between well-known engineering systems and phenomena and those found in biological situations are always of interest because if we know how to solve the engineering problem, we may be able to pick up a clue about how to solve the analogous biological problem. At the very least we may learn what biological measurements should be taken next. We are still a long way from curing atherosclerosis, but each new grain of understanding is a step in that direction.

The heart itself is a four-chambered pump. The mechanisms that control the rate of pumping and the volume of each stroke are not yet completely understood. They are sufficiently well known, however, to enable physicians

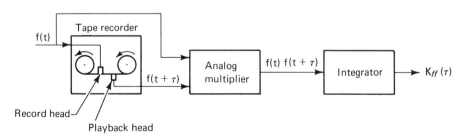

Figure 10.6
Equipment for measuring autocorrelation function.

and engineers to work together designing machines that temporarily take over the functions of the heart: either completely (the *heart-lung machine*) or partially (the *left-ventricular bypass*).

Because of its relative simplicity in comparison to other biological systems such as the CNS, the hormonal system, and the kidney, the heart has been the most popular object of cooperative endeavor between engineers and physicians. These endeavors have resulted recently in the need for engineers to be present in the actual operating room for certain cardiovascular operations.

Let us examine the basic physiology of the heart in order to get some idea of how it performs its function, and then, in Sections 10.5 and 10.6, we shall be prepared to look at some of the ways bioengineers have made real contributions to this area of cardiovascular medicine. Figure 10.7 indicates the relationships among the chambers of the heart and shows how the blood flows through these chambers.

The heart beats in a manner similar to the way one would wring out a wet towel. A twisting motion reduces the volume inside the chambers and

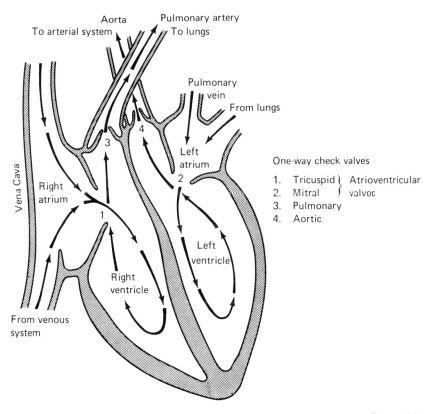

Figure 10.7
Blood flow in the heart.

blood takes the path of least resistance. The contracting (squeezing out) phase of the stroke is called *systole* (sis-toe-lee) and the relaxing (filling) phase is called *diastole* (die-ass-toe-lee). The important events that occur during a single stroke are the following: At the beginning of diastole, the atria both fill with blood; at the end of diastole, the atria begin to contract (the onset of systole), forcing blood down into the ventricles; as ventricular systole continues, the ventricular pressure rises, forcing the atrioventricular (AV) valves shut and the aortic and pulmonary valves open, releasing high-pressure blood to the arteries; and at the end of systole, the ventricles relax to the point where increasing ventricular volume is responsible for decreasing ventricular pressure to the extent that the aortic and pulmonary valves are forced shut by higher arterial pressure, and the AV valves are pushed open by blood flowing into the atria and ventricles from the venous system.

When a stethoscope or microphone is placed on the chest the familiar "lub-dub" sounds are heard. The first sound, lub, is due to the AV valves snapping shut, and the second sound, dub, is from the closure of the pulmonary and aortic valves. It can be noted that systole is shorter than diastole and that if the heart rate increases (because of exercise or excitement, for example), the time duration of diastole gets shorter and that of systole remains constant.

The normal heart goes through over 100,000 such beats each day. The muscular contraction that produces each beat is electrical in nature; that is, muscle contraction is due to electrical activity of the muscle fibers. If one were to connect a voltmeter such that one terminal of the voltmeter were connected to a wire that went inside a normal cardiac muscle fiber and the other terminal of the voltmeter were connected to the interstitial fluid surrounding the fiber, one would observe a voltage of approximately -80mV ($-.08$ V). If a strong enough electrical shock were applied to the fiber, a wave of depolarization would spread throughout the fiber, and the entire fiber would contract. This is the cause of all muscular contraction, and, therefore, the initiation and spread of such a wave of depolarization through the heart is what causes the heartbeat to occur in the manner in which it does. In humans there is an area in the right atrium called the sinoatrial node, sometimes called the *pacemaker*. This area of the heart depolarizes periodically of its own accord, without outside stimulus. The resulting wave of depolarization then spreads out over the upper half of the heart, causing it to contract and systole to begin. If the sinoatrial node becomes damaged for some reason, very often another region of the right atrium will take over and initiate the onset of the beat. If sufficient damage occurs and no other area is capable of initiating the depolarization, an artificial pacemaker may have to be inserted. As the wave of depolarization spreads over the entire right and left atria it eventually impinges on the *atrio ventricular node*. The AV node is located at the beginning of a special system of cardiac muscle fibers called the Purkinje system. These special fibers have a speed of conduction approximately six times the conduction speed of normal cardiac fiber. The wave of depolarization normally pauses for a few hundredths of a second at the AV node and then speeds down the Purkinje system and around both ventricles, causing ventricular contraction. The traveling impulse depolarizes the cardiac muscle temporarily as it goes so that the heart will not pass another wave until repolarization occurs. The depolarization period is approximately .3 sec in duration. The heart muscle is said to be *refractory* during this time. It is this refractory period that sets an upper limit on the heart-beat frequency of approximately 200 strokes/min. If electrodes are placed on the chest and connected to highly sensitive and specially constructed dc amplifiers, the effects of all these traveling waves of depolarization may be seen and recorded outside the body. Such a recording is called the electrocardiogram (EKG).

The left ventricle is the chamber of the heart that does the most work. It is this chamber that must develop enough pressure to force blood out into the entire arterial system of the body. It is on this chamber, then, that a great deal of research effort has been focused in attempts to make either temporary or permanent assist devices. One of the first of these devices was an air-driven artificial left ventricle jointly invented by the famous surgeon Dr. Michael E. DeBakey and by Dr. William W. Acres, Chairman of the Department of Chemical Engineering of Rice University. Another was a device jointly developed by surgeon Adrian Kantrowitz of Maimonides Hospital (Brooklyn, N.Y.) and his brother, Arthur Kantrowitz, Vice-President of Avco-Everett Research Laboratory. These devices and all others like them are essentially sophisticated, small, and efficient pumps that are controllable by the natural contraction of the heart. As of this date, no one has developed a successful, complete, artificial human heart. Efforts toward this end, however, are now being made. Complete artificial hearts have been more or less successfully operated in dogs and calves. These have not been sufficiently reliable as of yet to install them in humans, however. One very serious problem at present is that of finding an energy source that will keep the device in operation inside the intact human thorax (chest). There are two potential solutions to this problem: The first is to install an internal energy source such as a nuclear generator. A second possibility is to transmit sufficient energy through the skin by electromagnetic radiation. Both these alternatives have sufficient drawbacks at the moment to make this area of investigation a crucial one if a complete artificial heart is to be developed. Perhaps the solution to this problem will lie in a third alternative, which is to utilize the excess energy of the body itself such as mechanical movement or the electrical activity of muscle and nerve cells. However, very little is known about how to capture and make efficient use of this energy at this time.

10.5 ELECTROCARDIOLOGY

In Section 10.4 we saw that if a set of electrodes were placed on the wall of the chest and connected to the proper amplifiers, the electrical activity of the heart could be observed on a recording device external to the body.

The tracing is termed the electrocardiogram. A typical electrocardiographic signal is shown in Figure 10.8. The various points in the signal are labeled with letters that have become standard medical terminology over the years. Each of these effects is due to some action of the wave of depolarization that is traveling within the cardiac muscle itself. Observation of the relative sizes, shapes, and durations of these various parts of the electrocardiographic signal tell the physician a great deal about what is occurring within the heart muscle. A listing of the relationships between the internal waves of depolarization and the externally observed cardiographic events is as follows:

P: Transmission of impulse over atrium (slow wave)
P-Q: Delay of impulse at AV node
QRS complex: Transmission through ventricle (fast wave) via Purkinje system (atrial repolarization at this same time—hidden)
S-T: Refractory period
T: Ventricular repolarization (in opposite direction to wave of depolarization)

One of the first medical researchers to make note of this externally observable electrical activity was Willem Einthoven (1860–1927). Not only did he observe the electrocardiogram but also he developed a system of locations for the electrodes that is in use even today, although more modern electrode placements have recently been developed. Einthoven assumed the heart to be located in the middle of an equilateral triangle formed by drawing imaginary lines through three points: the right shoulder, left shoulder, and apex of the trunk. Since it was found that recordings from the shoulders and trunk did not differ very much from those obtained in using the wrists and ankles, the latter more convenient points were used instead. Thus, Einthoven's recording conventions were summarized in what became known as the Einthoven triangle (see Figure 10.9). In this system of electrode placement the right arm, the left arm, and the left leg are used as active electrode locations. The right leg is used as a ground point. Recording the voltage at the left arm with respect to the voltage at the right arm is known as the first Einthoven lead. If the left leg is recorded with respect to the right arm,

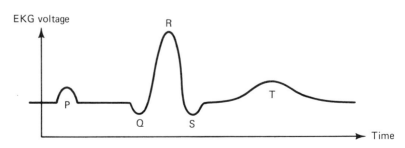

Figure 10.8
Typical electrocardiogram.

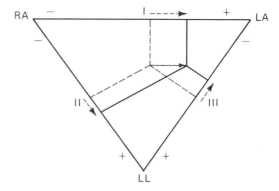

Figure 10.9
Einthoven's triangle of electrode locations.

this is known as the second Einthoven lead, and if the voltage of the left leg with respect to the left arm is recorded, this is known as the third Einthoven lead.

We are used to plotting functions on orthogonal axes, that is, axes that are at right angles to one another. The Einthoven triangle is an example of axes that are not orthogonal. We may think of each of the Einthoven electrocardiographic signals as being the projection (shadow) of a moving point on each of the Einthoven axes. The three axes themselves define a vertical plane. Since the three electrocardiographic signals vary in time, the point is obviously wandering in time on this vertical plane. Since the electrocardiographic signals are periodic, the point will repeat the same path each time the heartbeat occurs. The locus (path) of the point on the plane defined by the Einthoven triangle is called a vectorelectrocardiographic (VEKG) loop, as shown in Figure 10.10. An engineering student will recognize that one can define the location of a point on a plane by knowing its projection

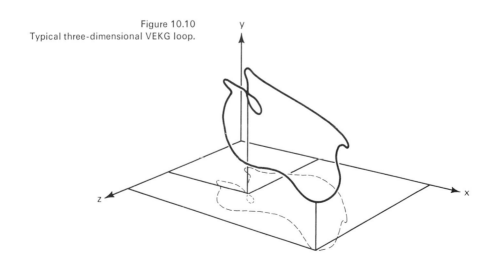

Figure 10.10
Typical three-dimensional VEKG loop.

on any two noncoincident axes. Since the Einthoven triangle supplies three axes, it is clear that one of these axes is redundant.

If an electrode is placed on the chest in the vicinity of the heart and another electrode on the back in the vicinity of the heart and if the resulting signal is amplified and recorded to produce a fourth electrocardiographic waveform, a three dimensional *vectorelectrocardiogram* may be obtained. A large number of bioengineers are presently involved not only in the design and manufacture of the sophisticated amplifying and recording systems necessary for use in this area but also in developing better methods of analysis and information retrieval for these electrocardiographic signals.

10.6 AUTOMATIC ELECTROCARDIOGRAPHIC ANALYSIS

At the present time one of the main thrusts of bioengineering is in making more readily available to the population at large the advances in biomedical research that have already been made. This area is called *health care delivery systems*. An example of such a system is the well-known chest X-ray. A mobile unit can drive into a neighborhood and, in the course of a few days, test almost every person in the local population for the presence of tuberculosis. Such tests, as you are no doubt aware, are in the nature of "screenings" rather than exact diagnoses. That is, a person is either told that he does not have tuberculosis or is told that he should see a physician. In the latter case it is not necessarily true that the person has tuberculosis but only that the tests showed something that should be brought to the attention of a physician, who will then make the final diagnosis. A similar type of screening could be accomplished in regard to heart disease if a mobile unit containing the proper amplifiers and a high-speed digital computer could be driven into a neighborhood. An extremely large amount of time and effort has been spent by bioengineers who have competence with digital computers to write the programs (*software*) that would accomplish almost instantaneous automatic analysis of electrocardiographic signals for such a screening unit. A very successful program for automatic analysis of EKG signals is presently available from the National Institute of Health. However, this program is a detailed diagnostic program and requires a large computer. Therefore, a great deal of effort still needs to be invested in the area of small machines, mobile units, and efficient health care delivery systems.

10.7 POPULATION GENETICS

An example of an area wherein the rigorous mathematical viewpoint has only recently come to be accepted is the field of population genetics. When the single name *genetics* is used, one is usually involved in a discussion of topics such as individual characteristics, dominant genes, recessive genes, mitosis, mutations, and genetic structure. If one is interested, rather, in the growth of some particular population and the effect of the environment and supply of resources on this population, the term *population genetics* is used.

Of prime importance in this area of science is the notion of *rate*. We might say that the population, X, of a particular species of fruit fly that we are studying numbers 100,000 members. Thus,

$$X = 100{,}000 \tag{10.2}$$

We could, at the same time, say that the rate of growth of this population is 20,000/day. We denote rate by placing a dot over the quantity, and so we write

$$\dot{X} = 20{,}000 \tag{10.3}$$

The reason that growth rates are important is twofold: First, they can be measured as simply or more simply than many other quantities (just count the population at two different times), and, second, one can predict what will happen in the near future if one knows the present value for the population, X, and the present growth rate, $\dot{X}$. The growth rate, $\dot{X}$, is the *slope* of the plot of $X(t)$. (See Figure 10.11.) Mathematically we say that $\dot{X} = \Delta X / \Delta t$ as the time change, Δt, becomes very small, where ΔX means a change in the value of X. This is no more complicated than saying "If I have traveled 120 miles and my present speed is 60 mph, I predict that 1 min from now I shall have gone 121 miles." In our mathematical notation,

$$X(0) = 120 \text{ miles}$$
$$\dot{X}(0) = 60 \text{ mph}$$
$$\Delta t = 1 \text{ min} = \tfrac{1}{60} \text{ hr}$$
$$\Delta X = 1 \text{ mile}$$

We can always write our predicted mileage as

$$X(\Delta t) = X(0) + \dot{X}(0)\, \Delta t \tag{10.4}$$

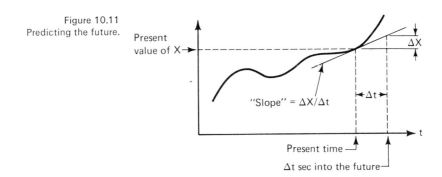

Figure 10.11 Predicting the future.

or, in English, "X at time Δt from now equals my present value of X plus the product of my present speed times Δt." (This assumes that Δt is a short enough interval so that my speed $\dot{X}$ does not change appreciably in that time.)

In biological work we often try to measure and/or guess the effect on a population's growth rate of all the other factors in that population's environment and then write down these effects in the form of an equation. Subsequently, we could use this equation to predict what the size of the population should be in the future. If our prediction turns out to have been correct, we know the effect of the outside influences on the population we are studying; if our prediction is wrong, we go back and modify our equations and try to predict again. This is the essence of *mathematical modeling*.

As an example, suppose that we had a population of turtles, x, and thought that we knew that the rate of growth of these turtles was numerically equal at any time to .001 times the square of the present population. We would write

$$\dot{x} = .001(x)^2 \tag{10.5}$$

and so we would predict that at a short time, Δt, in the future

$$x(\Delta t) = x(0) + .001x^2(0)\ \Delta t \tag{10.6}$$

Suppose that $x(0) = 2$ and that we choose Δt to be .1. Then we can use Equation (10.6) iteratively (again and again) as follows:

$$x(.1) = 2 + .001(2)^2 \times .1 = 2.0004 \tag{10.7}$$

$$x(.2) = 2.0004 + .001(2.0004)^2 \times .1 = 2.00080016 \tag{10.8}$$

$$x(.3) = 2.0008 + .001(2.0008)^2 \times .1 = 2.00120048 \tag{10.9}$$

$$\vdots$$

We should then plot $x(t)$ over a suitably long interval of time and see if our prediction matches the actual turtle population growth over that interval. If it does not, we must go back and modify Equation (10.5) and try again.

As a more complicated (and more realistic) example, let us consider a situation where two interdependent populations are growing in the same general area, each one affecting and being affected by the other.

An example of such a system is a particular seaweed and the limpets that make it the primary item of their diet. The limpet is a shellfish that adheres to rocks or timbers in shallow waters. The limpet population is a function of the available seaweed food supply; their feeding represents a continual "pruning" effect on the seaweed.

Let S represent the seaweed population in a particular acre along the coastal shelf. Let L represent the limpet population per acre in the same area. The interaction of the two species has been represented by the following equations:

$$\dot{S} = k_1 S - k_2 S^2 - k_3 SL \tag{10.10}$$

$$\dot{L} = bk_3 SL - k_4 L - k_5 L^2 \quad \text{(time unit is the day)} \tag{10.11}$$

where $k_1 = 1.1 =$ reproduction of the plant
$k_2 = 1.0 \times 10^{-5} =$ plant density factor
$k_3 = 1.0 \times 10^{-3} =$ the effect of limpet feeding
$k_4 = .9 =$ death of old age
$k_5 = 1.0 \times 10^{-4} =$ limpet density factor
$b = .02 =$ a birth rate coefficient

Normally, the ecological system is near equilibrium, and the casual observer is unaware of the interrelationship. A disaster such as that which occurred in English coastal waters in which an oil tanker, the *Torrey Canyon*, spilled vast quantities of oil can completely disrupt the balance of nature.

Assuming that such a disaster has reduced the limpet population to 10 limpets/acre and the seaweed population to 100/acre, it is our task to develop plots of both populations for a period of 30 days after the disaster.

The solution to this problem is found using the same method as in the previous example but using our new expressions for the growth rates given by Equations (10.10) and (10.11):

$$\dot{S}(0) = 1.1(100) - 10^{-5}(100)^2 - 10^{-3}(100)(10) = 108.9 \tag{10.12}$$

$$\dot{L}(0) = (.02)(10^{-3})(100)(10) - .9(10) - 10^{-4}(10)^2 = -8.99 \tag{10.13}$$

Next we use these results to predict a new value for S and for L at a time, say, .01 day later:

$$\begin{aligned} S(.01) &= S(0) + \dot{S}(0)\,\Delta t \\ &= 100 + (108.9)(.01) \\ &= 101.089 \end{aligned} \tag{10.14}$$

and

$$\begin{aligned} L(.01) &= L(0) + \dot{L}(0)\,\Delta t \\ &= 10 + (-8.99)(.01) \\ &= 9.91 \end{aligned} \tag{10.15}$$

Now we go back and use Equations (10.10) and (10.11) again to find the new growth rates:

$$\dot{S}(.01) = 1.1(101.089) - 10^{-5}(101.089)^2 - 10^{-3}(101.089)(9.91) \tag{10.16}$$

$$\dot{L}(.01) = (.02)(10^{-3})(101.089)(9.91) - .9(9.91) - 10^{-4}(9.91)^2 \tag{10.17}$$

Now we can predict the values for S and L at time $t = .02$ day in the same way as in Equation (10.14):

$$S(.02) = S(.01) + \dot{S}(.01)\,\Delta t$$
$$L(.02) = L(.01) + \dot{L}(.01)\,\Delta t$$

we use these values to find $\dot{S}(.02)$ and $\dot{L}(.02)$ from Equations (10.10) and (10.11) and then predict $S(.03)$ and $L(.03)$. This process can be continued for as long as we like. It is obviously a lot of work to do all these calculations, but since we are really only using four equations [(10.10), (10.11), (10.14), and (10.15)] over and over, this can be done easily on a digital computer. A program written in the FORTRAN IV language is shown below. The results are plotted together in Figure 10.12 for several different sets of initial values of L and S. We notice that, after about 20 days, both populations reach a stable, steady value of approximately

$S = 48,000$

$L = 620$

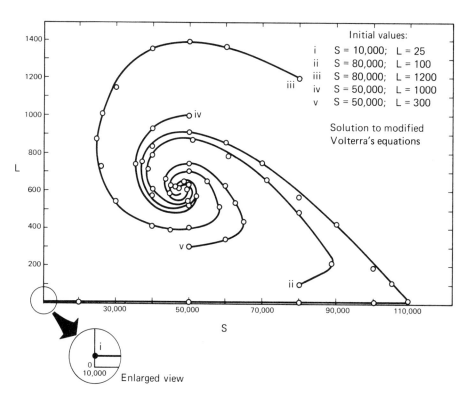

Figure 10.12
State trajectories of the limpet-seaweed ecosystem described by Equations (10.10) and (10.11).

regardless of the initial values, L(0) and S(0). This is clearly a most important point in Figure 10.12. Such points are called *singular points* and may be stable (as this one is) or unstable. They are very interesting from a mathematical point of view. A great deal of effort is usually spent in identifying them in any given problem and in determining whether each is stable or unstable and how sensitive its location is to changes in the coefficients of the governing equations.

```
LIST
  10  PROGRAM MODEL
  20  DIMENSION X(2),XDOT(2)
  30  DELT=0.01
  40  X(1)=100.
  50  X(2)=10.
  59  T=0.
  60  CONTINUE
  61  CONTINUE
  70  XDOT(1)=1·1*X(1)−.00001*(X(1)**2)−.001*X(1)*X(2)
  80  XDOT(2)=.00002*X(1)*X(2)−.9*X(2)−.0001*X(2)*X(2)
  90  X(1)=X(1)+XDOT(1)*DELT
 100  X(2)=X(2)+XDOT(2)*DELT
 110  T=T+DELT
 120  PRINT 121,T,X(1),X(2)
 121  FORMAT(1X,F5.3,F10.1,10X,F10.1)
 130  GO TO 60
 140  END
 150  ENDPROG
```

The formal mathematical name given to equations of the form of Equations (10.10) and (10.11) is *state equation*, because the variables S and L completely describe the *state* of the system being studied. The future action of the system (future values of S and L) may be predicted by the use of these equations given any initial system state. Bioengineers make heavy use of state equations in all sorts of varied situations and applications. If you decide to enter the bioengineering field, these methods will soon become "old hat" to you.

10.8 SUGAR BALANCE

One area that is typical of many in which the mathematical point of view of the bioengineer is useful is the study of the regulatory system in the body that controls the sugar level in the blood. *Sugar* is an imprecise word that is used to describe many different substances. When one eats, "sugar" in the form of *glucose* is normally taken into the stomach. The function of the liver is to store some of this sugar for later use. The liver can store sugar only in the form of *glycogen*, and so excess glucose is converted into glycogen for storage and, later, reconverted back to glucose for use by the cells of the body. Therefore, we may say that, normally, glucose or *blood sugar* is added to the blood in one or more of the following ways:

1. Digestive absorption (eating food).
2. Subcutaneous or intravenous injection.
3. Breakdown of liver glycogen (glycogenolysis) into glucose.
4. Absorption of glucose from tissue fluids.
5. Formation of glucose from other substances such as amino acids and glycerol. This occurs principally in the liver but also in the kidneys (gluconeogenesis).

Glucose leaves the blood by the following mechanisms:

1. Diffusion into the tissue fluids (for use by the cells of the body).
2. Storage as glycogen in the liver.
3. Excretion by the kidneys into the urine when the blood sugar concentration rises above a certain level (called the *renal threshold*—approximately 170 mg/100 cc).
4. Conversion into fat and tissue.

There are at least two chemical substances that aid or inhibit one or more of the mechanisms listed above. These substances are the hormones called *insulin* and *glucagon*, both of which are normally produced in the pancreas. A very approximate description of the action of these hormones is as follows:

Glucagon: When the blood sugar concentration falls to about 70 mg/100 cc the pancreas starts to secrete this hormone. Glucagon causes the liver to reconstruct usable glucose from its stored glycogen, thus raising the blood sugar concentration.

Insulin: This hormone aids the transport of glucose from the interstitial body fluid into the interior of the cardiac and skeletal muscle cells. It also is capable of abolishing release of glucose from the liver almost instantaneously.

The disease called diabetes is the result of a shortage of one or both of the above hormones. Periodic injections of insulin are the normally prescribed way of controlling this malfunction, since without insulin the ability of the cells to absorb their needed glucose from the interstitial fluid is dangerously reduced. Thus, blood sugar concentration rises and the body cells are starved for glucose. If allowed to continue, this condition (hyperglycemia) results in coma and death.

An excellent example of how bioengineers are making contributions to man's knowledge and betterment is to be found in this area. The work of several researchers has led to the formulation of the state equations for this biological system. These equations are indeed as complicated as one would expect and so are not given here, yet their solution lies well within the capabilities of large, modern digital and analog computers. The results may be summarized easily if one makes a plot of the *state-space* response of the system. This consists of a graph where the blood sugar concentration, C, is plotted on the horizontal axis and the rate of change of this concentration, $\dot{C}$, is plotted vertically, as shown in Figure 10.13. In other words, a point in the upper half of this *plane* (graph) tells us that the patient's blood sugar concentration is increasing. A point below the horizontal axis corresponds to a decreasing concentration. A point lying anywhere to the right of the vertical dashed line passing through the point N indicates a blood sugar concentration above normal, and a point to the left of this line indicates a lower than normal concentration. It has been determined that, within a reasonable degree of accuracy, regions may be defined on this graph that indicate the need for insulin or glucose for a diabetic patient. If, for example, the concentration of blood sugar is large and still increasing (as at point A), an insulin injection is needed. However, an equally large blood sugar level that is decreasing ($\dot{C}$ is negative), such as point B, indicates no need for injection because the diabetic system is in the process of correcting itself toward the "normal" point. The same is true for point D. Additional glucose is needed if the system state is located at a point such as E. The latter situation is particularly dangerous because sustained low blood sugar levels can cause irreversible damage to the brain and nervous system, and so immediate glucose injection is needed to prevent these effects and to move the system state toward the upper right.

This area is another field of medical biological science where the mathematical, rigorous, systems engineering approach is bearing fruit. There remains much to be done, however. For example, it might be possible to cut down on the number of insulin injections needed by a diabetic patient

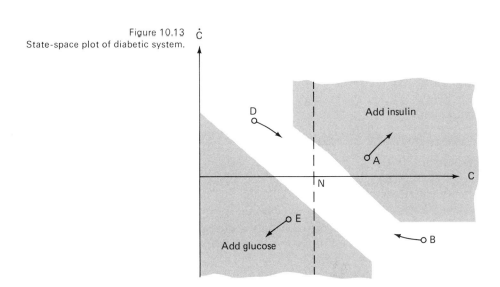

Figure 10.13 State-space plot of diabetic system.

if he could constantly monitor (perhaps by means of a small instrument that could fit into a brief case) his sugar balance system's state location in Figure 10.13. Then he would know when an insulin injection was needed (position *A*) and when it was not (positions *B* or *D*). Such a portable instrument is, as yet, unavailable, but no doubt will be available some day through the work of bioengineers.

10.9 THE KIDNEY

An organ that has caused almost as much bioengineering interest as the heart is the renal system (which consists of the urinary system, the kidneys, and their blood supply system). A gross simplification of this system is sometimes made when people say that it is a "filter" that removes waste products from the blood. This is a misleading description for several reasons. First, the kidneys have other functions than just that of removing waste products of cell metabolism from the blood. They also regulate the pH (acid-base chemical balance), osmotic pressure, volume, water content, and chemical makeup of blood and all other body fluids. In addition, one usually thinks of a filter as being some sort of fine screen that holds back certain contents of a fluid and allows the fluid itself to pass through. This is not at all the way the kidney functions.

The renal process is essentially one of almost totally removing everything from the blood (everything except blood cells and plasma proteins) and then selectively reinserting proper amounts of needed substances such as certain chemical ions, hormones, vitamins, and amino acids. The remaining *filtrate* is dumped out of the body into the urine. The refilling process, wherein the necessary substances are placed back into the blood stream, consists of two chemical mechanisms: dialysis and osmosis. Dialysis is the selective passage of ions and other chemical substances through a membrane. Osmosis is the process that takes place through a semipermeable membrane separating two solutions of differing solute concentration, wherein the water (or *solvent*) moves through the membrane so as to reach equilibrium between the two solutions. *Osmotic pressure* is the one-way pull exerted on water by that side of the system having the higher ion population or particle concentration.

If this were all that took place in the kidney, the engineering problems involved in building an artificial kidney (or *dializer*) would not be as large as they are. In the early stages of water reabsorption into the blood capillaries in the kidney, osmosis as described above takes place. But later on, as urine becomes more and more concentrated and therefore builds up its osmotic pressure, the membrane continues to withdraw water from the urine, despite the urine's greater osmotic pull for water. This is known as *negative osmosis* and is a process that is not yet understood well enough to manufacture a synthetic membrane that will accomplish it. Therefore, artificial dialysis—a process that is presently a life and death necessity for many hundreds of people in this country whose kidneys have ceased to function—is still far from perfected. There are many substances present in the blood in almost microscopically small concentrations whose functions and interrelationships are unknown. The search for membranes that will more closely duplicate the actual living nephrotic (kidney) membrane is now being pursued with great vigor by many companies and the bioengineers that work for them.

The following are some approaches that have been tried: to adsorb on activated carbon the waste products from the blood as they come through the synthetic membrane in an artificial kidney machine; this trick keeps the concentration gradient (difference between the waste product concentrations on either side of the membrane) as high as possible so as to shorten the length of time a patient must remain connected to the machine. A similar technique being experimented with is to attract the waste ions by means of an electric field placed across the membrane. When these and many other problems are finally solved, a portable artificial kidney (or even an implantable one) may become possible. The opportunities in this field for bioengineers and medical scientists to help mankind are unbounded.

10.10 THE BALANCE ORGANS

Not all biomedical engineers are working on problems related to disease and/or organ malfunction. An example of this is the research effort now going on in investigating the human balance control system. Basically, the human is able to maintain his balance and knowledge of his spatial orientation by means of five main sensory mechanisms:

1. Balance organs located in the inner ear:
   a. Semicircular canals
   b. Otolithic organs
2. Pressure receptors in the skin
3. Stretch receptors in the skeletal muscles
4. Movement of the internal organs that can be felt
5. Visual observation

Although there is presently an increasing number of investigators interested in malfunctions of this complicated biological control system, the greatest part of recent research in this area has been undertaken because of the occurrence of some unusual and unexpected illusions in high-speed airplanes and spacecraft. One example is the *oculogravic illusion*, consisting of apparent motion and displacement of a visual object in the direction of the resultant between the direction of a vehicle's acceleration and the direction of gravity. It has been observed to occur on the human centrifuge and in flight when a fixed target was being viewed by an observer.

Obviously, it is a matter of concern if a pilot or astronaut becomes disoriented or violently motion-sick when at the controls of his vehicle. Very little is known, presently, about the interactions among the five balance-sensing mechanisms. However, it is well accepted that such interactions are probably responsible for many of the illusions that are observed and are almost certainly responsible for much of the motion sickness that occurs.

The primary balance organs are the semicircular canals. There is a system of three interconnecting circular tunnels within the bony structure of each inner ear. Inside each of these tunnels is a thin-walled duct or canal, each of which is full of a fluid (*endolymph*). These ducts are connected at both ends to a sac called the utricle. There are three such ducts (set almost perpendicular to each other) in each ear. Near one end of each duct there is a structure (the *cupula*) much like a tiny swinging door that sticks out from the wall of the duct. The cupula blocks off the duct; but if by some means the endolymphatic fluid is caused to move against this swinging-doorlike structure, it will swing open and allow the fluid to flow. Neurons are connected to the hingelike part of the cupula, and so information is constantly being passed to the brain concerning the cupulas—whether each is open, in which direction, and to what degree. Typically, the canals respond to a sudden rotation of the head in the following way. If the rotation is sudden enough, the skull rotates, carrying the walls of the ducts and the cupulae with it but leaving the relatively heavy inertial fluid fixed in position. So, because the walls of the tunnel move and the endolymph does not, there is a relative motion between the fluid and the swinging cupula that forces the cupula to swing open. It is important to note that a *constant* angular velocity of the skull (for instance, such as we are all experiencing as the earth on which we stand rotates) produces no swing of the cupula. Only a *change* in angular velocity (such a change is called *angular acceleration*) will produce a cupula displacement. Therefore, we see that the semicircular canals are able to sense angular acceleration and transmit this information to the brain. We call such devices (either biological or artificial) angular acceleration *transducers*.

There is at least one other set of organs concerned with the determination of the spatial orientation of the skull that is located within the inner ear. These are the *otolithic organs* and are very much like miniature pendulums that usually hang straight down and that therefore tell the brain how the angular position of the head is fixed with respect to the direction of *apparent gravity*. We should realize that the direction of apparent gravity can be vastly different from the direction of true gravity. For example, if a person is riding in a centrifuge, in an airplane, or, for that matter, in almost any vehicle, apparent and actual gravity are different. We do not have space to describe all the experimental work presently being done in various attempts to discover how the messages transmitted from the balance sensors to the brain are processed and used by the CNS. It should be noted, however, that all these scientific efforts involve not only complex experimental hardware, much of which has been designed and built by bioengineers working with biological scientists, but also engineering-physics-type mathematics, which provides a basis for constructing mathematical models. And again we should emphasize that mathematical models, if they are continually updated and revised in the light of new experimental evidence, provide unique insight into the functioning of such biological systems.

10.11 SUMMARY

In this chapter we have obviously discussed only a very few areas wherein bioengineers are working. In addition, in the areas that have been mentioned here, we have only scratched the surface in an attempt to show you some of the kinds of things that bioengineers do and some of the techniques they typically use. Without question, an infinite number of opportunities lie ahead for biomedical engineers to contribute to the longevity and good health of mankind.

REFERENCES

1. BERNARD, E. E., and M. R. KARE (eds.), *Biological Prototypes and Synthetic Systems*. New York: Plenum Publishing Corporation, 1962.
2. BLESSER, W. B., *A Systems Approach to Biomedicine*. New York: McGraw-Hill Book Company, 1969.
3. GALAMBOS, R., *Nerves and Muscles*. Garden City, N.Y.: Anchor Books, Doubleday & Company, Inc., 1962.
4. GUCCIONE, E., "Biomedical Engineering," *Chemical Engineering*, Jan. 30, 1967, 107–124.
5. HARMON, L. D., "Studies with Artificial Neurons: Properties and Functions of an Artificial Neuron," *Journal of Acoustical Society 35*, Dec. 1963.
6. MARTIN, P. J., "Analysis of Myocardial Performance in the Blood Pressure Control System," *Automatika, G*, 1970, 175–191.
7. MILSUM, J. H., *Biological Control Systems Analysis*. New York: McGraw-Hill Book Company, 1966.
8. PLONSEY, R. (with Fleming, D. G.), *Bioelectric Phenomena*. New York: McGraw-Hill Book Company, 1969.
9. RUCH, T. C., M. D. PATTON, J. W. WOODBURY, and A. L. TOWE, *Neurophysiology*. Philadelphia: W. B. Saunders Company, 1964.
10. SCHWANN, H. P., *Biological Engineering*. New York: McGraw-Hill Book Company, 1969.
11. WOLAVER, L. E., "Mathematical Model for Blood Sugar Regulation of the Human Body and Its Control in Diabetes Mellitus," *5th Journees internationales de calcul analogique*, Dayton, Ohio: Applied Math Research Laboratory, Aerospace Research Laboratory.
12. WOOLDRIDGE, D.E., *The Machinery of the Brain*. New York: McGraw-Hill Book Company, 1963.

EXERCISES

1. What is a taxonometric viewpoint?
2. What are the advantages afforded by a mathematical model of a biological system?
3. What is the Lillie neuron?
4. What is the difference between the words *afferent* and *efferent* when applied to nerves?
5. What is bionics?
6. What is lateral inhibition? In what animal was it originally discovered?
7. Name two engineering techniques useful in EEG analysis.
8. What is the vena cava?
9. What is another name of the sinoatrial node? Where is it found?
10. Briefly describe the major contribution of Willem Einthoven.
11. In the seaweed-limpet problem discussed in this chapter, if the number of seaweed plants is 300/acre and there are 50,000 limpets/acre, what will be the maximum number of limpets per acre that will ever occur?
12. Use program MODEL (listed in this chapter) to obtain the state-space trajectory of the seaweed and limpet populations. Then plot each of these populations separately as a function of time.

11
introduction to experimental technique

G. Boothroyd

11.1 INTRODUCTION

This chapter on experimental technique is designed to establish a proper approach to the techniques and principles involved in conducting laboratory experiments.

Laboratory work provides an ideal opportunity for the consolidation and amplification of the topics studied during lecture periods; it also provides an opportunity to study the way in which engineering materials and equipment behave in practice. The development of a proper experimental approach, the experience of using instruments, and the ability to report adequately in good technical English on the work carried out in the laboratories form an important part of the student's education. In the senior year, many students undertake a small research project. It is here that the training in laboratory technique and the general ability to write technical reports, comprising literature surveys, general procedure, discussion of results, comments, conclusions, etc., are finally tested.

11.1.1 The Object of Laboratory Work

Laboratory experiments in an undergraduate course may serve several purposes. In the laboratory, a student may observe the physical behavior of engineering materials and engineering equipment for himself. An experiment may be conducted in order to measure certain properties of materials or other constants that are commonly used in engineering practice. This will acquaint the student with the standard methods of measuring these properties and constants. Laboratory work has other important uses: it provides an opportunity for informal contact between student and faculty, and it allows the student to gain experience in making experimental observations, in planning experiments, in keeping records of the work, and in technical report writing.

11.1.2 The Formal Technical Report

The purposes of a technical report on a piece of experimental work are threefold: (1) it forms a record of the work that may be used subsequently by the experimenter, (2) it forms a means of communicating the results of the work and the conclusions drawn to other engineers, and (3) it acts as a reference for future workers.

For these purposes a technical report will, in general, include the following information:

1. The date on which the experiment was performed.
2. The name of the experimenter.
3. *Title:* The title of the work.
4. *Object:* A clear and considered statement of the object of the work.
5. *Introduction:* This section should include a short statement giving the reasons for conducting the experiment and may include a statement of the findings of previous workers on the subject of the experiment.
6. *Theory:* This section should include the background theory to the experiment.
7. *Apparatus:* An adequate description of the equipment used and its calibration and estimated accuracy should be provided. The description should be sufficiently detailed so that another individual could repeat the experiment.
8. *Procedure:* This section should contain a description of the experimental procedure and techniques used.
9. *Results:* The results of the experiment should be presented clearly in either tabular or graphical form or both.
10. *Discussion of results:* In this section the results obtained in the experiment should be analyzed and discussed in the light of the theoretical background or the results obtained by previous workers or both.
11. *Conclusions:* This section should be a short concise statement of the main conclusions reached as a result of the experiment. The conclusions should have a direct bearing on the object of the experiment stated under 4 above.

A technical report is intended to be read by either a person who may not necessarily have the same engineering knowledge as the experimenter or by the experimenter himself at some future date. It is most important, therefore, that considerable thought be put into the presentation. For example, graphs and figures should be clearly numbered and headed and be referred to in the text. Also, in the writing of the report it should not be assumed that the reader will have an intimate understanding of the work carried out.

Clearly the writing of a technical report is a difficult and time-consuming task, and it will certainly not be possible for such a report to be written while the experiment is being conducted. It is therefore necessary to keep an adequate record of the work carried out in the laboratory while the experiment is being performed, and this record will be quite different in nature from the final technical report. For these purposes an *informal* laboratory notebook should be used. This notebook should contain a complete record of the essential information recorded while the experiment is being conducted and from which a formal technical report can be written if required.

Informal notebooks usually contain pages of graph sheets interspaced with two pages of ruled paper with columns superimposed. These books are an excellent arrangement for the purposes of recording experimental data during the experiment, the columns providing a means of tabulating results without

11.1.3 The Laboratory Notebook

the need for any further ruling of lines by the student. The information to be included in the notebook for each experiment includes the following:

1. The date on which the experiment was performed.
2. The title of the experiment.
3. The object of the experiment.
4. Essential information regarding the apparatus and instrumentation used and its calibration and estimated accuracy.
5. Experimental procedure (brief) together with any necessary definitions of the quantities measured.
6. The observations made during the experiment, together with calculations, graphs, and estimated accuracy of results.

These items should be inserted in ink or pencil during the experimental period. Either at the end of the experimental period or shortly afterwards an analysis of the results together with a brief conclusion based on the results of the experiments should be inserted. These reports should be an adequate and legible record of the experiment, containing all the relevant information that may be required in order to write up a formal report at some future date. It is emphasized that the laboratory notebook should be written up, in the main, in the laboratory while the experiment is in progress. The writing up of one experiment should be completed before the next experiment is undertaken.

11.1.4 Conducting the Experiment

The experimenter should first write the title, date, and object in his notebook, calibrate the equipment if necessary, and then insert headings in the columns for the purposes of recording his readings and calculations. He should then determine the range of results he expects to obtain and prepare the necessary axes for the graph as appropriate. These last steps will necessitate the complete planning of the experiment (which readings to take, what range of readings to take, etc.) before commencing with the experimental observations.

Undoubtedly the best approach, if possible, is to take one set of readings, note them down in the table, perform the necessary calculations, and mark the point on the graph before going on to the next set of readings. Points on graphs should be made with a dot to mark the exact position and then outlined with a circle, square, or other suitable shape to distinguish between sets of readings taken under different conditions.

Plotting the graph as one proceeds has the advantage that the readings for points that are obviously in error can be repeated immediately and the error corrected before proceeding with the next set of readings. A further advantage is that the experimental points can be spaced along the graph at regular intervals to cover the whole range of values of interest.

After completing the experiment, it will be necessary to write down briefly the procedure used in obtaining the experimental results and to estimate the errors involved. The results may now be discussed briefly and the main conclusions drawn, which again should be directly related to the object of the experiment.

Conducting an experiment in the manner outlined above is undoubtedly an art in itself and can be acquired only after considerable practice. It is very important therefore that the student adopt this approach from the outset.

11.2 ESTIMATION OF ERRORS

All observations are subject to error, so that what may be called the *true value* of a magnitude, such as a length, a time interval, or a temperature, cannot be ascertained. However, it is necessary to postulate that a true value exists and then to estimate the limits, called confidence limits, between which this value lies. These limits will give an indication of the precision with which the measurement has been made, closer estimated limits indicating a more precise measurement. It will be shown later how the confidence limits may be estimated.

The difference between the observed value of a magnitude and the true value is called the *error of observation*. Such errors may be classified under two main headings: *systematic* and *random* errors.

With systematic errors, the observed value deviates from the true value in a systematic way. For example, if the gain or amplification in an instrument is not adjusted correctly, then the observed values will deviate from the true value by a constant proportional amount if other errors present in the observations are negligible. Alternatively, if the needle on an indicating scale is bent, then the readings obtained will always deviate from the true value by a fixed amount, again assuming that other errors are negligible. The important point to remember is that systematic errors may be reduced or eliminated only by calibration of the equipment and method of observation against a standard and by applying the necessary corrections to subsequent readings. Systematic errors and instrument calibration will be dealt with separately.

Random errors are revealed when repeated observations of a fixed quantity are made; they are variable-in-magnitude, with positive and negative values occurring in repeated observations in no ascertainable sequence. Random errors may be dealt with by statistical analysis as follows.

11.2.1 Random Errors

It is not possible to estimate random errors from a single observation. There may be a large personal or accidental error, as well as that caused by the lack of precision of the particular instrument used. Suppose that a set of 10 repeated observations yields the following values:

9.8, 10.3, 10.0, 10.1, 9.7, 9.9, 10.2, 10.4, 9.6, 10.0

These observations suggest that the measured quantity lies between 9.6 and 10.4 (which is the range of these observations), and it is necessary to consider what is the "best" estimate of this quantity.

A well-known principle in statistical work is the *principle of least squares*. This states that the best estimate of a magnitude is given when the sum of the squares of the deviations of this value from the observed values is a minimum, or, in mathematical terms, when $\sum (x_i - \bar{x})^2$ is a minimum, where x_i is the observed value and $\bar{x}$ is the best estimate. Differentiating this expression and equating to zero we obtain

$$\frac{d \sum (x_i - \bar{x})^2}{d\bar{x}} = -2 \sum (x_i - \bar{x})$$

$$= 0 \text{ for a minimum} \quad (11.1)$$

Therefore $2 \sum x_i - 2 \sum \bar{x} = 0$ or $\sum x_i - n\bar{x} = 0$, where n is the number of observations made. Thus,

$$\bar{x} = \frac{\sum x_i}{n} \quad (11.2)$$

This means that the best estimate of a magnitude is the arithmetic mean of the n observations.

In the above example, the best estimate is 10.0. However, if further observations were made, they might be outside the range of $10.0 \pm .4$ or in the example quoted, say, 9.5 or 10.5. We can only say at this stage that many such further observations would probably be within the range and that the best estimate of the magnitude being measured is 10.0. Clearly, in scientific work it will be necessary to specify limits between which the true value of the magnitude lies, and a statement of the best estimate together with the range (10.0 ± 0.4) would be inadequate.

Observations will always be subject to random errors. Statistical work has shown that these errors follow a definite pattern; this is known as the *normal distribution*, mathematically defined as

11.2.2 Normal Distribution

$$y = \frac{1}{\sigma \sqrt{2\pi}} e^{-(x-\mu)^2/2\sigma^2} \quad (11.3)$$

where μ is the true value and σ is known as the *standard deviation*. The general form of the normal distribution curve is shown in Figure 11.1. This curve may be interpreted by considering that the area under the curve between x_1 and x_2 is equal to the probability of any observation x_i falling between x_1 and x_2. For example, when x_1 and x_2 are given the values $-\sigma$ and $+\sigma$, respectively, the probability becomes $\int_{-\sigma}^{\sigma} y \, dx$, which is equal to .6826.

Thus, a property of the normal distribution curve is that 68.26% of all observations fall within the range $\mu \pm \sigma$. Further, it is found that 95.4% of all observations fall within the range $\mu \pm 2\sigma$ and 99.73% within the range of $\mu \pm 3\sigma$.

Thus, if limits $\pm 2\sigma$ are placed on the true value, approximately 95% of all the observations will fall within these limits.

In experimental work, the number of individual observations that can be made is strictly limited, and therefore it is necessary to use a group of ob-

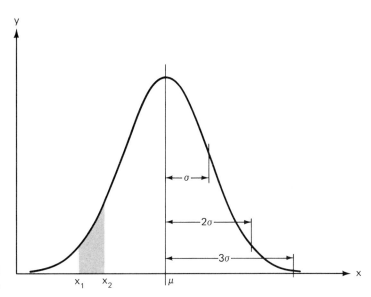

Figure 11.1
Curve showing normal distribution.

servations to estimate the actual distribution of the complete population of observations. For this purpose, the standard deviation, s, for a sample of n observations is defined as follows:

$$s = \left[\frac{1}{n-1}\sum_{i=1}^{n}(x_i - \bar{x})^2\right]^{1/2} \tag{11.4}$$

where s is the standard deviation of the individual observations from the mean $\bar{x}$. Clearly $\bar{x}$ and s are only estimates of μ and σ, respectively, since another set of observations of the same quantity will yield different values of $\bar{x}$ and s. The object in experimental work is usually to place limits on the mean and state with a specified degree of confidence that the true value lies between these limits.

11.2.3 Confidence Limits for the Mean

Suppose that an experimenter has derived from a number of observations the best estimate (mean of the sample) of the magnitude he is measuring. He will not be prepared to assert that the value obtained is the true value. He may feel confident that his value is quite close to the true value, although he is not in a position to say how close. He may overcome this difficulty by giving two limits, asserting with specified confidence that within these limits lies the true value. Such limits are called *confidence limits*.

It is usual for these purposes to adopt the 95% confidence limits, which are the limits corresponding to 95% probability. Thus, the experimenter who consistently uses 95% limits and asserts that they include the true value sought will, in the long run, be right 19 times out of 20.

The confidence limits for the mean of n readings are obtained by multiplying the estimate of the standard deviation, s, by a factor $t/\sqrt{n}$, where t is a quantity dependent on the number of observations on which the standard deviation is based.

Table 11.1 gives values of t for various values of n. For values of n greater than 20, t may be taken as approximately equal to 2.0.

Table 11.1
Table of t for 95% Confidence

n	t	n	t
2	12.71	12	2.20
3	4.30	13	2.18
4	3.18	14	2.16
5	2.78	15	2.14
6	2.57	16	2.13
7	2.45	17	2.12
8	2.36	18	2.11
9	2.31	19	2.10
10	2.26	20	2.09
11	2.23		

Referring to the figures given earlier, for a set of 10 readings, the 95% confidence limits may be calculated for the arithmetic mean as follows:

11.2.4 Example

Observed Value, x_i	Derivation from Arithmetic Mean, $x_i - \bar{x}$	Square of the Deviation, $(x_i - \bar{x})^2$
9.8	−.2	.04
10.3	.3	.09
10.0	.0	.00
10.1	.1	.01
9.7	−.3	.09
9.9	−.1	.01
10.2	.2	.04
10.4	.4	.16
9.6	−.4	.16
10.0	.0	.00
$100.0 = \sum x_i$		$.60 = \sum (x_i - \bar{x})^2$ (sum of the squares of the deviations)

$$\bar{x} = \frac{\sum x_i}{n} = 10 \text{ (arithmetic mean)}$$

Now

$$s^2 = \frac{1}{9} \times .6$$

and

$$s = \frac{\sqrt{.6}}{3} = \frac{.774}{3} = .258 \quad \text{(standard deviation)}$$

From Table 11.1, for $n = 10$, $t = 2.26$ for 95% confidence. Thus, the confidence limits are given by

$$\pm \frac{2.26}{\sqrt{10}} \times .258 = \pm \frac{2.26}{3.17} \times .258 = \pm .184$$

Thus, it can be stated with a confidence of 95% that the true value of the quantity lies between $10.0 \pm .2$.

11.3 SYSTEMATIC ERROR AND INSTRUMENT CALIBRATION

It has already been stated that errors in observations are a combination of random errors (which may be treated by statistical analysis) and systematic errors. Systematic errors can be revealed only by calibration of the measuring system or instrument against a standard.

When calibrating an instrument, the object is to determine the correction or the amount that must be added to the indicated value, due regard being paid to sign, to obtain the "true" value of the quantity measured. It is important to keep in mind the various factors or instrument characteristics that give rise to errors.

11.3.1 Zero Error

Zero error is defined as the indication when the magnitude of the physical quantity presented to the instrument is zero. Many instruments, particularly electrical instruments, have a control giving zero adjustment, and the first step when using the instrument would be to adjust the zero setting.

11.3.2 Datum Error

Datum error is defined as the indication error when a physical quantity of a magnitude prescribed for the purpose of verifying the setting is presented to the instrument. With many instruments an adjustment of the gain or amplification is provided that may be set by presenting a known magnitude of a physical quantity to the instrument and then adjusting for datum error. With instruments having linear characteristics, adjustment of the zero and datum settings would be sufficient for a quick and simple calibration. This is illustrated in Figure 11.2, where curve *A* is the initial characteristic before adjustment, curve *B* is the characteristic after adjustment of zero setting, and curve *C* is the required characteristic obtained finally by adjustment of the datum setting. Facilities for this simple form of calibration are usually provided for standardizing the instrument immediately prior to making a series of measurements. It should be remembered that the characteristics of an instrument will vary and that regular calibration of the instrument against a standard may be required in addition to the zero and datum setting procedures.

The following factors describe the performance of instruments. If they are not investigated, large errors in measurement can occur.

11.3.3 Stability

Stability is the reproducibility of the mean readings of an instrument when tested under defined conditions of use, repeated on different occasions separated by intervals of time that are long compared to the time of taking a reading.

Electrical measuring instruments are prone to instability, particularly during the period immediately after switching on. For this reason the manufacturers will usually specify that the instrument must be left switched on (warmed up) for a certain period before the instrument is used.

11.3.4 Constancy

Constancy is the reproducibility of the (uncorrected) reading of an instrument over a period of time, when the quantity to be measured is presented continuously and the conditions of the test are allowed to vary within specified limits. With many instruments the indication will be affected by changes in ambient conditions, and a test for constancy will reveal this characteristic. Such a test will indicate the range of conditions under which the instrument

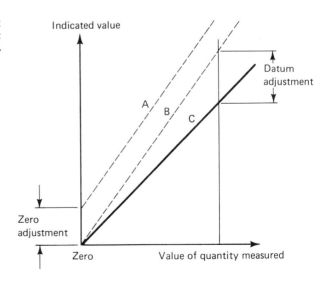

Figure 11.2
Zero and datum adjustments for an instrument with linear characteristics.

will give acceptable performance or, alternatively, the corrections that should be applied to subsequent readings to allow for changes in these conditions.

11.3.5 Discrimination

Discrimination is the smallest change in the quantity measured that produces a perceptible movement of the index. One of the main causes of poor discrimination in instruments is hysteresis, resulting from friction or backlash in moving parts. This effect is most strikingly revealed when a series of readings is taken both with the physical quantity increasing in magnitude and decreasing in magnitude; Figure 11.3 illustrates this effect.

In addition to the factors described above, there will always be an observation error when reading instruments. This is the error committed by the observer when reading the indication of the instrument and may be due, for example, to parallax, to faulty estimation of the fractional part of a scale interval, or to simple misreading of the indication. Instruments are often provided with a means of reducing observation errors (e.g., vernier scales and mirrors behind indicators to reduce parallax), and some instruments are designed to present information in numerical form, in which case the only observation error possible is due to misreading the numerical information. An example of the latter is a digital clock.

11.3.6 General Terms Relating to Performance of Instruments

The *precision* of an instrument is measured by the spread of the indications obtained when a physical quantity of a constant magnitude is presented to the instrument, the measurements being taken over a period of time and under various conditions of test. With an instrument having high precision, the indications obtained would be contained within narrow limits.

The *accuracy* of an instrument describes the degree of closeness with which the indications of an instrument approach the true values of the

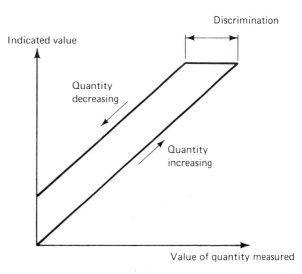

Figure 11.3
Effect of hysteresis and backlash on instrument performance.

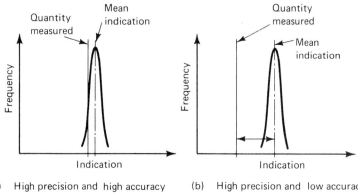
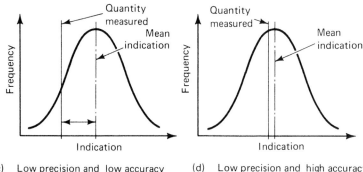

Figure 11.4
Instrument characteristics.

quantities measured (this means that the mean indication error is small in an accurate instrument). To illustrate these terms, Figure 11.4 shows typical frequency distributions for the indications of an instrument having (a) high precision and high accuracy, (b) high precision and low accuracy, (c) low precision and low accuracy, and (d) low precision and high accuracy.

11.4 PRESENTATION OF RESULTS

Often, the most convenient form for the presentation of the results of an experiment is in the form of a graph. A graph displays relationships in such a way that the reader is able to see at a glance the trends occurring. For example, Figure 11.5 shows the results of an experiment presented in both tabular and graphical form. Comparison of the two methods illustrates the relatively large effort required on the part of the reader to appreciate the nature of the relationship from the tabulated results. It is also easier to draw conclusions from data presented in graphical form. For instance, the small peak in the relationship can readily be observed in the graph in Figure 11.5, whereas this would be most difficult to deduce from the table.

In many experiments the results can be presented in the form of a straight-line graph even when the relationship is basically nonlinear. The following discussion first explains the features of a linear relationship and then shows how some nonlinear relationships can be converted to a linear form.

One of the advantages of a linear relationship is the relative ease with which a straight line can be drawn through a series of experimental points. Further, a simple statistical method is available and will be described later that allows the experimenter to calculate the "best" straight line through a series of points.

The first step is to choose the appropriate scales and the form of the variables; this will depend on the relationship that it is desired to illustrate. In general, however, relationships will be one of the two forms described below.

11.4 presentation of results

11.4.1 Properties of Linear Relationships

Linear relationships may be written in the form

$$y = mx + C \tag{11.5}$$

In this case if y is plotted against x, then the relationship will form a straight line that has a slope of m and an intercept on the y axis of C (Figure 11.6). With this kind of experimental curve the results could show that the rela-

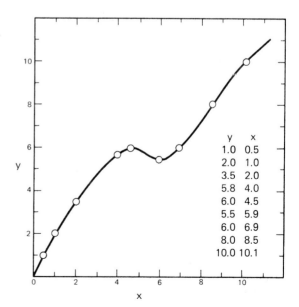

Figure 11.5
Graphical and tabular presentation of data.

y	x
1.0	0.5
2.0	1.0
3.5	2.0
5.8	4.0
6.0	4.5
5.5	5.9
6.0	6.9
8.0	8.5
10.0	10.1

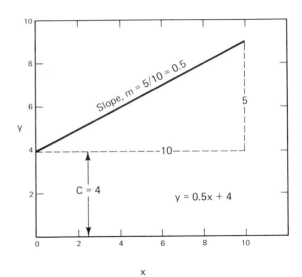

Figure 11.6
Linear relationship.

tionship between y and x is indeed linear; the slope of the line and the intercept could be used to determine the values of m and C.

11.4.2 Conversion of Nonlinear Relationships to Linear Form

A relationship of the form $y = ae^{b/x}$ may be converted to a linear relationship by taking logarithms of both sides of the equation. Hence,

$$\log_e y = \log_e a + \frac{b}{x} \tag{11.6}$$

In this case, $\log_e y$ would be plotted against $1/x$, the slope of the straight line resulting would be equal to b, and the intercept on the $\log_e y$ axis would give $\log_e a$.

An alternative method is to plot the relationship on log-linear graph paper (sometimes known as semilog graph paper), in which case y would be plotted on the logarithmic scale and $1/x$ on the linear scale (Figure 11.7). It can be seen from the figure that on the logarithmic scale the distances are marked off so as to be proportional to the logarithm of the number indicated on the scale.

Another form of nonlinear relationship is

$$y = ax^m \tag{11.7}$$

This again may be converted to a linear expression by taking logarithms. Hence,

$$\log y = \log a + m \log x \tag{11.8}$$

In this case $\log y$ would be plotted against $\log x$, and the slope of the resulting straight line would be equal to m and the intercept on the $\log y$ axis

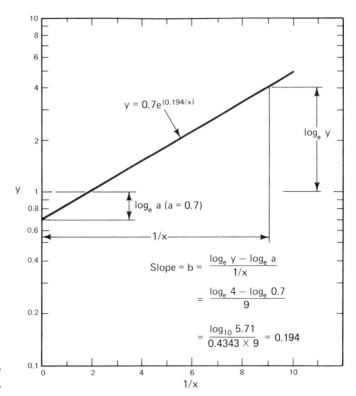

Figure 11.7
Relationship $y = ae^{b/x}$.

would give $\log a$. Alternatively, the results may be plotted on log-log graph paper as shown in Figure 11.8.

11.4.3 Drawing the Curve

Since there will usually be uncertainties in the measurements obtained, the true curve would not be expected to pass through all experimental points. The question then arises as to what method or criteria should be used to draw the curve in its best position. Before making this decision, it is first necessary to consider for what purpose the curve is to be used.

A curve is an easy way of visualizing the qualitative nature of the dependence of one variable on another. If this is the purpose, then the accuracy of the curve is not too important, and it should be drawn so that it fairly represents the nature of the relationship taking into account the experimental scatter.

It may be desired to estimate the value of one variable at given values of the other variable between measured values (interpolation) or outside the range of measured values (extrapolation). Particularly in the case of extrapolation, the accuracy of the curve is most important, as errors in drawing the curve will be magnified in the final result. Extrapolation can normally be carried out only with a linear relationship, and in this case the slope of the straight line is the most important factor and the least-squares method described below should be used.

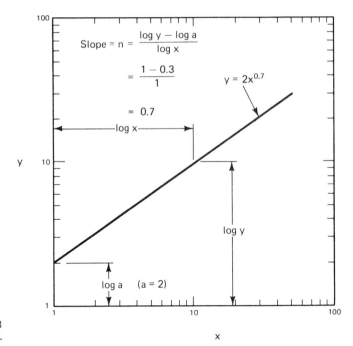

Figure 11.8
Relationship $y = ax^n$.

The purpose of a curve may be to eliminate insofar as possible the effects of random errors, or, in other words, to smooth out the data. This, of course, is one of the main advantages in presenting experimental results in the form of a graph where a reading having an unusually large error may be neglected whereas in a statistical analysis the reading would normally be taken into account.

Derivatives or integrals may be derived from an experimental relationship by plotting the slope of the curve or the area under the curve. In this case, accuracy is all-important since errors introduced in drawing the curve will be magnified in the subsequent derivations or calculations. For this reason, it is always preferable to plot the relationship desired rather than to plot the variables noted during the experiment and then to derive the desired relationship from the resulting curve.

It may be necessary to determine a mathematical relationship between two variables. The technique of dimensional analysis, which will be dealt with separately, can help in the choice of appropriate scales.

It is nearly always assumed when drawing the curve that one variable is a continuous function of the other. The smoothest curve is usually that curve with the least number of points of inflection. It is also helpful sometimes to keep the number of points above and below the line equal, although in most cases it is sufficient to draw the curve by eye. The drawing of the curve is easier, of course, when the general relationship between the variables is known.

Obviously, greater accuracy can be obtained with a greater number of experimental points. In general, it may be said that with a linear relationship at least six points should be plotted and in the case of a more complicated relationship a greater number of points would be necessary.

11.4.4 Best Straight Line

One way to obtain the best straight line through a series of points is by the method of least squares. In this method, it is assumed that all the scatter in the points is due to errors in measuring only one of the variables, for example, y in Figure 11.9. The best straight line can now be defined as that which minimizes the errors in the y direction.

Assuming that the best straight line has the equation

$$y = mx + c \qquad (11.9)$$

the problem is now to determine the values of the constants m and c. If point i, with coordinates x_i and y_i, does not lie on this line, then the y coordinate of the point is in error by a distance e_{yi}, given by

$$y_i = mx_i + c + e_{yi} \qquad (11.10)$$

The best straight line is therefore that which minimizes $\sum e_{yi}^2$. Now

$$\sum e_{yi}^2 = \sum (y_i - mx_i - c)^2 \qquad (11.11)$$

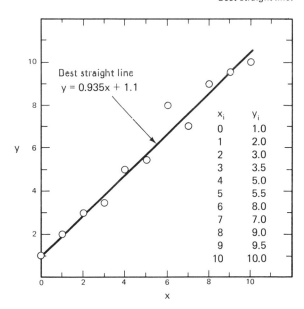

Figure 11.9
Best straight line.

Differentiating the right-hand side of this equation with respect to both m and c, the unknown constants, and equating to zero gives

$$\sum x_i(y_i - mx_i - c) = 0 \tag{11.12}$$

and

$$\sum (y_i - mx_i - c) = 0 \tag{11.13}$$

These equations are applied to the data in Figure 11.9 as follows: first, the terms in parentheses in Equations (11.12) and (11.13) are written with the original data points:

$$
\begin{aligned}
1.0 &- 0m - c \\
2.0 &- 1m - c \\
3.0 &- 2m - c \\
3.5 &- 3m - c \\
5.0 &- 4m - c \\
5.5 &- 5m - c \\
8.0 &- 6m - c \\
7.0 &- 7m - c \\
9.0 &- 8m - c \\
9.5 &- 9m - c \\
10.0 &- 10m - c
\end{aligned}
$$

Summing these and equating to zero will satisfy Equation (11.13):

$$63.5 = 55m + 11c \tag{11.14}$$

Multiplying each term by the appropriate value of x_i, summing the result, and equating to zero will satisfy Equation (11.12):

$$
\begin{aligned}
0.0 &- 0m - 0c \\
2.0 &- 1m - 1c \\
6.0 &- 4m - 2c \\
10.5 &- 9m - 3c \\
20.0 &- 16m - 4c \\
27.5 &- 25m - 5c \\
48.0 &- 36m - 6c \\
49.0 &- 49m - 7c \\
72.0 &- 64m - 8c \\
85.5 &- 81m - 9c \\
100.0 &- 100m - 10c
\end{aligned}
$$

and

$$420.5 = 385m + 55c \tag{11.15}$$

Solution of Equations (11.14) and (11.15) now gives the required values of m and c as follows: Multiplying (11.14) by 5 gives

$$317.5 = 275m + 55c \qquad (11.16)$$

Subtracting (11.16) from (11.15),

$$103.0 = 110m$$

and

$$m = \tfrac{103}{110} = .935$$

Substitution of this value of m in (11.14) gives

$$63.5 = 51.4 + 11c$$

Hence,

$$c = \frac{12.1}{11} = 1.1$$

Thus, the best straight line is given by

$$y = .935x + 1.1 \qquad (11.17)$$

and is shown in Figure 11.9.

11.5 REDUCTION OF VARIABLES

11.5.1 Dimensional Analysis

One of the most powerful tools in experimental work is the technique of dimensional analysis. Application of this technique allows a problem having several variables to be reduced to one having fewer groups of these variables. This can often save a considerable amount of experimental effort.

Dimensional analysis is based on the fact that the dimensions of the terms in an equation must be the same. Clearly, one cannot equate feet to seconds or add minutes to degrees Fahrenheit. It is not necessary for the purposes of dimensional analysis to know whether a length is being measured in inches or miles or whether a speed is being measured in miles per hour or feet per second. The important point is whether it is a distance or a speed, etc.

It is usual to represent the various dimensions as follows: mass $= M$; length $= L$; time $= T$. It now follows, for example, that

$$\text{velocity} = \frac{\text{length}}{\text{time}} = \frac{L}{T} \qquad (11.18)$$

$$\text{acceleration} = \frac{\text{velocity}}{\text{time}} = \frac{L}{T^2} \tag{11.19}$$

$$\text{force} = \text{mass} \times \text{acceleration} = \frac{ML}{T^2} \tag{11.20}$$

For the purposes of this chapter, we shall concern ourselves with the three basic dimensions M, L, and T since all variables except those involving temperature and heat may be derived in terms of these. [*Note*: In some problems it is more convenient to work in terms of the three basic dimensions, force (F), length (L), and time (T). In this case mass $= FT^2/L$].

Many experimental problems may be stated in the following form: "We wish to know how the variable x is affected by the independent variables y and z". This may be written in mathematical terms as

$$x = \text{function }(y, z) \tag{11.21}$$

and the problem is to determine the nature of the function or relationship. However, before proceeding to set up the experiment, it often pays to check by dimensional analysis whether the number of variables [in Equation (11.21) there are three] can be reduced.

Suppose that we guess that the final relation is in the form

$$x = \text{constant} \times y^3 \times z^2 \tag{11.22}$$

and that we know the dimensions of x are those of velocity (L/T), the dimensions of y are those of length (L), and the dimensions of z are those of acceleration (L/T^2).

Substitution of the dimensions in Equation (11.22) will show that the relationship cannot possibly be of this form because the dimensions of the right-hand side are not equal to those of the left-hand side.

Expressing our guess as to the final relationship in more general terms, we may write

$$x = \text{constant} \times y^a \times z^b \tag{11.23}$$

and substituting the dimensions in each side of the equation we obtain

$$\frac{L}{T} = L^a \times \left(\frac{L}{T^2}\right)^b \tag{11.24}$$

It is now possible to write two equations in terms of a and b, which will ensure that the dimensions of length (L) and time (T) will balance in Equation (11.24).

Thus, for length

$$1 = a + b \tag{11.25}$$

and for time

$$-1 = -2b \tag{11.26}$$

From (11.25) and (11.26)

$$b = \tfrac{1}{2}$$

and

$$a = \tfrac{1}{2}$$

Inserting these values in Equation (11.23) we obtain

$$\frac{x^2}{yz} = \text{constant} \tag{11.27}$$

and x^2/yz can be considered as one variable in the experiment. Thus, the experimental problem has been simplified to a consideration of the single value of x^2/yz simply by analysis of the dimensions of the variables.

Another useful result of this analysis is that it is now possible to state the relative effects of changing the original variables without performing an experiment. For example, Equation (11.27) shows that the effect of doubling the magnitude of y would be the same as that of doubling the magnitude of z, or that doubling the magnitude of x would have the same effect as that of reducing y or z by a factor of 4.

For dimensional analysis, it is not necessary to assume a relation between variables. The problem above would have been solved in a more general way by raising each variable in the original expression (11.21) to an unknown index. Thus,

$$x^{a_1} = \text{function}(y^{a_2}, z^{a_3}) \tag{11.28}$$

Substitution of the dimensions of the variables now gives

$$\left(\frac{L}{T}\right)^{a_1} = \left(L^{a_2}, \left(\frac{L}{T^2}\right)^{a_3}\right) \tag{11.29}$$

For the dimension L we may now write

$$a_1 = a_2 + a_3 \tag{11.30}$$

and for T

$$-a_1 = -2a_3 \tag{11.31}$$

Thus,

$$a_3 = \frac{a_1}{2}$$

and

$$a_2 = \frac{a_1}{2}$$

Substituting these values in (11.28) we obtain

$$x^{a_1} = \text{function } (y^{a_1/2}, z^{a_1/2})$$

Collecting terms with a common index and squaring, we obtain

$$\frac{x^2}{yz} = \text{constant} \tag{11.32}$$

which is identical to Equation (11.27).

11.5.2 Knowledge of Linear Relationships

In some experimental situations it is possible to deduce that a linear relationship exists between two of the variables, and this will allow these variables be combined into one.

For example, if a mass m is swinging on a string and the resulting maximum pull on the string is p, then an identical mass swung in an identical way will also give a pull p. If the two situations are now joined together, a mass of $2m$ will result in a total pull of $2p$, and, hence, the pull p is directly proportional to the mass m.

In any solution to the problem, therefore, the variables p and m must always appear as a ratio p/m, or the pull per unit mass, and it is now possible to consider the ratio p/m as one variable.

A further example would be a simple coiled spring where the deflection, y, of the spring is proportional to the load applied, w. Here, the ratio y/w, which is the deflection per unit load, can be used as one variable. This ratio is usually referred to as the spring stiffness.

Summarizing, it can be stated that any reduction in the number of variables to be considered in a problem will reduce the experimental effort, and such reductions should be sought by the application of dimensional analysis and by any existing knowledge of the relationships between individual variables.

However, before the problem can be tackled in this way, it is necessary to decide which variables are important in the problem. Inclusion of variables that do not affect the problem can lead to an incorrect answer.

11.5.3 Choice of Variables

Taking a simple example, suppose that we wish to know how long it takes an object to fall to the ground from a given height under conditions where we can neglect air resistance.

The dependent variable is the time, t, taken for the object to fall. A variable that will affect the result is clearly the height, h, from which the object is dropped. An object falls because of the pull of gravity, and, hence, although it cannot be readily varied, we must include g, the acceleration due to gravity in our analysis. Finally, it is necessary to decide whether we should include the mass or weight of the object dropped. Reasoning will help us to decide this.

If two 1-lb masses were dropped at the same moment and from the same height, they would strike the ground simultaneously. Thus, they could be considered joined together as a 2-lb mass. It is clear, therefore, that without resorting to experiment we have proved that the time taken for a 2-lb mass to fall is the same as that for a 1-lb mass to fall, and hence t in the problem is not affected by the mass of the object. Hence, the mass should not be included in the analysis.

Another example would occur in deciding whether the width of the straight blade of a bulldozer should be included as a variable in considering the height of the pile of earth created in front of the blade when the bulldozer is removing a thin strip of earth from a field. Since two bulldozers operating side by side would have identical piles of earth in front of the blade, it follows that the width of the blade does not affect the variable under consideration and should not be included in an analysis.

11.5.4 Example of Reduction of Variables

Figure 11.10 shows a straight slender horizontal beam of rectangular cross section, firmly fixed at one end and having a vertical load applied at its free end. The beam, with this kind of support, is known as a cantilever beam, and the effect of the load will be to bend the beam in the manner shown. Often, in practice, it is desirable to know the deflection (or vertical displacement) of the beam at the point of loading.

Clearly, the shape of the beam will affect this deflection, and thus the length l, the breadth b, and the depth d, will be important variables in the problem. The load w on the beam will also affect the deflection, as will the material from which the beam is manufactured. If we concern ourselves with situations where the beam is distorted only elastically (not permanently deformed by the load) and if we decide that we are interested in ordinary engineering materials, we can say that the deflection will always be in proportion to the load and that the important characteristic of the material can be expressed by Young's modulus of elasticity E, which has the dimensions F/L^2 (i.e., tons per square inch or pounds per square inch, etc.).

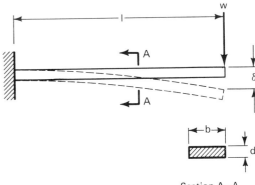

Figure 11.10
Cantilever beam of rectangular cross-section.

Section A–A

We shall suppose that the object of the experiment is to obtain a relationship from which it will be possible to predict the elastic deflection δ of any cantilever beam of rectangular cross section.

The problem may be stated initially as follows:

$$\delta = \text{function}\,(w, l, b, d, E) \tag{11.33}$$

Thus, we have a problem with six variables; any change in one of the five variables on the right-hand side of the expression will affect the magnitude of δ.

Without first considering the reduction of variables it might be thought necessary to design an experiment such that each of the variables on the right-hand side of Equation (11.33) can be varied and its effect on δ measured.

In this case, the value of δ could be measured for various loads w, various lengths l, various breadths b, etc. If all possible combinations of conditions are investigated and effects are plotted on graphs using five points on each graph, the total number of experiments required would be $5^5 = 3{,}125$, requiring the manufacture of 125 different beams and presenting the results on 125 different graphs. This clearly would be a considerable undertaking, and, further, the results could be used only to predict the behavior of beams within the range studied experimentally.

We shall now see how this problem can be reduced to one where only five individual tests are made requiring only one experimental beam and the results are presented on one graph. It will also be shown that this graph can be used to predict the behavior of beams of different size and material to that used in the experiment.

Since it has been stated that, for elastic beams, the deflection δ is proportional to the load w, then any solutions to the problem will contain the ratio of these variables w/δ, i.e., the load per unit deflection or stiffness. This allows us immediately to eliminate one variable from our considerations by dealing with this combination of w and δ rather than to treat these variables separately.

Further consideration will show that the deflection of the beam is always inversely proportional to its breadth b or, alternatively, that the stiffness w/δ is proportional to the breadth b. Two beams arranged side by side with the same load on each would be subjected to the same deflection. These beams could, therefore, be joined together as one, giving rise to the situation where a beam of twice the breadth would have twice the stiffness. It follows from this argument that stiffness is proportional to breadth and that the two variables can be combined as one, or $w/\delta b$, which can be referred to as the stiffness per unit width of beam.

Equation (11.33) can now be rewritten as

$$\frac{\delta b}{w} = \text{function}\,(l, d, E) \tag{11.34}$$

and dimensional analysis can be used to obtain further reduction of variables. Following the procedure outlined earlier, we can express Equation (11.34) as follows:

$$\left(\frac{w}{\delta b}\right)^{a_1} = \text{function } (l^{a_2}, d^{a_3}, E^{a_4}) \tag{11.35}$$

Substituting the dimensions of the variables $\delta b/w = L^2/F$, $l = L$, $d = L$, $E = F/L^2$, where F = force and L = length, we obtain

$$\left(\frac{F}{L^2}\right)^{a_1} = L^{a_2} \times L^{a_3} \times \left(\frac{F}{L^2}\right)^{a_4} \tag{11.36}$$

For the dimensions of length to be consistent

$$-2a_1 = a_2 + a_3 - 2a_4 \tag{11.37}$$

and for the dimensions of force to be consistent

$$a_1 = a_4 \tag{11.38}$$

and from (11.37) $a_3 = -2a_1 - a_2 + 2a_4$ or

$$a_3 = -a_2 \tag{11.39}$$

Substitution of Equations (11.38) and (11.39) in (11.35) gives

$$\left(\frac{w}{\delta b}\right)^{a_4} = \text{function } (l^{a_2}, d^{-a_2}, E^{a_4}) \tag{11.40}$$

Collecting terms with a common index,

$$\left(\frac{\delta E b}{w}\right)^{-a_4} = \text{function}\left[\left(\frac{l}{d}\right)^{a_2}\right]$$

or

$$\frac{\delta E b}{w} = \text{function}\left(\frac{l}{d}\right) \tag{11.41}$$

We have now reduced the original problem containing six variables to one having only two dimensionless groups of variables, and in the experiment it is necessary only to study how changes in l/d affect $\delta Eb/w$ to obtain the unknown functional relationship of Equation (11.41).

This can be accomplished by varying either l or d for a cantilever beam, and since it is a simple matter to vary l in the manner shown in Figure 11.11, this procedure would be chosen in practice.

11.5.5 An Experiment

Suppose, in an experiment on a cantilever steel beam of rectangular cross section, that the results shown in Table 11.2 were obtained. Young's modulus

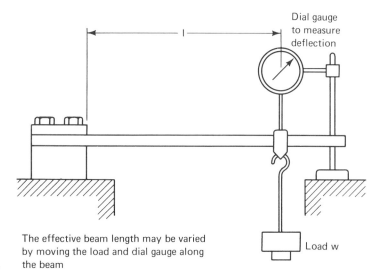

Figure 11.11
Experiment on cantilever beam.

The effective beam length may be varied by moving the load and dial gauge along the beam

for the steel beam was $E = 30 \times 10^6$ lb/in², the beam depth $d = .25$ in., and the beam width $b = 1.0$ in.

Table 11.2
Results of an Experiment on a Cantilever Beam

Length of Beam (in.)	Deflection under 1-lb Load (in.)
4	.0005
6	.0020
8	.0045
10	.0090
12	.0150

The relationship to be investigated is given by Equation (11.41), and it is therefore necessary to calculate for each experimental result the values of the dimensionless quantities $\delta Eb/w$ and l/d. It is best to arrange this calculation in tabular form where each step is kept to a simple arithmetical operation, as shown in Table 11.3. Trial plots of the results will show that the relationship is of the form

$$\frac{\delta Eb}{w} = a\left(\frac{l}{d}\right)^n \qquad (11.42)$$

and thus a linear relationship exists between log ($\delta Eb/w$) and log (l/d), where the intercept will be equal to log a and the slope will give n.

Table 11.3
Calculation of Dimensionless Variables

$$\frac{\delta Eb}{w} = \frac{\delta \times 30 \times 10^6 \times 1}{1} = \delta \times 30 \times 10^6$$

$$\frac{l}{d} = \frac{l}{.25} = 4.0 \times l$$

(1) l	(2) δ	(3) $\delta Eb/w = (2) \times 30 \times 10^6$	(4) $l/d = 4.0 \times (1)$
4	.0005	15,000	16
6	.0020	60,000	24
8	.0045	135,000	32
10	.0090	270,000	40
12	.0150	450,000	48

The appropriate values to be plotted on a linear graph are shown in Table 11.4. The best straight line through these points is found by the least-squares method to be

$$\log_{10}\left(\frac{\delta Eb}{w}\right) = 3.084 \log_{10}\left(\frac{l}{d}\right) + .487 \tag{11.43}$$

or

$$\frac{\delta Eb}{w} = a\log_{10} .487 \left(\frac{l}{d}\right)^{3.08}$$

$$= 3.07\left(\frac{l}{d}\right)^{3.08} \tag{11.44}$$

Table 11.4
Values To Be Plotted

$\log_{10}\left(\frac{\delta Eb}{w}\right)$	$\log_{10}\left(\frac{l}{d}\right)$
4.1761	1.2041
4.7781	1.3802
5.1303	1.5051
5.4314	1.6021
5.6532	1.6812

Figure 11.12 shows a plot of the values of $\delta Eb/w$ and l/d together with the best straight line.

It can be seen that considerable extrapolation was needed to obtain the intercept on the vertical axis and that a large measure of uncertainty must be associated with this value.

Clearly, if reasonable accuracy is required, the experimental curve shown in Figure 11.12 should not be used to predict the deflection for beams having l/d ratios outside the limits of the experimental results (i.e., $16 < l/d < 48$). It should be noted that the important parameter here is the l/d ratio and not the size of the beam. The curve could be used to predict the deflection for a beam 20 ft long and 12 in. in thickness quite satisfactorily. It can also be used to predict the behavior of a beam of any elastic material for which the value of E is known.

Apart from the errors caused by extrapolation, there is a further more important reason the behavior of beams having l/d ratios outside the experimental range cannot be predicted. There is no evidence to show that the relation obtained between $\delta Eb/w$ and l/d holds outside the experimental range. Indeed, if experiments were to be conducted for beams having l/d ratios of the order of unity or less, it would be found that the results deviate from the linear relation of Figure 11.12. This example has shown, however,

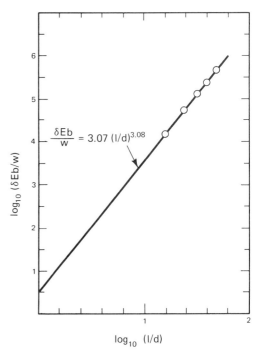

Figure 11.12
Results of experiment on cantilever beam.

how experimental effort can be reduced considerably by proper analysis of the variables involved.

REFERENCES

1. KAPP, R. O., *The Presentation of Technical Information*. London: Constable & Co. Ltd., 1948.
2. SCHENK, H., JR., *Theories of Engineering Experimentation*. New York: McGraw-Hill Book Company, 1961.
3. WILSON, E. B., *An Introduction to Scientific Research*. New York: McGraw-Hill Book Company, 1952.

EXERCISES

1. A weighing machine is observed to read 10 lb when no weight is placed on it. When a weight of 200 lb is applied, the reading is 190 lb. Assuming that the relation between the observed reading and the value of the weight applied is always linear, estimate the weight of a person who obtains a reading of 150 lb when he stands on the machine.
2. In an experiment the following repeated observations were made: .233, .228, .230, .223, .217, .221, .223, .220, .218, and 0.219.
   a. What is the best estimate of the value?
   b. Calculate the range within which you would be prepared to assert with 95% confidence that the true value lies.
3. In an experiment where the effect of x on the value of y was studied, it was expected that a linear relation between $\log_e y$ and x would exist. The following results were obtained in the experiment:

x	$\log_e y$
3.5	2.1
7.0	3.2
10.0	4.3
12.0	5.5

   a. Obtain the equation for the best straight line of the form $\log_e y = mx + c$.
   b. Rewrite the experimental relationship between y and x in the form $y = ae^{mx}$.
4. In a transient heat transfer experiment, the temperature θ was expected to depend on the time t according to the relation

   $$\theta = at^n$$

   or

   $$\log \theta = \log a + n \log t$$

   The following observations were made during the experiment:

θ	t
.1	.0
3.2	2.0
10.0	2.0
31.7	4.0
100.0	5.0
317.0	8.0

By the least-squares method, obtain the values of a and n in the equation that best fit the results of the experiment.

5. The stiffness (deflection per unit load S) for a cantilever beam of circular cross section depends on the diameter of the beam d, its length l, and Young's modulus of elasticity for the material E.
   a. Show by dimensional analysis that

   $$SdE = \text{function}\left(\frac{l}{d}\right)$$

   b. In an experiment on cantilever beams of various diameters it is found that a plot of $\log_{10}(SdE)$ against $\log_{10}(l/d)$ has a slope of 3 and an intercept on the $\log_{10}(SdE)$ scale of .83. From this information derive an expression giving the deflection of the beam δ in terms of the load W, the length l, the modulus of elasticity E, and the second moment of area of the beam cross section I ($I = \pi d^4/64$).

12

chemical and thermal processes

W. L. Short

J. R. Kittrell

12.1 INTRODUCTION

In this chapter, we will briefly describe some of the chemical and thermal concepts used in the practice of engineering and show typical examples of their applications. Much of the material contained in this chapter often falls in the areas of chemical and/or mechanical engineering. The application of these principles can be found in such diverse areas as air pollution control, power generation, heat engine design, oil refining, and manufacturing.

12.2 CONSERVATION OF MATTER

The engineer calls the law of conservation of matter a material balance. The law of conservation of matter states simply that matter (mass) cannot be created or destroyed. There is, of course, an exception to the previous statement. In atomic fission, the simple material balance does not hold, since in this process matter is transformed into energy.

For an example of the application of a material balance let us consider a power plant. We are interested in calculating the quantity and composition of the stack gas, i.e., in particular how much sulfur dioxide and particulate matter is produced by a typical power plant. Let us assume, for purposes of this example, that we are going to use coal, and that we will burn 560,000 pounds per hour (as shown later, this is *approximately* the amount required for *one* 800 megawatt power plant).

In the combustion process the coal is burned in the presence of air liberating heat and producing combustion products. Coal consists chemically of carbon, hydrogen, oxygen, nitrogen and inerts (ash). During combustion the carbon is converted to either carbon dioxide or carbon monoxide

$$C + O_2 \longrightarrow CO_2$$
$$2C + O_2 \longrightarrow 2CO$$

and the hydrogen to water vapor.

$$2H_2 + O_2 \longrightarrow 2H_2O$$

Essentially all of the carbon (more than 99.9%) is converted to CO_2 and very little to CO.

As an example of our material balance calculation consider the coal described in the following table.

Constituent	Weight Percent
Carbon	77
Hydrogen	5
Sulfur	8
Water	3
Ash	7

Let us suppose that the combustion takes place with 20% excess air (20% more oxygen than required for burning all the combustibles). Power plant furnaces seldom run at very low or zero excess air because of the tendency for smoke formation. Typical values of excess air are 10-20% for fuel oil and 20-40% for coal. We wish to calculate the amounts and percentages of CO_2, SO_2, and ash in the combustion gases when 560,000 lbs/hr. of coal are burned. First, we must calculate the number of pounds of carbon, hydrogen, and sulfur that are burned each hour, using the analyses above and the coal firing rate (560,000 lbs/hr).

$$\text{Weight carbon burned} = \left(0.77 \frac{\text{lb carbon}}{\text{lb coal}}\right)\left(560,000 \frac{\text{lbs coal}}{\text{hr}}\right)$$

$$\text{Weight carbon burned} = 431,000 \text{ lbs/hr}$$

$$\text{Weight hydrogen burned} = \left(0.05 \frac{\text{lb hydrogen}}{\text{lb coal}}\right)\left(560,000 \frac{\text{lbs coal}}{\text{hr}}\right)$$

$$\text{Weight hydrogen burned} = 28,000 \text{ lbs/hr}$$

$$\text{Weight sulfur burned} = \left(0.08 \frac{\text{lb sulfur}}{\text{lb coal}}\right)\left(560,000 \frac{\text{lbs coal}}{\text{hr}}\right)$$

$$\text{Weight sulfur burned} = 44,800 \text{ lbs/hr}$$

Now, we must calculate the number of moles of these components burned in order to calculate the air requirements (using the molecular weights of the species involved).

$$\text{Moles carbon burned} = \frac{431,000 \text{ lbs carbon/hr}}{12 \text{ lbs/lb mole}} \quad \text{(molecular weight of carbon is 12)}$$

$$\text{Moles carbon burned} = 35,900 \text{ lb moles/hr}$$

Moles hydrogen (H_2) burned $= \dfrac{28{,}000 \text{ lbs/hr}}{2 \text{ lbs/lb mole}}$ (molecular weight of hydrogen is 2)

Moles H_2 burned $= 14{,}000$ lb moles/hr

Moles sulfur burned $= \dfrac{44{,}800 \text{ lbs/hr}}{32 \text{ lbs/lb mole}}$ (molecular weight of sulfur is 32)

Moles sulfur burned $= 1{,}400$ lb moles/hr

The theoretical amount of oxygen required to burn each of these components is calculated as follows:

For carbon, which burns to form CO_2,

$$C + O_2 \longrightarrow CO_2$$

one mole of oxygen is required for each mole of carbon burned. Hence, 35,900 lb moles of carbon require 35,900 lb moles of oxygen. Furthermore, each lb mole of any gas occupies 379 ft³ measured (at 14.7 psia and 60°F); thus for burning carbon, the oxygen requirement is

$$O_2 \text{ required for carbon} = (35{,}900 \text{ lb moles/hr})\left(379 \dfrac{\text{ft}^3}{\text{lb mole}}\right)$$

$$= 13{,}600{,}000 \text{ ft}^3/\text{hr}$$

Similarly, for hydrogen, which burns to form water vapor,

$$H_2 + 1/2\, O_2 \longrightarrow H_2O$$

one-half mole of oxygen is required for each mole of hydrogen burned. Hence, 14,000 lb moles of hydrogen require 0.5 (14,000) = 7,000 lb moles of oxygen. Converting to cubic feet,

$$O_2 \text{ required for hydrogen} = (7{,}000 \text{ lb moles/hr})\left(379 \dfrac{\text{ft}^3}{\text{lb mole}}\right)$$

$$= 2{,}650{,}000 \text{ ft}^3/\text{hr}$$

For combustion of sulfur, which burns to form SO_2,

$$S + O_2 \longrightarrow SO_2$$

one mole of oxygen is required for each mole of sulfur burned. Hence, 1,400 lb moles of sulfur require 1,400 lb moles of oxygen. Converting to cubic feet,

$$O_2 \text{ required for sulfur} = (1{,}400 \text{ lb moles/hr})\left(379 \dfrac{\text{ft}^3}{\text{lb mole}}\right)$$

$$= 530{,}000 \text{ ft}^3/\text{hr}$$

The theoretical O_2 required to burn carbon, hydrogen and sulfur, then, is

$$\text{Theoretical } O_2 \text{ requirement} = 13{,}600{,}000 + 2{,}650{,}000 + 530{,}000$$
$$= 16{,}780{,}000 \text{ ft}^3/\text{hr}$$

Since air contains only 21% oxygen, then the amount of air required to provide 16,780,000 ft³ of oxygen is

$$\text{Theoretical air requirement} = \frac{16{,}780{,}000 \text{ ft}^3 O_2/\text{hr}}{0.21 \text{ ft}^3 O_2/\text{ft}^3 \text{ air}}$$

$$\text{Theoretical air requirement} = 79{,}700{,}000 \text{ ft}^3 \text{ air}/\text{hr}$$

However, this furnace is to operate with 20% excess air. Hence, the actual air requirement is the theoretical air requirement plus 20% of the theoretical air requirement

$$\text{Actual air requirement} = 1.20 \, (79{,}700{,}000 \text{ ft}^3)$$
$$= 95{,}600{,}000 \text{ ft}^3/\text{hr}$$

Note that nearly 100 million ft³/hour of air per hour is required to burn the coal for an 800 megawatt power plant.

Let us now calculate the emission rates of the products of combustion leaving the furnace stack. All of the nitrogen (N_2) entering the furnace with the combustion air will leave as N_2 except for whatever amount reacts to form nitrogen oxides. Operating data and theoretical calculations indicate that the nitrogen oxides concentration in a stack gas is typically less than 1,000 ppm (parts per million) and often less than 500 ppm (1,000 ppm is 0.1%). Hence, for simplicity, we can assume that all the entering N_2 leaves as N_2 without serious error in the *total* stack gas flow rate. Thus, N_2 in the combustion gases can be calculated from the N_2 content of the combustion air (70%)

$$N_2 \text{ in combustion gas} = \left(0.79 \frac{\text{ft}^3 N_2}{\text{ft}^3 \text{ air}}\right)(95{,}600{,}000 \text{ ft}^3/\text{hr air})$$
$$= 75{,}500{,}000 \text{ ft}^3/\text{hr}$$

The total amount of oxygen entering with this combustion air can similarly be calculated (air contains 21% oxygen)

$$O_2 \text{ entering with combustion air} = \left(0.21 \frac{\text{ft}^3 O_2}{\text{ft}^3 \text{ air}}\right)(95{,}600{,}00 \text{ ft}^3 \text{ air})$$
$$= 20{,}100{,}000 \text{ ft}^3/\text{hr } O_2$$

However, we know that 16,780,000 ft³ of O_2 is required to burn all of the carbon, hydrogen, and sulfur in the coal, so the remainder passes out the stack with the combustion (stack) gases.

O_2 in combustion (stack) gases = 20,100,000 − 16,780,000
= 3,320,000 ft³/hr

Note that the Law of Conservation of Mass only requires that *mass* be conserved, not volume (ft³).

Since one mole of sulfur dioxide is formed from every mole of sulfur burned, the combustion of 1,400 moles of sulfur will produce 1,400 moles of sulfur dioxide in the stack gas. Converting this to cubic feet

SO_2 in combustion gases = 1,400 lb moles/hr (379 ft³/lb mole)
= 530,000 ft³/hr

The water vapor in the stack gas results from two sources, water in the coal (3%) and water formed by burning hydrogen. Since one mole of water is formed from every mole of hydrogen burned, the combustion of 14,000 moles of hydrogen will produce 14,000 moles of water vapor in the stack gas. Converting this to cubic feet,

H_2O from combustion = (14,000 lb moles/hr) (379 ft³/lb mole)
= 5,300,000 ft³/hr

Water inherent in the coal (3% water) is

weight of water in coal = $\left(0.03 \frac{\text{lb water}}{\text{lb coal}}\right)\left(560,000 \frac{\text{lbs coal}}{\text{hr}}\right)$
= 16,800 lbs/hr

Since the molecular weight of water (H_2O) is 18 lbs/lb mole, this represents

moles of water in coal = $\frac{16,800 \text{ lbs } H_2O/\text{hr}}{18 \text{ lbs/lb mole}}$
= 933 lb moles/hr

Since, when vaporized, there will be 379 ft³ for each lb mole,

H_2O from coal = (933 lb moles/hr)(379 ft³/lb mole)
= 356,000 ft³/hr

The total water in the combustion gases, then, is the water from the coal plus the water from burning hydrogen:

H_2O in combustion gases = 5,300,000 + 356,000
= 5,656,000 ft³/hr

The emission levels and the percentage of each component in the combustion gases, then, can be summarized as follows.

COMPONENT	EMISSION LEVEL, FT³/HOUR (MEASURED AT 14.7 PSIA, 60°F)	PERCENTAGE
N_2	75,500,000	76.56
O_2	3,320,000	3.37
CO_2	13,600,000	13.79
SO_2	530,000	0.54
H_2O	5,656,000	5.74
Total	98,606,000	100.00

Let us now consider what happens to the ash. The ash content of the coal is

$$\left(.07 \frac{\text{lb ash}}{\text{lb coal}}\right)\left(560,000 \frac{\text{lbs coal}}{\text{hr}}\right) = 39,200 \text{ lbs/hr}$$

This quantity of ash is produced in the combustion process. However, not all the ash is emitted to the atmosphere, and in fact only a very small portion reaches the atmosphere even from a so called "polluting" furnace. A portion of the ash is deposited directly in the combustion zone of the furnace and is collected, much as the ashes might be collected and removed from a home fireplace. Another substantial portion is removed in the fly ash removal equipment. This equipment might consist of a cyclone, electrostatic precipitator or scrubber. (Not all plants have all this equipment and in fact many plants have only a cyclone for fly ash removal.) For a well designed system, it is not unreasonable that 99% or more of the fly ash formed will be removed. Hence, in our example, the emitted fly ash becomes

$$\text{Particulate Emission} = (0.01)(39,200)$$
$$= 392 \text{ lbs/hr}$$

The percentage removal varies from case to case and the above number is only suggested as being perhaps typical. It is certainly not the maximum possible removal for all types of ash.

The alert reader will note that we have not calculated the quantity of nitrogen oxides (NO_x) formed. When the fuel is burned, nitrogen is present, and at high temperatures nitrogen oxide is readily formed.

$$N_2 + O_2 \longrightarrow 2NO$$

The amount formed cannot be determined from a material balance *alone*, but requires in addition a knowledge of thermodynamics and kinetics, discussed in subsequent sections.

12.3 CONSERVATION OF ENERGY AND THE FIRST AND SECOND LAWS OF THERMODYNAMICS

In many ways, the principles and techniques developed from a thorough understanding of thermodynamics are required in other different areas of chemical engineering—energy balances, stage operations. The science of thermodynamics originated almost simultaneously with the development of the steam engine—in fact the word *thermodynamics* means "heat in motion."

Almost all engineering disciplines use thermodynamics. Thermodynamics is a study of the application of three basic "laws." The first law is an energy balance and is a statement of the conservation of energy. The second law concerns availability of energy—a simple example arising from it is the well known fact that heat will flow *spontaneously* from a high temperature region to a low temperature region. The third law states that at absolute zero temperature (about $-460°F$) molecular motion ceases.

From the three laws of thermodynamics one can derive a host of useful ideas and concepts that will enable us to predict directions of chemical reactions, determine the feasibility of new processing routes, predict thermochemical data (e.g., heat capacities), and so forth. The ideas most pertinent to the purposes of this chapter are the energy balance (a use of the first law), the concept of an ideal power cycle (a use of the second law), and vapor-liquid equilibrium (a use of the second law).

The energy balance simply states that energy is neither created nor destroyed. This brief statement ignores the possibility of nuclear fission reactions in which matter is converted to energy. In order to apply the energy balance, *all* forms of energy must be accounted for and expressed in the same units (Btu's, foot pounds, etc.). Different types of energy that are commonly encountered include heat, work, potential energy, kinetic energy, and internal energy. An example of a less common form of energy would be surface energy.

The Law of Conservation of Energy may be generally written as shown in the following equation.

Rate of accumulation of internal, potential, and kinetic energy in system	=	Net rate of flow of internal, potential, and kinetic energy into system	+
		Net rate of heat addition to system	−
		Net rate of work done by system on surroundings	(12.1)

This transport of energy is shown schematically in Figure 12.1.

This equation asserts there are three different types of energy that must be considered: internal energy, potential energy, and kinetic energy. The Law

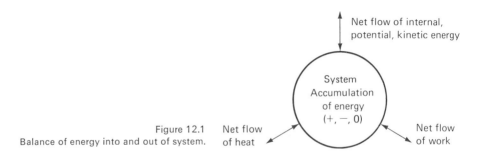

Figure 12.1 Balance of energy into and out of system.

of Conservation of Energy also asserts an interchangeability between energy and work, that is, that work is a form of energy. Note that the last two terms in the equation (the heat being transferred to the system and the work being done by the system) are states of energy in *transition*. This energy cannot be accumulated by the system in this fashion, but rather must be stored as either internal energy, potential energy, or kinetic energy.

Energy is a very essential factor in carrying out processes important to man, including chemical processing and environmental applications. The combustion process (e.g., burning coal) can transform fuel into light and heat. The heat can then be transferred to water, and this water can store energy at a higher internal energy level as steam. The stored energy from the fuel (steam) may be used to drive mechanical equipment (releasing energy as work) or it may be used to heat buildings through steam radiators (releasing energy as heat). The input of heat into a river would be stored as an increased temperature of the water; consequently, any fish in the river may be adversely affected by this stored energy (that is, by the higher temperature of the stream). If the combustion process is required to provide some specific amount of heat, this heat can be related to the amount of fuel that is required. Thus, the amount of pollutants that will be entering the atmosphere from this combustion process can be calculated and pollution control devices selected accordingly.

12.3.1
Internal Energy

Energy may be stored in a system in the form of internal energy. To understand how this energy is stored, we must recall that there are several sub-particles making up any substance; for example, a glass of water contains many individual water molecules and these water molecules in turn contain sub-particles (neutrons, protons, and electrons). The internal energy of this glass of water is determined in part by the movement of the sub-particles within the water. The molecular constituents are continually moving in translational motion, rotational motion, and vibrational motion. If the temperature of the water is increased, this component of the internal energy, represented by the movement of the molecules, increases. At high temperatures the molecular particles are moving much more rapidly than at low temperatures. Other forms of internal energy are also present. There are electrical and magnetic interactions between the sub-particles which contribute to this other component of internal energy. The total internal energy of a system is generally not known; however, for our purposes we can simply consider the internal energy of the substance relative to a reference state. For example, we can fully characterize the internal energy of the glass of water by defining its temperature, say 100°F, relative to some standard of reference temperature, let us say 0°F. Because of the internal energy term arising from the interaction of the molecules, we must also specify the state of the substance in the reference condition; for example, it may be water at 0° or solid crystalline ice at 0°.

12.3.2 Kinetic Energy

Another important form of energy (called kinetic energy) is that due to movement of the mass as a whole, rather than the molecular constituents of that substance described in Section 12.3.1. Kinetic energy is generally defined on a relative basis also, using the motion of one mass relative to another. We normally assume that the motion of a body may be considered relative to that of the motion of the earth, which is taken to be zero on a relative basis. Consequently, fundamentals of physics can define the kinetic energy of a moving body by the following equation

$$\text{K.E.} = 1/2 \, mv^2 \tag{12.2}$$

where m is the mass of the system and v is the velocity of the system. This suggests that a baseball thrown by a pitcher possesses internal energy (which is measured in part by the temperature of the baseball) and kinetic energy (which is measured in part by its velocity relative to the earth). If this baseball has been thrown at 60 mph (2680 cm/sec) and the baseball has a mass of 400 gms, then the kinetic energy of this baseball would be shown by the following equation

$$\text{K.E.} = 1/2 \, (400 \text{ gms}) \left(2680 \frac{\text{cm}}{\text{sec}}\right)^2 = 14.3 \times 10^8 \text{ g-cm}^2/\text{sec}^2 \tag{12.3}$$

12.3.3 Potential Energy

The potential energy of a substance is defined as the potential of that mass for doing work. Any two masses are known to exert an attraction on one another. A baseball thrown straight up in the air is attracted back towards the earth by the gravitational force of the earth. Consider this baseball when it reaches the apex of its travel up into the air. At this instant of time, the baseball contains a potential energy which is measurable relative to that of the earth. The potential energy of any mass is generally expressed by the following equation:

$$\text{P.E.} = mZ \tag{12.4}$$

where the distance Z is the height of that mass, m, above the surface of the earth. In the case of the baseball, this potential energy is achieved as a result of the expenditure of energy. Consequently, when a pitcher throws a baseball at a velocity of 60 mph as it leaves the pitcher's hand (and if there is no frictional work done on the atmosphere by the baseball's passing through the atmosphere), then we can calculate the ultimate height the baseball will achieve by combining equations (12.3) and (12.4).

Let us consider a few simple examples.

Example 12.1:

A 500 pound block is raised 50 feet. What is the change in potential energy?

$$P.E. = (500)(50) \text{ foot-pounds}$$
$$= \frac{(500)(50)}{778} \text{ Btu}$$
$$= 32.1 \text{ Btu}$$

Example 12.2:

If the block from Example 12.1 is allowed to drop from rest 50 feet to the ground, what is its kinetic energy just before striking the ground (neglecting any air friction)?

$$K.E. = \frac{mv^2}{2gc}$$

To use the above relationship, we need to know the final velocity. Fundamental physics relationships tell us that

$$v^2 = V^2 + 2as$$
$$V = \text{initial velocity} = 0$$
$$a = \text{acceleration}$$
$$s = \text{distance}$$
$$v^2 = (2)(32.2)(50)$$
$$K.E. = \frac{(2)(32.2)(50)(500)}{(2)(32.2)(778)} \text{ Btu}$$
$$K.E. = 32.1 \text{ Btu}$$

Note that we can obtain the same answer by simply equating the K.E. and P.E. terms as Equation (12.1) suggests. (Compare results of Examples 12.1 and 12.2.)

The second law of thermodynamics concerns itself with the availability of energy, and was first conceived in terms of power cycles (steam engines, boiler plants, etc.). The first law does not tell us anything about the direction in which energy is transferred, the minimum work required to perform separations, the maximum theoretical work from a power plant cycle, etc. . . The second law is a statement of these limitations. There are many ways of expressing the second law—for our purposes we will use the following.

1. A device (machine, etc.) cannot convert all heat solely into work (some of the heat must be wasted).

2. It is impossible for heat to flow spontaneously from a low temperature to a high temperature (without, for example, the addition of work).

To illustrate the second law, let us consider a simple power plant cycle which is sketched in Figure 12.2. Water is pumped into a boiler at a relatively high pressure (say 1000 psi) where it is vaporized into superheated steam. This steam leaves the boiler at a high pressure and is used to drive a turbine by allowing the steam pressure to decrease to some low value near atmospheric pressure; the steam at this point is called exhaust steam. The turbine operates basically on the same principle as a paddle wheel on a boat, the steam pressure is used to turn the turbine blades which are connected to a rotating shaft. This shaft drives an electric generator. The exhaust steam leaving the turbine is condensed in a water (or air) cooled condenser. The condensed water is then pumped back to the boiler for reuse.

The diagram shows the pressures, temperatures and heats of vaporization and condensation for the appropriate points in the process. Note from Figure 12.2 that the heat of vaporization of water at 1000 pounds per square inch is much less than at atmospheric pressure.

If we call the boiler temperature T_H and the condenser temperature T_C, the second law of thermodynamics tells us that the efficiency of the power cycle is approximately

$$\text{Eff} = \frac{T_H - T_C}{T_H}$$

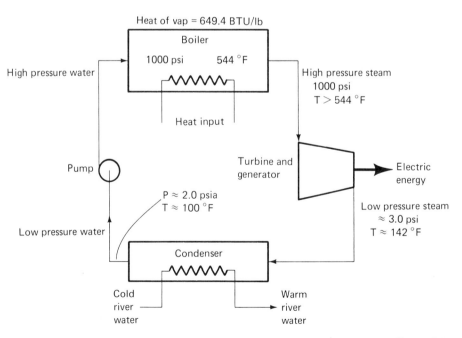

Figure 12.2
Schematic diagram of a power plant.

The above expression is strictly true only for a Carnot cycle (a special form of a power cycle), but we will use it as an estimate of actual power plant efficiency. The high temperature depends upon the pressure at which the steam is generated—the higher the pressure, the higher the temperature. Table 12.1 shows the vapor pressure of water as a function of temperature.

Table 12.1
Vapor Pressure of Water

Temperature, °F	Pressure, psia
32	0.0886
40	0.1217
60	0.2561
80	0.5067
100	0.9487
150	3.716
200	11.525
212	14.696
250	29.82
300	67.01
400	247.25
500	680.80
600	1543.2
700	3094.1

The low temperature depends upon the temperature at which the heat can be rejected in the condenser, i.e., either the air or water (river) temperature. Let us assume that the two temperatures are 500°F (680 psia) and 90°F respectively. Then the efficiency is

$$\text{Eff} = \frac{(500 + 460) - (90 + 460)}{(500 + 460)}$$

$$\text{Eff} = \frac{960 - 550}{960} = 0.427 \text{ or } 42.7\%$$

Note that degrees absolute are used for the temperatures—degrees absolute are degrees Fahrenheit plus 460 or degrees Centigrade plus 273. The efficiency of 42.7% tells us that (100 − 42.7) or 57.3% of all heat produced will be rejected into the river.

Let us now use the concept of an energy balance and the second law of thermodynamics to calculate the fuel requirements for a 800 megawatt power plant. A power plant of this size is typical of those currently under construction, although there are plans to build much larger plants. What we wish to calculate is the amount of fuel required and the amount of heat rejected to the river (thermal pollution). In the previous section on material balances we used this same example to estimate the SO_2 and particulate emissions.

We require 800 megawatts of power or

$$\text{Power} = \frac{(800,000,000) \text{ watts}}{746 \text{ watts}} \times \text{HP} \times \frac{42.4 \text{ Btu}}{\text{HP min}} \times \frac{60 \text{ min}}{\text{hr}}$$

$$\text{Power} = 2.73 \times 10^9 \text{ Btu/hr}$$

But this figure is the power generated and does not take into account the efficiency of converting heat to work.

The actual heat release in the boiler must be, for 42.3% efficiency,

$$\text{Heat} = \frac{2.73 \times 10^9}{0.423}$$

$$\text{Heat} = 6.45 \times 10^9 \text{ Btu/hr}$$

If we assume that the heating value of the coal is 11,500 Btu/lb, then we require

$$\text{Coal} = \frac{6.45 \times 10^9}{1.15 \times 10^4} = 560,000 \text{ lbs/hr of coal}$$

or 280 tons per hour of coal

The heat release to the river is

$$\text{Thermal Pollution} = (6.45 - 2.73) \times 10^9$$
$$= 3.72 \times 10^9 \text{ Btu/hr}$$

or 3.72 billion Btu/hr of heat

One Btu is the amount of heat required to raise the temperature of one pound of water one degree Fahrenheit, so this heat will raise the temperature of 372 million pounds per hour of water about 10 degrees Fahrenheit. These heat quantities are truly stupendous, yet they refer to only one 800 megawatt (approximately) power plant. The annual electrical energy consumption in the United States is approximately 1640 million megawatts per year (1970).[2] It is estimated that consumption will increase to 3240 million megawatts in 1980 and to 6072 million megawatts by 1990.

12.4 KINETICS AND CHEMICAL EQUILIBRIUM

Kinetics may be considered as a study of the rate and mechanism by which one chemical species is converted to another. The terms "rate" and "mechanism" are discussed below in some detail. To illustrate the discussion of kinetics, we will once again talk in terms of an air pollution problem, viz. the removal of SO_2 and NO_x from power plant stack gases.

It has been estimated that there are some 60 to 70 different processes or methods under investigation for the removal of SO_2 and/or NO_x from stack gases. One possible process utilizes a reaction of carbon monoxide with each pollutant (sulfur dioxide and nitric oxide) to produce harmless compounds

$$2CO + SO_2 \rightleftarrows 2CO_2 + 1/2\, S_2$$
$$CO + NO \rightleftarrows CO_2 + 1/2\, N_2$$

There is also a side reaction that can occur between sulfur (one of the products) and CO to form carbonyl sulfide (COS)

$$CO + 1/2\, S_2 \rightleftarrows COS$$

There are three questions we wish to answer concerning the above reactions.

1. Given the stack gas composition and temperature (i.e., percent SO_2, CO, CO_2, N_2, etc.) what is the maximum theoretical conversion of the SO_2 to S_2?

    Suppose we have a vessel at 1000°F to which is added gas of composition typical of a power plant stack—76% N_2, 14% CO_2, 1000 ppm SO_2, 500 ppm CO, and the balance H_2O and O_2. If this mixture is allowed to react and we let it sit sufficiently long in the vessel (at 1000°F) so that there eventually is no further change in composition, what will be the final concentrations of CO, SO_2, COS, S_2, etc.? This resulting point is called *chemical equilibrium* and may be calculated using thermodynamic principles.

2. How long, starting from the same initial composition as in (1), will it take to reach chemical equilibrium? (The answer may be years.)

    The equilibrium concentration of SO_2 might be 800 ppm (for an initial SO_2 concentration of 1000 ppm). We may also wish to know how long it will take to reach 80% of the equilibrium conversion (80% × (1000 − 800)) or 840 ppm SO_2. In other words, we want to know the rate of chemical reaction.

3. By what reaction path does the SO_2 react with the CO to form CO_2 and S_2, i.e., what is the mechanism? For example, do both species adsorb on a catalyst and react on its surface, are there intermediate products, e.g., radicals, formed?

Both these latter two questions pertaining to reaction rates and mechanisms fall in the domain of kinetics. Typically the chemical engineer is more concerned with reaction rates than with reaction mechanisms, because the rate data are of vital importance in the design and operation of a chemical reactor to carry out a given conversion. The subject of kinetics is also intimately involved with the subject of catalysis. Many chemical reactions proceed very slowly unless a catalyst is present in which case the *rate* of reaction may be increased manyfold.

For purposes of illustration we will use the reaction

$$CO + NO \rightleftharpoons CO_2 + \tfrac{1}{2}N_2$$

and assume that there are no other reactions to consider. Suppose the initial concentrations of the reactants are:

CO	0.1 moles
NO	0.2 moles

and the temperature is 1000°F. We want to find the chemical *equilibrium* composition of the mixture. Thermodynamics allows us to define an equilibrium constant for the reaction which, for the present illustration, is taken to be

$$K_{eq} = \frac{a_{CO_2} a_{N_2}^{1/2}}{a_{CO} a_{NO}}$$

where K_{eq} is the equilibrium constant and a_{N2}, a_{CO2}, a_{SO2}, a_{CO} are the "activities" of the chemical species. For gases at low pressures the activity is nearly always equal to what is called the partial pressure. The partial pressure is defined as

$P_i = y_i P_{tot}$

P_i = partial pressure of component i

y_i = mole fraction of component i

P_{tot} = total pressure

The equilibrium constant is found both theoretically and experimentally to vary with temperature as indicated in Figure 12.3.

Suppose we let x denote the number of moles of NO which react. Then the moles of NO remaining will be $(0.2 - x)$. Similarly, the equilibrium concentrations of the other components will be

NO	$(0.2 - x)$
CO	$(0.1 - x)$
CO_2	x
N_2	$0.5x$
The total is	$(0.3 - 0.5x)$

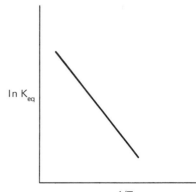

Figure 12.3

and the partial pressures become

$$P_{NO} = \frac{(0.2 - x)}{(0.3 - 0.5x)} P_{tot}$$

$$P_{CO} = \frac{(0.1 - x)}{(0.3 - 0.5x)} P_{tot}$$

$$P_{CO_2} = \frac{x}{(0.3 - 0.5x)} P_{tot}$$

$$P_{N_2} = \frac{(0.5x)}{(0.3 - 0.5x)} P_{tot}$$

The expression for the equilibrium constant is then

$$K_{eq} = \frac{P_{CO_2} P_{N_2}^{1/2}}{P_{NO} P_{CO}}$$

$$= \frac{\frac{x}{(0.3 - 0.5x)} P_{tot} \sqrt{\frac{0.5x}{0.3 - 0.5x} P_{tot}}}{\frac{0.2 - x}{(0.3 - 0.5x)} P_{tot} \frac{0.1 - x}{0.3 - 0.5x} P_{tot}}$$

Using a value for K_{eq}, we then determine the equilibrium mixture, since only x is unknown.

12.4.2 Reaction Kinetics

Whereas thermodynamics provides us with information as to how far a reaction could go if given an infinite time to react, reaction kinetics provides information concerning the actual rate at which a reaction proceeds. We are familiar with a variety of very rapid reactions, for example the reaction of hydrogen and oxygen in the inverted beaker observed in high school chemistry. In the presence of a flame, this reaction proceeds so rapidly that a loud pop is heard. Thermodynamics says that this reaction will go completely to water at room temperature. However, it is not until a burning ember is inserted into the mixture that the reaction takes place. A slower reaction with which we are familiar is the rusting of iron, particularly in the presence of water. Most reactions that we are concerned with in industrial practice occur at rates between those of the hydrogen-oxygen reaction and the iron-oxygen reaction.

The applications of kinetics in the United States economy are very broad. Chemical reaction kinetics play a major role in producing goods worth more than 100 billion dollars per year. This represents products accounting for $\frac{1}{6}$ of the value of all goods manufactured in the United States. In general, these reactions are utilized to convert a low value raw material into a high value product. Examples wherein reactions are used to produce a valuable product are tabulated below.

1. Gasoline from crude oil
2. Jet fuel from crude oil

3. All drugs and pharmaceuticals
4. All heavy chemicals (acetic acid, fertilizers, etc.)
5. All plastics (telephone, wire coating, intrauterine devices, hula hoops, etc.)
6. All synthetic fibers for clothing (nylon, dacron, etc.)
7. Food stuffs (cheese, beer, clarified apple juice, ethanol, etc.)
8. Solution to environmental problems (removal of nitric oxide, sulfur dioxide, etc.)

One of the great efforts in the general field of chemical engineering kinetics of the last few decades has been to find new catalysts which cause a reaction to proceed at relatively less severe conditions and to proceed more completely to the desired product, but without themselves being consumed by the reaction. These catalysts are characterized by being able to form a chemical intermediate which is not readily formed in the absence of the catalyst and which then can react to easily form the desired product. For example, without a catalyst, normal butane cannot readily be transformed to isobutane. Conversion of normal butane to isobutane is desirable because the isobutane is more useful in the production of gasoline.

We can force this reaction to proceed if we raise the temperature high enough. In such a case, the high temperature would transmit such high energies to the normal butane molecule that it would ultimately rearrange and fly apart as shown below.

$$\text{normal butane} \rightarrow \text{isobutane} + \text{methane} + \text{ethane} + \text{propane} + \text{ethylene} + \cdots$$

It is apparent that we do get some isobutane, as desired, but we also may obtain many other nondesired products due to the severe reaction conditions required.

In the presence of a catalyst, however, which, in this case, is simply a solid material that is contacted with the normal butane at a moderately high temperature, the following reaction sequence proceeds:

The normal butane becomes attached to the catalyst.

The catalyst removes a hydrogen atom from the normal butane leaving a positive charge in its place on the normal butane. This positively charged ion then readily becomes rearranged to a more stable form of an isobutane ion.

This isobutane ion then recovers the hydrogen picked up by the catalyst to produce isobutane and the fresh regenerated catalyst. This catalyst can then repeat the reaction sequence by picking up additional normal butane. Hence, we see a reaction sequence that more efficiently gives isobutane through the action of the catalyst, and, in addition, the catalyst is unchanged as a result of the reaction.

We must also be able to predict the rate of reaction—the rate at which the reactants disappear, the rate at which products form, and the rate of reaction between products formed and other reactants to form various by-products. This information will enable us to calculate the size of the reactor needed to remove a specified amount of reactants or to form a specified amount of products.

For the reaction

$$CO + NO \rightleftarrows CO_2 + 1/2\, N_2$$

the expression for the rate of formation of CO_2 could be

$$\Gamma = k[\text{concentration NO}][\text{concentration CO}]$$

where k is the rate constant. We do not know without experimental data what the rate expression will be. For example, the rate of formation of CO_2 could be given by

$$\Gamma = k[\text{concentration NO}]$$

i.e., independent of the CO concentration.

To illustrate the magnitude of chemical reactor required for our stack gas problem, let us assume that we calculate from the rate equations that 90 percent of the SO_2 and NO_x will be removed in 0.1 seconds. This percent removal is about what is required to meet typical air pollution standards.

From the section on material balances, the volume of stack gas produced is 98,163,000 cubic feet per hour, or 27,200 cubic feet per second. The volume of the reactor is then

$$27{,}200 \frac{\text{ft}^3}{\text{sec}} \times 0.1 \text{ sec}$$
$$= 2720 \text{ ft}^3$$

If we were to assume that the height is three times the diameter then the reactor size would be

$$\left(\frac{D^2}{4} \times 3D\right) = 2720$$

$$D = 15.1 \text{ ft}$$

and the dimensions of the reactor are 15 feet diameter by 45 feet high.

12.5 HEAT TRANSFER

The transfer of heat occurs in practically every engineering process and is encountered in everyday life in countless situation—e.g., car radiators, air conditioners, home furnaces, etc. There are three modes of heat transmission: conduction, convection, and radiation.

Conduction is the transfer of heat from one part of a body to another or between two bodies in actual physical contact.

Convection is the transfer of heat from one point within a fluid to another point within a fluid. Convection can be further subdivided into natural and forced convection. In natural convection the heat transfer occurs solely as a result of density differences resulting from temperature differences. In forced convection, the fluid is in motion.

Radiation is the transfer of heat by radiant energy. Electromagnetic waves are radiated by all bodies in all directions at all temperatures. These electromganetic waves carry energy. When this energy strikes (contacts) another body, a part of it will be absorbed, a part reflected, and a part transmitted.

In many actual cases, heat transfer will occur by more than one of these means simultaneously. The rate of heat transfer (amount of heat transferred per unit time) is proportional to a driving force (temperature difference),

rate $\propto \Delta T$

or

$$\frac{dQ}{dt} \propto \Delta T$$

The proportionality constant is known as the heat transfer coefficient—in the case of pure conductivity the coefficient is known as the thermal conductivity of the material. Excellent heat transfer materials such as copper have high thermal conductivities; excellent insulators, such as fiber glass, air, and asbestos, have low thermal conductivities.

As an example of the use of heat transfer concepts, combined with energy balances, let us consider what happens to the heat discharged from a power plant to a river. The reader will recall that not all of the energy produced from the fuel combustion is converted into electrical energy (in fact most of it is not).

In writing an energy balance for a thermally polluted river, some of the terms which must be considered are discussed below.

Q_{sn} net short wave radiation flux delivered through the water surface air interface after losses by absorption and scattering in the atmosphere and by reflection at the surface.

Q_{at} net atmospheric long wave radiation flux delivered through the interface.

Q_{BR} long wave water surface back radiation to the atmosphere.

Q_E energy loss by evaporation.

Q_C convective energy flux between the water surface and the overlying air mass. This energy may either flow from the air to the water or from the water to the air depending upon the temperatures of each.

We will briefly describe the various terms above, their significance and relative magnitudes.

Net Short Wave Radiation

The net short wave radiation passing through the air-water interface may be described generally as the extraterrestial radiation flux entering at the top of the atmosphere less losses incurred by scattering and absorption in the atmosphere and by reflection from the water surface. The rate is obviously a complicated function of many variables such as time of year, time of day, material in the atmosphere which can cause scattering and/or absorption, cloud cover, weather conditions (rain, snow), etc.

Atmospheric Radiation

A part of the long wave radiation which is emitted by solid or liquid bodies at the earth's surface is absorbed by water vapor and ozone in the surrounding atmosphere. These constituents, in turn, radiate back to the ground or into space.

Long Wave Back Radiation

This is essentially self-explanatory and is the radiation leaving the river surface.

Evaporation

Evaporation is one of the most important terms in the particular energy balance—its presence attests to the fact that cooling ponds may be used as a method of eliminating thermal pollution. When water evaporates from the surface of the river (or lake) into the air, it changes from a liquid to a vapor; for each pound of water vaporized approximately 970 Btu of heat are required. The only source of heat is the river itself and hence, the temperature of the river water will tend to decrease as evaporation is taking place. Water can evaporate into the air as long as the air is not saturated—or in perhaps more familiar terms as long as the relative humidity is less than 100%. The rate of evaporation depends upon the driving force which depends upon the relative humidity of the air. Also cold air can hold less water than warm air, i.e., one pound of cold air will contain less water at 100% relative humidity than one pound of warmer air also at 100% relative humidity.

Convection

In the case of thermal pollution heat transfer the term advection is often used rather than convection. Advection refers to the same mechanism of heat transfer, but implies that the transfer is in a horizontal direction, i.e., that the wind blows parallel to the river surface.

In order to carry out the energy balance for the thermal pollution of a river, we must be able to estimate the magnitude of each of the above terms. In addition, there are other means by which energy might leave the river, viz. conduction to the river bed or to the river banks. It has also been reported that rocks in the river are capable of absorbing significant quantities of radiant energy and can act as "hot spots," particularly in late afternoon and evening.

What are the magnitudes of each of the above terms? Some reasonably careful estimates have been made of each and the numbers shown below in Figure 12.4 are typical of those either measured or generally accepted as being reliable estimates.

Jimeson and Adkins[1] have presented a typical heat balance for a 1000 megawatt power plant, which is reproduced below in Figure 12.5.

The authors state that a 1,000 megawatt power plant operating with a 15°F increase in water temperature in the condenser requires about 1,400 cubic feet per second of water. If a cooling tower is employed for an 800 megawatt power plant, the tower will be approximately 400 feet in diameter and 450 feet high. These towers cool the water by allowing a portion of it to

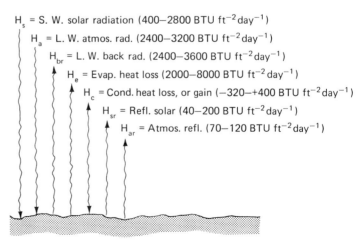

Figure 12.4 Mechanisms of heat transfer across a water surface.

H_s = S. W. solar radiation (400–2800 BTU ft^{-2} day^{-1})
H_a = L. W. atmos. rad. (2400–3200 BTU ft^{-2} day^{-1})
H_{br} = L. W. back rad. (2400–3600 BTU ft^{-2} day^{-1})
H_e = Evap. heat loss (2000–8000 BTU ft^{-2} day^{-1})
H_c = Cond. heat loss, or gain (−320–+400 BTU ft^{-2} day^{-1})
H_{sr} = Refl. solar (40–200 BTU ft^{-2} day^{-1})
H_{ar} = Atmos. refl. (70–120 BTU ft^{-2} day^{-1})

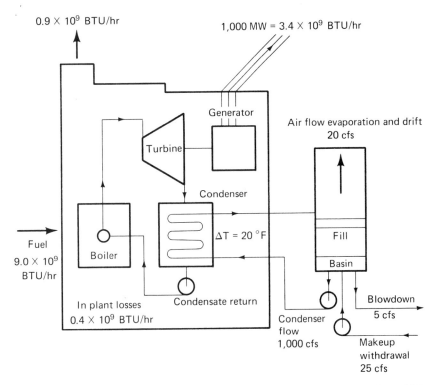

Figure 12.5
Heat balance of 1,000 megawatt fossil steam plant operating at full load.

evaporate into the air passing through the tower. The heat required to evaporate the water will result in a decrease in water temperature.

Jimeson and Adkins[1] have also given the following costs of cooling water systems for steam electric plants:

Type of System	Investment Cost $/kw
Once through	2.00—3.00
Cooling Pond	4.00—6.00
Evaporative Cooling Tower	
Natural Draft (no fans)	6.00—9.00
Mechanical Draft (fans)	5.00—8.00

The cooling costs correspond roughly to 0.2 to 0.4 mils/kw hr (one mil is 0.1 cent).

12.6 MASS TRANSFER AND DIFFUSION

A chemical engineer is often concerned with the changing of the composition of characteristics of mixtures and solutions, through methods not involving chemical reactions. Some of these required changes may be mechanical in nature, for example filtration or settling to segregate large and small particles. A large number of these separations involve changes in the composition of solutions and are known as diffusional or mass transfer operations.

A simple example of diffusion is that shown in Figure 12.6. A container is divided into two regions, A and B, containing two different pure gases, for example nitrogen and oxygen. The two regions A and B are initially separated by a membrane. When the membrane is removed, the two gases will diffuse (mix) until the concentration is uniform throughout. The gases diffuse initially because there is a concentration gradient—this gradient is referred to mathematically as (dc/dx). The gas A will diffuse or migrate to the left and B to the right until the concentration of A and B is essentially uniform throughout.

In an analogous way, we can make use of the same principle of mass transfer—chemical species migrate from a high concentration region to a low concentration region—to separate or purify chemicals by distillation. To discuss the principles of mass transfer as applied to distillation we first need to look at the concepts of vapor pressure and vapor liquid equilibrium.

Vapor pressure is familiar to all of us via the statement "water boils at 212°F." If we were to enclose pure water in a container as shown in Figure 12.7 and measure the pressure in the vapor space, the pressure measured

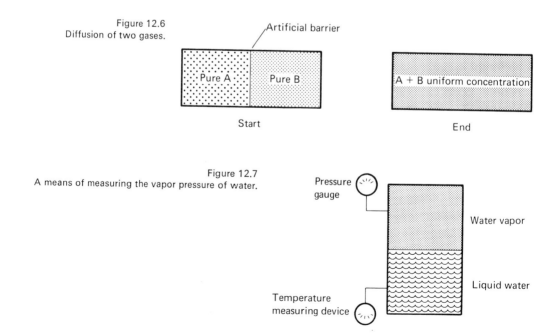

Figure 12.6
Diffusion of two gases.

Figure 12.7
A means of measuring the vapor pressure of water.

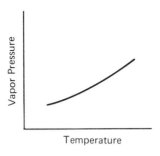

Figure 12.8
Graph of vapor pressure of water.

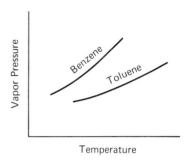

Figure 12.9
Sketch of vapor pressures of benzene and toluene.

would be the vapor pressure of water at the temperature of the container.

We could measure the vapor pressure of water at several temperatures and plot the data as sketched in Figure 12.8 (see also Table 12.1).

Suppose we were to measure the vapor pressure of a binary (two component) mixture, for example benzene and toluene. Because there is more than one component present, the additional variable of composition must now be considered. The shape of vapor pressure curve of the two pure chemicals benzene and toluene are shown in Figure 12.9. Note that at any one temperature the vapor pressure of benzene is larger than that of toluene. Benzene is said to be more volatile than toluene. Suppose we place a mixture of benzene and toluene in a vessel and allow the liquid and vapor to come to equilibrium, as shown in Figure 12.10. If the liquid contains 50% benzene and 50% toluene, the vapor composition will not be the same, but will in fact contain more benzene than toluene, i.e., at equilibrium composition the vapor will normally contain a higher concentration of the more volatile component, than the liquid.

The concept of vapor liquid equilibrium is used for designing and operating distillation columns, which are essentially a number of equilibrium stages each similar to the type described above. The subject area in chemical engineering is often called staged operations, the principles for which come from the principles of mass transfer and thermodynamics.

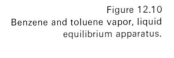

Figure 12.10
Benzene and toluene vapor, liquid equilibrium apparatus.

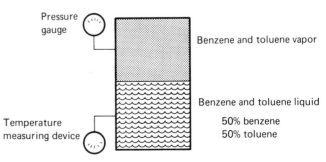

Staged operations is a title given to some of the unit operations which consist of or include what is known as equilibrium stages. These unit operations include distillation, extraction, and absorption. We shall discuss stage operations in terms of distillation because this is particularly applicable in the pollution control field, in the use of what are called stripping towers, e.g., ammonia strippers, hydrogen sulfide strippers, to remove these pollutants from water.

The equipment used often (but not always) contains plates or trays; hence the name stage. The word equilibrium connotes thermodynamic equilibrium as was illustrated by our example of benzene-toluene. One equilibrium stage corresponds to one "box" of vapor and liquid in equilibrium.

Figure 12.11 shows a flow diagram for a typical ammonia stripper. The stripper consists of a column with trays, a reboiler and a condenser. The reboiler is used to generate vapor which passes upward in the column. The condenser is used to condense the vapor leaving the top of the column that is collected as liquid and partly returned to the column (as a liquid). The trays are used to provide mixing of the vapor and liquid so that the vapor leaving each tray will be in "equilibrium" with the liquid leaving each tray, much as described above for our benzene-toluene example. In the case of NH_3-H_2O, ammonia is more volatile and will be enriched in the vapor relative to the liquid.

The interior of the stripper contains special attachments (internals) to separate the liquid and vapor leaving each tray and conduct it to the next tray. Also special internals are provided to generate the vapor in the reboiler.

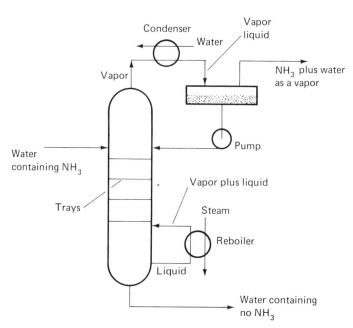

Figure 12.11
Schematic diagram of a stripper that removes ammonia from water.

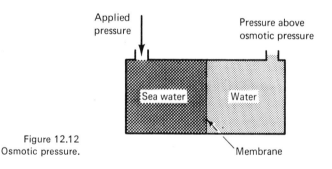

Figure 12.12
Osmotic pressure.

Details of these internals are discussed in many texts on chemical engineering, but are beyond the scope of this chapter.

An interesting example of the importance of diffusion arises in the desalting of sea water by the process of reverse osmosis, which is a membrane separation technique. Osmosis is a phenomenon that has been known for many years. When an aqueous solution of solute is separated from pure water by a semi-permeable membrane (a membrane that permits the passage of water but not the solute), the water will always pass through the membrane into the solution thereby diluting it. This phenomenon is called osmosis and was first reported by Nollet in 1748. We may visualize the osmotic process by considering Figure 12.12. The mechanical pressure required to prevent osmosis of the water into the solution is called the osmotic pressure. If the mechanical pressure exceeds the osmotic pressure, water will flow from solution to the pure water side (the membrane is assumed impermeable to the salt). This process is known as reverse osmosis and is used for the production of drinking water from sea water. The osmotic pressure of sea water, which contains about 3.5% salts, is approximately 350 psi. Reverse osmosis units operate at pressures of 1000 to 1500 psi.

A new and exciting area of mass transfer in chemical engineering is the use of artificial membranes to effect separations by reverse osmosis. In the early 1960s, Loeb and Sourirajan[2] discovered that cellulose acetate membranes could be used to make pure water from salt water. These membranes are very similar in appearance to a plastic film such as that used to wrap fruits and vegetables in a grocery store.

The experimental apparatus used by Loeb[3] is sketched in Figure 12.13. The membrane is placed on a porous backing and placed inside a container. The salt water is added and nitrogen gas used to apply pressure. In the case of sea water about 1500 psi is required before significant amounts of water pass through the membrane. The water produced is suitable for drinking. Several demonstration plants employing this principle have been built recently and are now in operation.

Other areas where chemical engineers are actively using the principles of membrane separation processes include the design of artificial kidneys, the use of membranes to make ultrapure water (used in manufacturing of magnetic tapes), waste treatment processes e.g., cheese whey industry, electroplating industry, etc.

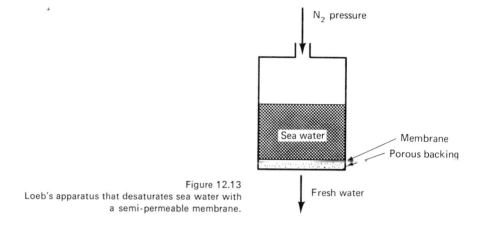

Figure 12.13
Loeb's apparatus that desaturates sea water with a semi-permeable membrane.

12.7 CONSERVATION OF MOMENTUM (MOMENTUM BALANCES)

Momentum of a body is defined, in physics, as the product of its mass (m) and its velocity (v) or momentum $= mv$. Because velocity is a vector quantity, momentum is also a vector quantity, i.e., direction must be specified when describing it. Momentum balances are based on Newton's second law.

$$F = ma \tag{12.5}$$

or

Force = (Mass)(Acceleration)

In terms of derivative expressions, this is written as

$$F = m\frac{d^2x}{dt^2} \tag{12.6}$$

or

$$F = m\frac{dv}{dt} \tag{12.7}$$

where x is position (distance) and t is time. Equation (12.7) may be used to express Newton's second law of motion as follows, that force which is proportional to the time rate of change of momentum

$$F \propto \frac{d}{dt}(mv) \tag{12.8}$$

Two simple examples will illustrate the concept of momentum. In order for water to flow down a pipe, a pump is required, for example the boiler feed water pump in the power plant. What does this pump do? It takes water at a low pressure and pumps it to a high pressure and a high velocity as illustrated in Figure 12.14.
Hence the momentum is increased. Another type of pump is the heart in our body, which is used to drive blood through our arteries. In both of these cases as we get farther away from the pump on the high pressure side [e.g., travel from (1) to (2)], we will note that the pressure decreases. It decreases because we are using momentum to overcome viscous shear, often called friction.

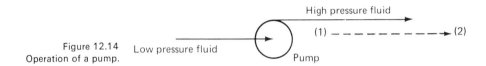

Figure 12.14 Operation of a pump.

The principles of momentum conservation are widely used by chemical engineers to design plastics extrusion equipment, to describe the flow of blood in the human body, to design artificial heart-lung machines used in open heart surgery, to design oil pipelines which will be ecologically safe but are economical to build, to describe the dispersion of pollutants in the atmosphere and the settling of particulate matter from the atmosphere, to describe the motion of aerosols and how they enter our lungs, to describe cloud formation using aerosol theory, and so on.

The principles of the conservation of momentum are also widely employed in other engineering disciplines—for example civil engineering, in the design of hydraulic conduits, canals, river channels, etc. The subject areas are often titled fluid mechanics or hydraulics and form a major component in the training of an engineer.

12.8 PROCESS DESIGN

In process design all knowledge of the fundamentals of chemistry and engineering, some of which were touched on previously, plus a knowledge of economics are brought to bear.

The process design function may take several forms.

1. The interpretation of experimental data from a chemistry lab to evaluate whether the idea is worth pursuing further. (Is it economic, competitive, or technically feasible in a commercial role?) If the idea is worth pursuing, the process design will be used to give guidance in the future experiments to be performed.
2. A process design for evaluation by top management relative to building a new plant.

3. A process design for actual construction. This design is the most detailed by far of the three types. The difference between a successful and an unsuccessful plant is often the amount of attention given to detail in this design stage.

12.9 PROCESS CONTROL

During the process design stage, and before the plant can be built and operated, the control instrumentation must be specified. Until a few years ago, this consisted simply of specifying level controls, pressure controls, flow controls, etc. Now with the advent of large high-speed computers, entire process plants may be under the control of a single computer, which uses continuous data input from the plant coupled with a mathematical model describing the process to correctly and most economically control the process. The entire area of process control, although exceedingly complex and outside the scope of this book, is certainly one of the most sophisticated and exciting frontiers of the chemical engineering profession.

12.10 SUMMARY

By now, the intent of a statement that a chemical engineer must be "a flexible person for a flexible career" should be obvious.

The challenges to the chemical engineer to use his knowledge and talents to help solve the problems confronting society are great, but his training and diversity of skills make him well suited to the tasks ahead.

REFERENCES

1. JIMESON, R. M., and G. G. ADKINS, "Waste Heat Disposal in Power Plants," *Chemical Engineering Progress 67*, No. 7, 64, 1971.
2. LOEB, S., and S. SOURIRAJAN, *Advan. Chem. Series 38*, 117, 1962.
3. LOEB, S., "Preparation and Performance of High Flux Cellulose Acetate Desalination Membranes," *Desalination by Reverse Osmosis*, V. MERTEN, Ed., MIT Press, 1966.

EXERCISES

1. Find out the flow rate of a fairly large river near your home. If its temperature is 70°F upstream of an 800 megawatt power plant, what will the average temperature be immediately downstream from the plant?
2. An example of catalysis not discussed in the chapter is the catalytic muffler. Briefly describe what it is and what it will do.
3. *One* of the reasons for removing lead from gasoline is that lead poisons the catalyst in the catalytic muffler described above. What does this mean?

4. From the figures given in the chapter for power consumption in the U.S., if only coal were burned as a fuel (with the same heating value as in the example) how many tons of coal would be required nationwide?
5. How many tons per year of ash are produced by an 800 megawatt power plant? If it has a density of 200 lbs/ft³ how much land will this cover to a depth of one foot?
6. How many railway cars of coal per day does an 800 megawatt power plant require? (You need to know the density of coal and the volume of a rail car.)

appendix A

Trigonometric Identities   445

Often-Used Trigonometric Functions   446

Approximations for Small Angles   447

appendix B

Prefixes and Symbols   448

appendix C

Conversion Factors   449

appendix A

TRIGONOMETRIC IDENTITIES

$$1 = \sin^2 A + \cos^2 A = \sin A \csc A = \tan A \cot A$$
$$= \cos A \sec A$$

$$\sin A = \frac{\cos A}{\cot A} = \frac{1}{\csc A} = \cos A \tan A$$
$$= \pm(1 - \cos^2 A)^{1/2}$$

$$\cos A = \frac{\sin A}{\tan A} = \frac{1}{\sec A} = \sin A \cot A$$
$$= \pm(1 - \sin^2 A)^{1/2}$$

$$\tan A = \frac{\sin A}{\cos A} = \frac{1}{\cot A} = \sin A \sec A$$

$$\sin A = \frac{e^{jA} - e^{-jA}}{2j}$$

$$\cos A = \frac{e^{jA} + e^{-jA}}{2}$$

$$\sin (A \pm B) = \sin A \cos B \pm \cos A \sin B$$

$$\cos (A \pm B) = \cos A \cos B \mp \sin A \sin B$$

$$\tan (A \pm B) = \frac{\tan A \pm \tan B}{1 \mp \tan A \tan B}$$
$$= \frac{\tan A \cot B \pm 1}{\cot B \mp \tan A}$$

$$\cot (A \pm B) = \frac{\cot A \cot B \mp 1}{\cot B \pm \cot A}$$
$$= \frac{\cot A \mp \tan B}{1 \pm \cot A \tan B}$$

$$\sin 2A = 2 \sin A \cos A$$

$$\cos 2A = \cos^2 A - \sin^2 A$$

$$\tan 2A = (2 \tan A)/(1 - \tan^2 A)$$

$$\sin 3A = 3 \sin A - 4 \sin^3 A$$

$$\cos 3A = -3 \cos A + 4 \cos^3 A$$

$$\sin \tfrac{1}{2}A = \pm \left[\frac{1 - \cos A}{2} \right]^{1/2}$$

$$\cos \tfrac{1}{2}A = \pm \left[\frac{1 + \cos A}{2} \right]^{1/2}$$

$$\tan \tfrac{1}{2}A = \frac{\sin A}{1 + \cos A} = \frac{1 - \cos A}{\sin A}$$

$$\sin A \pm \sin B = 2 \sin \tfrac{1}{2}(A \pm B) \cos \tfrac{1}{2}(A \mp B)$$

$$\cos A + \cos B = 2 \cos \tfrac{1}{2}(A + B) \cos \tfrac{1}{2}(A - B)$$

$$\cos B - \cos A = 2 \sin \tfrac{1}{2}(A + B) \sin \tfrac{1}{2}(A - B)$$

$$\tan A \pm \tan B = \frac{\sin (A \pm B)}{\cos A \cos B}$$

$$\cot A \pm \cot B = \frac{\sin (B \pm A)}{\sin A \sin B}$$

$$\sin^2 A - \sin^2 B = \sin (A + B) \sin (A - B)$$

$$\cos^2 A - \sin^2 B = \cos (A + B) \cos (A - B)$$

$$\tan \tfrac{1}{2}(A \pm B) = \frac{\sin A \pm \sin B}{\cos A + \cos B}$$

$$\cot \tfrac{1}{2}(A \mp B) = \frac{\sin A \pm \sin B}{\cos B - \cos A}$$

$$\cos^2 A = \tfrac{1}{2}(\cos 2A + 1)$$

$$\cos^3 A = \tfrac{1}{4}(\cos 3A + 3 \cos A)$$

$$\cos^4 A = \tfrac{1}{8}(\cos 4A + 4 \cos 2A + 3)$$

$$\sin^2 A = \tfrac{1}{2}(-\cos 2A + 1)$$

$$\sin^3 A = \tfrac{1}{4}(-\sin 3A + 3 \sin A)$$

$$\sin^4 A = \tfrac{1}{8}(\cos 4A - 4 \cos 2A + 3)$$

$$\sin A \cos B = \tfrac{1}{2}[\sin (A + B) + \sin (A - B)]$$

$$\cos A \cos B = \tfrac{1}{2}[\cos (A + B) + \cos (A - B)]$$

$$\sin A \sin B = \tfrac{1}{2}[\cos (A - B) - \cos (A + B)]$$

OFTEN-USED TRIGONOMETRIC FUNCTIONS

Angle (degs)	Sine	Cosine	Tangent
0	0	0	0
30	$\frac{1}{2}$	$\frac{1}{2}\sqrt{3}$	$\frac{1}{3}\sqrt{3}$
45	$\frac{1}{2}\sqrt{2}$	$\frac{1}{2}\sqrt{2}$	1
60	$\frac{1}{2}\sqrt{3}$	$\frac{1}{2}$	$\sqrt{3}$
90	1	0	$\pm\infty$
180	0	-1	0
270	-1	0	$\pm\infty$
360	0	1	0
0–90	$+$	$+$	$+$
90–180	$+$	$-$	$-$
180–270	$-$	$-$	$+$
270–360	$-$	$+$	$-$

APPROXIMATIONS FOR SMALL ANGLES

$$\sin\theta = \left(\theta - \frac{\theta^3}{6} \cdots\right)$$

$$\tan\theta = \left(\theta + \frac{\theta^3}{3} \cdots\right) \quad \text{radians}$$

$$\cos\theta = \left(1 - \frac{\theta^2}{2} \cdots\right)$$

$\sin\theta = \theta$
  With less than 1% error up to $\theta = .24$ radian $= 14.0°$
  With less than 10% error up to $\theta = .78$ radian $= 44.5°$

$\tan\theta = \theta$
  With less than 1% error up to $\theta = .17$ radian $= 10.0°$
  With less than 10% error up to $\theta = .54$ radian $= 31.0°$

appendix B

PREFIXES AND SYMBOLS

Multiple	Prefix	Symbol
10^{12}	Tera	T
10^{9}	Giga	G
10^{6}	Mega	M
10^{3}	Kilo	k
10^{2}	Hecto	h
10	Deka	da
10^{-1}	Deci	d
10^{-2}	Centi	c
10^{-3}	Milli	m
10^{-6}	Micro	μ
10^{-9}	Nano	n
10^{-12}	Pico	p
10^{-16}	Femto	f
10^{-18}	Atto	a

appendix C

CONVERSION FACTORS

To Convert	Into	Multiply by	Conversely, Multiply by
Acres	Square feet	4.356×10^4	2.296×10^{-5}
	Square meters	4,047	2.471×10^{-4}
	Square yards	4.84×10^3	2.066×10^{-4}
Ampere-hours	Coulombs	3,600	2.778×10^{-4}
Angstroms	Nanometers	10^{-1}	10
Atmospheres	Bars	1.0133	.9869
	Millimeters of mercury at 0°C	760	1.316×10^{-3}
	Feet of water at 4°C	33.90	2.950×10^{-2}
Bars	Newtons per square m	10^5	10^{-5}
Btu	Foot-pounds	778.3	1.285×10^{-3}
	Joules	1,054.8	9.480×10^{-4}
	Kilogram-calories	.2520	3.969
	Horsepower-hours	3.929×10^{-4}	2,545
Bushels	Cubic feet	1.2554	.8036
Calories (I. T.)	Joules	4.1868	.238
Calories (thermochem)	Joules	4.184	.239
Celsius (centigrade)	Fahrenheit	$°C \times \frac{9}{5} = °F - 32$	
		$(°C + 40) \times \frac{9}{5} = (°F + 40)$	
Cubic feet	gallons (liq. U.S.)	7.481	.1337
	Liters	28.32	3.531×10^{-2}
Cubic inches	Cubic centimeters	16.39	6.102×10^{-2}
	Cubic feet	5.787×10^{-4}	1,728
	Cubic meters	1.639×10^{-5}	6.102×10^4
	Gallons (liq. U.S.)	4.329×10^{-3}	231
Cubic meters	Cubic feet	35.31	2.832×10^{-2}
	Cubic yards	1.308	.7646
Degrees (angle)	Radians	1.745×10^{-2}	57.30
Dynes	Pounds	2.248×10^{-6}	4.448×10^5
	Newtons	10^{-5}	10^5
Electron volts	Joules	1.602×10^{-19}	$.624 \times 10^{19}$
Ergs	Foot-pounds	7.376×10^{-8}	1.356×10^7
Ergs	Joules	10^{-7}	10^7
Fathoms	Feet	6	.16667
	Meters	1.8288	.5467

appendix C

To Convert	Into	Multiply by	Conversely, Multiply by
Feet	Centimeters	30.48	3.281×10^{-2}
Footcandles	Lumens per square meter	10.764	.0929
Footlamberts	Candelas per square meter	3.4263	2.919×10^{-1}
Foot-pounds	Horsepower-hours	5.050×10^{-7}	1.98×10^{6}
	Kilogram-meters	.1383	7.233
	Kilowatt-hours	3.766×10^{-7}	2.655×10^{6}
Gallons (liq U.S.)	Cubic meters	3.785×10^{-3}	264.2
	Gallons (liq. Br Imp) (Canada)	.8327	1.201
Grams	Dynes	980.7	1.020×10^{-3}
	Grains	15.43	6.481×10^{-2}
	Ounces (avoirdupois)	3.527×10^{-2}	28.35
	Poundals	7.093×10^{-2}	14.10
Grams per centimeter	Pounds per inch	5.600×10^{-3}	178.6
Grams per cubic centimeter	Pounds per cubic inch	3.613×10^{-2}	27.68
Grams per square centimeter	Pounds per square foot	2.0481	.4883
Horsepower (boiler)	Btu per hour	3.347×10^{4}	2.986×10^{-5}
Horsepower (550 ft-lb/sec)	Btu per minute	42.41	2.357×10^{-2}
Horsepower	Foot-pounds per minute	3.3×10^{4}	3.030×10^{-5}
	Kilowatts	.745	1.342
	Kilogram-calories per	10.69	9.355×10^{-2}
Inches	Centimeters	2.540	.3937
	Feet	8.333×10^{-2}	12
	Miles	1.578×10^{-5}	6.336×10^{4}
	Mils	1,000	.001
	Yards	2.778×10^{-2}	36
Joules	Foot-pounds	.7376	1.356
	Ergs	10^{7}	10^{-7}
Kilogram-calories	Kilogram-meters	426.9	2.343×10^{-3}
	Kilojoules	4.186	.2389
Kilogram force	Joules	9.81	.102
Kilogram-meters	Newtons	.102	9.81
Kilograms	Pounds (avoirdupois)	2.205	.4536
Kilograms per kilometer	Pounds (avdp) per mile (stat)	3.548	.2818
Kilograms per square meter	Pounds per square foot	.2048	4.882
Kilometers	Feet	3281	3.048×10^{-4}
Kilowatt-hours	Btu	3413	2.930×10^{-4}
	Foot-pounds	2.655×10^{6}	3.766×10^{-7}
	Joules	3.6×10^{6}	2.778×10^{-7}
	Kilogram-calories	860	1.163×10^{-3}
	Kilogram-meters	3.671×10^{5}	2.724×10^{-6}

appendix C

To Convert	Into	Multiply by	Conversely, Multiply by
Knots* (nautical miles per hour)	Feet per second	1.688	.5925
Knots	Meters per minute	30.87	.03240
	Miles (stat) per hour	1.1508	.8690
Leagues	Miles (approximately)	3	.33
Links	Inches	7.92	.1263
Liters	Cubic centimeters	1,000	.001
	Cubic meters	.001	1,000
	Cubic inches	61.02	1.639×10^{-2}
	Gallons (liq. U.S.)	.2642	3.785
	Pints (liq. U.S.)	2.113	.4732
$\log_e$ or ln	$\log_{10}$	.4343	2.303
Lumes per square foot	Foot-candles	1	1
Meters	Yards	1.094	.9144
Microns	Meters	10^{-6}	10^6
Miles (nautical)*	Feet	6,076.1	1.646×10^{-4}
	Meters	1,852	5.400×10^{-4}
	Miles (statute)	1.1508	.8690
Miles (statute)	Feet	5,280	1.894×10^{-4}
	Kilometers	1.609	.6214
Miles per hour	Kilometers per minute	2.682×10^{-2}	37.28
	Feet per minute	88	1.136×10^{-2}
	Kilometers per hour	1.609	.6214
Millibars	Inches of mercury (32°F)	.02953	33.86
Millibars (10^3 dynes/cm²)	Pounds per square foot	2.089	.4788
Mils	Meters	2.54×10^{-5}	3.94×10^4
Nepers	Decibels	8.686	.1151
Newtons	Dynes	10^5	10^{-5}
	Kilograms	.1020	9.807
	Poundals	7.233	.1383
	Pounds (avoirdupois)	.2248	4.448
Ounces (fluid)	Quarts	3.125×10^{-2}	32
Ounces (avoirdupois)	Pounds	6.25×10^{-2}	16
Pints	Quarts (liq. U.S.)	.50	2
Pounds of water (dist)	Cubic feet	1.603×10^{-2}	62.38
	Gallons	.1198	8.347
Quarts	Gallons (liq. U.S.)	.25	4
Rods	Feet	16.5	6.061×10^{-2}
Slugs (mass)	Pounds (avoirdupois)	32.174	3.108×10^{-2}
Square feet	square meters	9.290×10^{-2}	10.76
Square inches	Circular mils	1.273×10^6	7.854×10^{-7}
	Square centimeters	6.452	.1550

*Nautical mile based on International Nautical Mile adopted by U. S. Dept. of Defense and U.S. Dept. of Commerce, July 1, 1954.

To Convert	Into	Multiply by	Conversely, Multiply by
Square miles	Square yards	3.098×10^6	3.228×10^{-7}
	Acres	640	1.562×10^{-3}
	Square kilometers	2.590	.3861
Square millimeters	Circular mils	1973	5.067×10^{-4}
Watts	Btu per minute	5.689×10^{-2}	17.58
	Ergs per second	10^7	10^{-7}
	Foot-pounds per minute	44.26	2.260×10^{-3}
	Horsepower (550 ft-lb/sec)	1.341×10^{-3}	745.7
	Horsepower (metric) (542.5 ft-lb/sec)	1.360×10^{-3}	735.5
	Kilogram-calories per minute	1.433×10^{-2}	69.77
Watt-seconds (joules)	Gram-calories (mean)	.2389	4.186
Yards	Feet	3	.3333

Note: Pounds are avoirdupois in every entry except where otherwise indicated.

index

A

accuracy of instrumentation, 393
acres, 449
action potential, 359
adiabatic, 167
advection, 434
afferent nerve, 358
aircraft, 119, 120
air pollution:
    and energy, 165
    and wind, 164
    clean air acts, 148, 190, 197
    definition of, 158
    primary standards, 197
    secondary standards, 197
    transportation, 101, 102, 103, 121
air pollution constraints, 159
air pollution emissions, 149
air pollution legislation, 189
air pollution sources, 175
air quality act, 191
air quality control regions, 191
air quality criteria, 191
alpha rhythm, 392
ambient air quality standards, 192, 195
amplifier, 310, 318
amplitude, 323, 351, 352
ampere-hours, 449
Amtrak, 92, 95, 113
analog, 330, 352
analytic engine, 273
angstroms, 449
antenna, 310, 312, 313
approximations, trigonometric, 447
arch, 33, 46, 60, 69, 74, 78
    natural, 34
    semicircular, 72
    shallow, 73
    true, 34
artificial organs:
    heart, 367
    kidney, 378
    nerve, 358
    pacemaker, 366
autocorrelation function, 363
automotive emissions, 151
automotive pollution, 178, 182, 185, 191, 192, 198
average value, 258

B

Babbage, Charles, 273
Baker, Benjamin, 81
balance organs, 379
bandwidth, 324, 349
bars, 449
BART, 107, 113
base, 283
beam, 33, 50, 54
beams:
    cross, 49, 69
    girders, 69
    simple, 34
    stiffening, 57, 60
bend test, 211
best straight line, 399
binary, 283
binary functions, 296
biological sciences, engineers in, 19
Bionics, 360
bit, 284
blood sugar, 375
breadboards, 25
Brooklyn Bridge, 75
BTU, 449
bushels, 449

C

cable, 311
    coaxial, 311
calculator, 275
calibration of instruments, 391
calories, 449
cantilever, 80, 81
capacitance, 262
capacitor, 260, 262
    energy storage in, 262
    symbol for, 263
    voltage-current relation, 263
carbon dioxide, 167, 177, 179, 189
carbon monoxide, 177, 182, 188
carrier, 343
catalysis, 427, 430
cathode-ray tube, 252
cell, 284
    electronic, 299
Celsius, 449
ceramics, 230
channel, 309, 310, 312, 325
charge, 236, 237
    on a capacitor, 263
chemical equilibrium, 427, 428
Chesapeake Bay Bridge-Tunnel, 70, 81
CIA, 17
circuit, 21
    integrated, 251

454

circuit (cont.)
  parallel, 244
  series, 243
  series-parallel, 246
  series resistor-capacitor, 267
  series resistor-inductor, 265
  two-loop, 247
circuit element:
  active, 238
  linear, 242
  non dissipative, 260
  passive, 238
  two terminal, 239
Civil Aeronautics Board (CAB), 96
clipping, 314
code, 309, 321, 341, 351, 352, 353
  computer, 283
  Morse, 315
coding, computer, 283
combustion, 141, 149
  products of, 149
combustion pollution, 177
commutator, 338
complementation, 292
composites, 228
compression test, 207
computer, 305, 309
  general purpose digital, 273, 276
conductance:
  definition, 238
  for parallel elements, 246
conductivity, definition, 238
conductor, 311
confidence limits in the mean, 389
conservation of energy, 420
conservation of mass, 136
conservation of matter, 414
conservation of momentum, 440
constancy of instruments, 392
consulting, 20
Consumers' Union, 5
convection, 434
conversion:
  A/D, 331
  D/A, 336
  units, 449, 450, 451, 452
counter, 299
covered bridges, 78, 79
creep test, 217
cross-talk, 317
cubic feet, 449
cubic inches, 449
cupula, 380
current, 236
  alternating, 251

current (cont.)
  definition, 237
  direct, 251
cycle, 323

D

daPonte, Antonio, 74
deformation, 41, 42, 44
degrees, angle, 449
demand-type system:
  transportation, 107, 108, 109
demodulation, 319, 344
demography, transportation, 105
DeMorgans' theorem, 295
Department of Transportation (D.O.T.), 93, 94, 95
derivative, 237
desalination, 439
design, 21, 202
design process, 22, 23, 24
destination, 309
development, 15
dialysis, 378
diastable, 366
differential equation, 266
diffusion, 436
digital, 330, 352
dimensional analysis, 401
discrimination in instruments, 393
dispersion, 314
distortion, 312
doctor of engineering, 2
Donora, Penn., 172
draftsman, 14
dry adiabatic lapse rate, 167
dynes, 449

E

echoes, 316
economy and computers, 275
education, engineers in, 20
effective value:
  of periodic signals, 258
  of sinusoidal signals, 259
efferent nerve, 358
efficiency of power plant, 424
effluent fee, 196
Einthoven, Willem, 368
electrocardiogram, 367
  computer analysis of, 370
electroencephalogram, 361
electromagnetic spectrum, 347
electromotive force, 238

electronics, 231
electron microscope, 221
electron volts, 449
emissions power plant, 419
    sulfur dioxide, 418
emission standards, 192, 195
endolymph, 380
energy:
    availability of, 423
    conversion factors, 135
    costs, 140
    environmental pollution, 147
    first law of thermodynamics, 137
    in electric field, 262
    in magnetic field, 260
    second law of thermodynamics, 138
    supply, 133
energy and air pollution, 165
energy consumption, 129
    breakdown, 133
    U. S., 131
    world, 131
energy conversion systems, 139
    automotive power plant, 143
    definition of, 134
    efficiency, 136, 138
    fuel cell, 145
    gas turbine engine, 144
    MHD, 145
    power output, 142
    Rankine cycle, 141
    thermionic, 146
energy utilization:
    transportation, 104, 122, 123
engineering, computer, 273
engineers:
    bioengineer, 356
    biomedical, 356
    chemical, 88
    civil, 88
    computer, 273
    industrial, 16, 28, 88
    mechanical, 28, 88
    self-employed, 20
Engineers' Joint Council for Professional Development, 6
ENIAC, 274
Environmental Protection Agency (EPA), 96, 197
epidemiologic studies, 192, 195
equalizer, 317
equilibrium, 36-38, 46, 48, 50, 56, 63, 64, 68
equilibrium equations, 40, 55, 62
ergs, 449

error:
    datum, 391
    detection, 322
    zero, 391
errors:
    estimation of, 387
    of obervation, 387
    random, 387
    systematic, 387, 391
estimate, best, 388
ethics, 6
evaporation, 433
experiments, conduction of, 386
exponential function, 254, 264
    derivative of, 264
    integral of, 265

F

fan-out, 293
Faraday's law, 255
fathoms, 449
fatigue test, 216
Federal Aviation Administration (FAA), 17, 94
Federal Bureau of Investigation (FBI), 17
Federal Communications Commission (FCC), 17, 349
Federal Highway Administration (FHWA), 94
Federal Railroad Administration (FRA), 93
feet, 450
filter, 326, 327, 348
flip-flop, 298
fluid entrained vehicle, 117
fluorides, 174
foot candles, 450
foot lamberts, 450
foot pounds, 450
force, 33, 36, 50
    components, 37, 50
    compression, 43, 66, 69
    concurrent, 39, 40, 64, 67
    coplanar, 36, 39
    moment, 36, 39, 52, 55
    reactions, 50, 54
    shear, 43, 52
    tensile, 43
    tension, 45, 57, 66
    thrust, 63, 73
foundations, 37
    abutments, 36, 50
    piers, 36
frequency, 322, 323, 351, 352
frequency discriminator, 347
function, logical, 293
functions, trigonometric, 446

G

gallons, 450
generator, 255
geometry, 47
    parabolic, 60
    semicircular, 72
    triangle, 47, 63
glucagon, 376
glucose, 376
glycogen, 376
GNP, 275
government, engineering in, 16
graduate school, 2
grams, 450
graph, 306
graphs, 394
gravity vacuum transit, 117
greenhouse effect, 167, 180
guitar tuner, 25, 26, 27

H

hardness, 209
harmonic, 323
Harvard Mark I, 274
heart, 364
heat, 135
heat transfer, 432
Hertz, 323
Hodgkin, A. L., 359
horsepower, 450
human implants, 232
Huxley, A.F., 359
hydrocarbons, 184, 186
hydrogen sulfide, 172
hydrophone, 320
hypoglycemia, 376

I

identities, trigonometric, 445, 446
impact test, 213
inches, 450
inductance, 261
inductor, 260
    energy storage in, 260
    symbol for, 261
    voltage current relation, 261
industry, 15
    computer, 275, 276
information, 305
input devices, 277
input-output characteristic, 314, 350, 352

integral, 363
integrated circuit (IC), 274
interference, 311, 317, 351
internal energy, 421
Internal Revenue Service, 18
Interstate Commerce Commission (ICC), 96
Interstate Highway System, 89
interurban systems, transportation, 112
intraurban systems, transportation, 107
instructions, computer, 287, 288
inventor, 21

J

jobs, 14
joules, 450

K

kidney, 378
kilogram, 450
kilometers, 450
kilowatt-hours, 450
kinetics, 426, 429
Kirchoff:
    current law, 244
    voltage law, 240
Kitty Hawk, 91
knots, 451

L

laboratory work, objective of, 384
land utilization, transportation, 103, 104
Large Scale Integration (LSI), 275
LASER, 312, 314
lateral inhibition, 360
law, engineering in, 18
leagues, 451
least squares, principle of, 387
Lenz's law, 261
Lillie nerve model, 358
limpets vs. seaweed, 372
linear induction motor, 117
links, 451
liters, 451
loads, 33, 36, 47, 84
    masonry, 61
    parabolic, 62
    traffic, 46, 57
    uniform, 57
logic, digital, 292
logic network, 297
lumes, 451

M

McCulloch, Warren S., 357
machine language, 283, 288
Mackinac Straits Bridge, 77
magnetic core, 278
magnetic field:
    constant, 255
    time varying, 261
magnetic suspended system, 118, 119
management, engineers in, 18
marketing, 16
mass transportation, history of, 90
material balance, 414
materials, 48, 202
    cast iron, 80
    concrete, 67, 70, 79
    iron, 79
    steel, 48, 79
    stone, 48
    wood, 48, 83
MBA, 19
mechanical testing, 203
medical sciences, engineers in, 19
membrane, 439, 440
memory:
    cell, 284
    computer, 278
    LSI, 280
    magnetic tape, 280
    random access, 279
    scratch pad, 281
    sequential, 282
    solid state, 281
message, 309
meters, 451
Meuse River Valley, 172
microscope:
    electron, 221
    optical, 220
microwaves, 312, 316
miles, 451
millibars, 451
mils, 451
MOBIDIC, 274
model:
    Hodgkin-Huxley, 359
    mathematical, 356
modulation, 310, 339, 343
    amplitude (AM), 343
    carrier, 343
    frequency (FM), 345
    phase, 343
    pulse amplitude (PAM), 340

modulation (cont.)
    pulse code (PCM), 340, 353
    pulse frequency, 342
    pulse position, 353
    pulse shape, 353
    pulse width, 342
multiplexing, 317, 338
    frequency division, 348
    time division, 338

N

Nader, Ralph, 5, 30
National Aeronautics and Space Administration (NASA), 17, 21
national air pollution problem, 160
National Bureau of Standards (NBS), 17
National Science Foundation (NSF), 16, 21
National Society of Professional Engineers, 6
natural resource depletion, transportation, 105
Naval Research Laboratory (NRL), 16
nepers, 451
nephrotic membrane, 379
nerve, 357
    afferent, 358
    efferent, 358
    electronic model, 360
    neuron, 357
    synapse, 359
neuron, 357 (also see nerve)
newtons, 451
Newton's second law, 440
Niagara Bridge, 75
nitric oxide, 178, 185
nitrogen oxide emissions, 419
nitrogen oxides, 178, 185, 189
node, 239
noise, 317, 331
    thermal, 318
noise pollution, transportation, 105, 121
normal distribution, 388
notebook, laboratory, 385
numbers:
    binary, 283, 286
    decimal, 283, 286
    hexadecimal, 285, 286
    octal, 285, 286

O

octave, 349
oculogravic illusion, 379
Office of High Speed Ground Transportation (OHSGT), 95

Ohm's Law, 241
operations code, 287
optical fiber, 312
oscillograph, 307
oscilloscope, 252, 307
osmotic pressure, 378
otolithic organs, 379
otto cycle, 144
ounces, 451
output devices, 282

P

Palladio, 78
pan, 158, 174, 187
Papineau-Leblanc Bridge, 71
particulates, 169, 178, 180, 189, 193, 419
patent, 18, 25, 27
patent law, 18
period, 323
peroxyacetyl nitrates, 158, 174, 187
phase, 323, 351, 352
phase angle, 259
Ph.D., 2
photochemical smog, 174, 186
photo oxidation, 159
pints, 451
plastics, 229
pleating machine, 28
pollutants:
    accumulation of, 163
    primary, 176
    secondary, 176
pollutant scavenging, 163, 187
polygraph recorder, 308
Pont d'Avignon, 73
Pont du Gard, 72
population genetics, 370
potential, 238
potential energy, 422
pounds, 451
power:
    average, 257, 260
    in resistors, 241
power spectral density, 361
Poza Rica, Mexico, 172
precision of instrumentation, 393
prefixes, 448
process control, 442
process design, 441
production engineering, 15
professional conduct, 6
professional engineer, 6
professional life, 6

program, computer, 289, 290
proposal, 11
prototype, 22
pulse, 310
pump operation, 441

Q

quantization, 332, 352
quarts, 451

R

radiation, 433
radiation inversion, 168
radix, 283
rail systems, 113, 114
rate, 371
reactance:
    capacitive, 264
    inductive, 262
reaction mechanism, 427, 430
reaction rate, 427
receiver, 309, 310
regeneration, 339
register, 284, 287
relations:
    clients and employees, 7
    public, 7
relationships, linear, 395
renaissance, 35, 71
renal threshold, 376
report, technical, 384
research, 15
resistance:
    definition, 238
    for copper wire, 239
    for parallel elements, 246
    for series elements, 244
resistivity, definition, 238
response:
    amplitude, 326
    phase, 326
results, presentation of, 394
reverse osmosis, 439
Rhine Bridge, 70
Rialto Bridge, 74
rods, 451
Roebling, John, 75
Romans, 72
root-mean-square value, definition, 258

S

sales, 16
sampling, 332, 352
satellite, 310, 313, 336
scheduled-type systems, transportation, 107, 109, 110, 111
science:
    applied, 13
    basic, 13
sciences, 13
second law of thermodynamics, 423
semicircular canals, 379
signal, 306, 309
    guided, 310
    logic, 291
signal corps, 16
sine wave (sinusoid), 323, 351, 352
singular point, 375
sluggishness, 315
slugs, 451
smog, 160
social issues, transportation systems, 106
social sciences, engineers in, 19
societies:
    honor, 13
    professional, 11
society, role of engineering in, 4
sound, underwater, 319
source, 309
span, 33, 69, 70
    length, 48, 50, 58, 74, 75, 83
    multiple, 33, 48
    single, 49
spectrum, 324, 325, 343, 347
square feet, 451
square inches, 451
square miles, 452
square millimeters, 452
stability of instruments, 392
standard deviation, 388, 389
state equations, 373
    computer program to solve, 375
stationary source, 198
Steinman, David, 77
strain, 203
stratosphere, 165
strength, 204, 205, 213, 218
stress, 52, 70, 203
subsistence inversion, 168
sulfur dioxide, 177, 183, 188, 193
superalloys, 228
superposition, 249

suspension, 33, 35, 45, 71, 75
    cable, 56, 57, 60
    cable stayed, 71
    suspenders, 57
symbol, 331
symbols, 448
synergism, 172, 193
systems, transportation, 88
systems concepts, mass transportation, 90
systole, 366

T

Tacoma Narrows Bridge, 75–77
taxonometric viewpoint, 356
technician, 14
telephone, 305, 310
teletypewriter, 321
television, 305, 309, 330
temperature inversion, 167
tension test, 203
test:
    bend, 211
    compression, 207
    creep, 217
    fatigue, 216
    impact, 213
    mechanical, 203
    tension, 203
testing, 15
thermal oxidation, 158
thermal pollution, 153, 433, 434
    water uses, 153
thermodynamics, 420
timber, 78, 79
time constant, 264
    of RC circuit, 268
    of RL circuit, 267
time-sharing, 320
toxicologic studies, 193, 195
Tracked Air Cushion Vehicles (TACV), 115
trains, 92
trajectories, 374
transducers, biological, 380
transfer function, 326, 352
transient, 254, 264
    in RC circuits, 267
    in RL circuits, 266
transistors, 274
transmission, multipath, 315
transmitter, 309
transportation:
    convenience, 99
    cost, 98

transportation (cont.)
   death rates, 101
   magazines and journals, 124, 125
   planning, 96
   reliability, 100
   safety, 100
   speed, 100
   use of modes, 89
transportation systems:
   criteria, 98
   modern criteria, 101
trigonometric approximations, 447
trigonometry, 445, 446, 447
troposphere, 165
true value of a magnitude, 387
truss, 33, 35, 47, 49, 63, 77
   joints, 47, 64, 84
truth table, 292, 293
tube flight, 117
Tube Vehicle Systems (TVS), 115, 116, 117

U

UMALAB, 287–290, 300
uniflow, transit system, 118
Urban Mass Transportation Administration (UMTA), 93, 95
U. S. Coast Guard (USCG), 94

V

vapor liquid equilibrium, 437
variables:
   choice of, 404
   reduction of, 101

vector electrocardiogram, 369
voltage, 238
   across a capacitor, 263
   across a resistor, 242
   across an inductor, 261
   drop, 240
   rise, 240
   sawtooth, 253
   sinusoidal, 252
   time-varying, 251
von Neumann, John, 274

W

water:
   ammonia removal, 438
   vapor pressure, 425
water pollution, transportation, 105
watts, 452
watt-seconds, 452
waveform (waveshape), 306
   periodic, 323, 351
waveguide, 312
wavelength, 343
work, 135
Wright brothers, 91

X

x-ray diffraction, 223

Y

yards, 452